T0296077

LONDON MATHEMATICAL SOCIETY LECTURE NOTE SERIES

Managing Editor: Professor M. Reid, Mathematics Institute,
University of Warwick, Coventry CV4 7AL, United Kingdom

The titles below are available from booksellers, or from Cambridge University Press at www.cambridge.org/mathematics

234 Introduction to subfactors, V. JONES & V.S. SUNDER
235 Number theory: Séminaire de théorie des nombres de Paris 1993–94, S. DAVID (ed)
236 The James forest, H. FETTER & B. GAMBOA DE BUEN
237 Sieve methods, exponential sums, and their applications in number theory, G.R.H. GREAVES *et al* (eds)
238 Representation theory and algebraic geometry, A. MARTSINKOVSKY & G. TODOROV (eds)
240 Stable groups, F.O. WAGNER
241 Surveys in combinatorics, 1997, R.A. BAILEY (ed)
242 Geometric Galois actions I, L. SCHNEPS & P. LOCHAK (eds)
243 Geometric Galois actions II, L. SCHNEPS & P. LOCHAK (eds)
244 Model theory of groups and automorphism groups, D.M. EVANS (ed)
245 Geometry, combinatorial designs and related structures, J.W.P. HIRSCHFELD *et al* (eds)
246 p-Automorphisms of finite p-groups, E.I. KHUKHRO
247 Analytic number theory, Y. MOTOHASHI (ed)
248 Tame topology and O-minimal structures, L. VAN DEN DRIES
249 The atlas of finite groups - Ten years on, R.T. CURTIS & R.A. WILSON (eds)
250 Characters and blocks of finite groups, G. NAVARRO
251 Gröbner bases and applications, B. BUCHBERGER & F. WINKLER (eds)
252 Geometry and cohomology in group theory, P.H. KROPHOLLER, G.A. NIBLO & R. STÖHR (eds)
253 The q-Schur algebra, S. DONKIN
254 Galois representations in arithmetic algebraic geometry, A.J. SCHOLL & R.L. TAYLOR (eds)
255 Symmetries and integrability of difference equations, P.A. CLARKSON & F.W. NIJHOFF (eds)
256 Aspects of Galois theory, H. VÖLKLEIN, J.G. THOMPSON, D. HARBATER & P. MÜLLER (eds)
257 An introduction to noncommutative differential geometry and its physical applications (2nd edition), J. MADORE
258 Sets and proofs, S.B. COOPER & J.K. TRUSS (eds)
259 Models and computability, S.B. COOPER & J. TRUSS (eds)
260 Groups St Andrews 1997 in Bath I, C.M. CAMPBELL *et al* (eds)
261 Groups St Andrews 1997 in Bath II, C.M. CAMPBELL *et al* (eds)
262 Analysis and logic, C.W. HENSON, J. IOVINO, A.S. KECHRIS & E. ODELL
263 Singularity theory, W. BRUCE & D. MOND (eds)
264 New trends in algebraic geometry, K. HULEK, F. CATANESE, C. PETERS & M. REID (eds)
265 Elliptic curves in cryptography, I. BLAKE, G. SEROUSSI & N. SMART
267 Surveys in combinatorics, 1999, J.D. LAMB & D.A. PREECE (eds)
268 Spectral asymptotics in the semi-classical limit, M. DIMASSI & J. SJÖSTRAND
269 Ergodic theory and topological dynamics of group actions on homogeneous spaces, M.B. BEKKA & M. MAYER
271 Singular perturbations of differential operators, S. ALBEVERIO & P. KURASOV
272 Character theory for the odd order theorem, T. PETERFALVI. Translated by R. SANDLING
273 Spectral theory and geometry, E.B. DAVIES & Y. SAFAROV (eds)
274 The Mandelbrot set, theme and variations, T. LEI (ed)
275 Descriptive set theory and dynamical systems, M. FOREMAN, A.S. KECHRIS, A. LOUVEAU & B. WEISS (eds)
276 Singularities of plane curves, E. CASAS-ALVERO
277 Computational and geometric aspects of modern algebra, M. ATKINSON *et al* (eds)
278 Global attractors in abstract parabolic problems, J.W. CHOLEWA & T. DLOTKO
279 Topics in symbolic dynamics and applications, F. BLANCHARD, A. MAASS & A. NOGUEIRA (eds)
280 Characters and automorphism groups of compact Riemann surfaces, T. BREUER
281 Explicit birational geometry of 3-folds, A. CORTI & M. REID (eds)
282 Auslander–Buchweitz approximations of equivariant modules, M. HASHIMOTO
283 Nonlinear elasticity, Y.B. FU & R.W. OGDEN (eds)
284 Foundations of computational mathematics, R. DEVORE, A. ISERLES & E. SÜLI (eds)
285 Rational points on curves over finite fields, H. NIEDERREITER & C. XING
286 Clifford algebras and spinors (2nd Edition), P. LOUNESTO
287 Topics on Riemann surfaces and Fuchsian groups, E. BUJALANCE, A.F. COSTA & E. MARTÍNEZ (eds)
288 Surveys in combinatorics, 2001, J.W.P. HIRSCHFELD (ed)
289 Aspects of Sobolev-type inequalities, L. SALOFF-COSTE
290 Quantum groups and Lie theory, A. PRESSLEY (ed)
291 Tits buildings and the model theory of groups, K. TENT (ed)
292 A quantum groups primer, S. MAJID
293 Second order partial differential equations in Hilbert spaces, G. DA PRATO & J. ZABCZYK
294 Introduction to operator space theory, G. PISIER
295 Geometry and integrability, L. MASON & Y. NUTKU (eds)
296 Lectures on invariant theory, I. DOLGACHEV
297 The homotopy category of simply connected 4-manifolds, H.-J. BAUES
298 Higher operads, higher categories, T. LEINSTER (ed)
299 Kleinian groups and hyperbolic 3-manifolds, Y. KOMORI, V. MARKOVIC & C. SERIES (eds)
300 Introduction to Möbius differential geometry, U. HERTRICH-JEROMIN
301 Stable modules and the D(2)-problem, F.E.A. JOHNSON
302 Discrete and continuous nonlinear Schrödinger systems, M.J. ABLOWITZ, B. PRINARI & A.D. TRUBATCH
303 Number theory and algebraic geometry, M. REID & A. SKOROBOGATOV (eds)
304 Groups St Andrews 2001 in Oxford I, C.M. CAMPBELL, E.F. ROBERTSON & G.C. SMITH (eds)
305 Groups St Andrews 2001 in Oxford II, C.M. CAMPBELL, E.F. ROBERTSON & G.C. SMITH (eds)
306 Geometric mechanics and symmetry, J. MONTALDI & T. RATIU (eds)
307 Surveys in combinatorics 2003, C.D. WENSLEY (ed.)
308 Topology, geometry and quantum field theory, U.L. TILLMANN (ed)
309 Corings and comodules, T. BRZEZINSKI & R. WISBAUER
310 Topics in dynamics and ergodic theory, S. BEZUGLYI & S. KOLYADA (eds)
311 Groups: topological, combinatorial and arithmetic aspects, T.W. MÜLLER (ed)
312 Foundations of computational mathematics, Minneapolis 2002, F. CUCKER *et al* (eds)

313 Transcendental aspects of algebraic cycles, S. MÜLLER-STACH & C. PETERS (eds)
314 Spectral generalizations of line graphs, D. CVETKOVIC, P. ROWLINSON & S. SIMIC
315 Structured ring spectra, A. BAKER & B. RICHTER (eds)
316 Linear logic in computer science, T. EHRHARD, P. RUET, J.-Y. GIRARD & P. SCOTT (eds)
317 Advances in elliptic curve cryptography, I.F. BLAKE, G. SEROUSSI & N.P. SMART (eds)
318 Perturbation of the boundary in boundary-value problems of partial differential equations, D. HENRY
319 Double affine Hecke algebras, I. CHEREDNIK
320 L-functions and Galois representations, D. BURNS, K. BUZZARD & J. NEKOVÁŘ (eds)
321 Surveys in modern mathematics, V. PRASOLOV & Y. ILYASHENKO (eds)
322 Recent perspectives in random matrix theory and number theory, F. MEZZADRI & N.C. SNAITH (eds)
323 Poisson geometry, deformation quantisation and group representations, S. GUTT *et al* (eds)
324 Singularities and computer algebra, C. LOSSEN & G. PFISTER (eds)
325 Lectures on the Ricci flow, P. TOPPING
326 Modular representations of finite groups of Lie type, J.E. HUMPHREYS
327 Surveys in combinatorics 2005, B.S. WEBB (ed)
328 Fundamentals of hyperbolic manifolds, R. CANARY, D. EPSTEIN & A. MARDEN (eds)
329 Spaces of Kleinian groups, Y. MINSKY, M. SAKUMA & C. SERIES (eds)
330 Noncommutative localization in algebra and topology, A. RANICKI (ed)
331 Foundations of computational mathematics, Santander 2005, L.M PARDO, A. PINKUS, E. SÜLI & M.J. TODD (eds)
332 Handbook of tilting theory, L. ANGELERI HÜGEL, D. HAPPEL & H. KRAUSE (eds)
333 Synthetic differential geometry (2nd Edition), A. KOCK
334 The Navier–Stokes equations, N. RILEY & P. DRAZIN
335 Lectures on the combinatorics of free probability, A. NICA & R. SPEICHER
336 Integral closure of ideals, rings, and modules, I. SWANSON & C. HUNEKE
337 Methods in Banach space theory, J.M.F. CASTILLO & W.B. JOHNSON (eds)
338 Surveys in geometry and number theory, N. YOUNG (ed)
339 Groups St Andrews 2005 I, C.M. CAMPBELL, M.R. QUICK, E.F. ROBERTSON & G.C. SMITH (eds)
340 Groups St Andrews 2005 II, C.M. CAMPBELL, M.R. QUICK, E.F. ROBERTSON & G.C. SMITH (eds)
341 Ranks of elliptic curves and random matrix theory, J.B. CONREY, D.W. FARMER, F. MEZZADRI & N.C. SNAITH (eds)
342 Elliptic cohomology, H.R. MILLER & D.C. RAVENEL (eds)
343 Algebraic cycles and motives I, J. NAGEL & C. PETERS (eds)
344 Algebraic cycles and motives II, J. NAGEL & C. PETERS (eds)
345 Algebraic and analytic geometry, A. NEEMAN
346 Surveys in combinatorics 2007, A. HILTON & J. TALBOT (eds)
347 Surveys in contemporary mathematics, N. YOUNG & Y. CHOI (eds)
348 Transcendental dynamics and complex analysis, P.J. RIPPON & G.M. STALLARD (eds)
349 Model theory with applications to algebra and analysis I, Z. CHATZIDAKIS, D. MACPHERSON, A. PILLAY & A. WILKIE (eds)
350 Model theory with applications to algebra and analysis II, Z. CHATZIDAKIS, D. MACPHERSON, A. PILLAY & A. WILKIE (eds)
351 Finite von Neumann algebras and masas, A.M. SINCLAIR & R.R. SMITH
352 Number theory and polynomials, J. MCKEE & C. SMYTH (eds)
353 Trends in stochastic analysis, J. BLATH, P. MÖRTERS & M. SCHEUTZOW (eds)
354 Groups and analysis, K. TENT (ed)
355 Non-equilibrium statistical mechanics and turbulence, J. CARDY, G. FALKOVICH & K. GAWEDZKI
356 Elliptic curves and big Galois representations, D. DELBOURGO
357 Algebraic theory of differential equations, M.A.H. MACCALLUM & A.V. MIKHAILOV (eds)
358 Geometric and cohomological methods in group theory, M.R. BRIDSON, P.H. KROPHOLLER & I.J. LEARY (eds)
359 Moduli spaces and vector bundles, L. BRAMBILA-PAZ, S.B. BRADLOW, O. GARCÍA-PRADA & S. RAMANAN (eds)
360 Zariski geometries, B. ZILBER
361 Words: Notes on verbal width in groups, D. SEGAL
362 Differential tensor algebras and their module categories, R. BAUTISTA, L. SALMERÓN & R. ZUAZUA
363 Foundations of computational mathematics, Hong Kong 2008, F. CUCKER, A. PINKUS & M.J. TODD (eds)
364 Partial differential equations and fluid mechanics, J.C. ROBINSON & J.L. RODRIGO (eds)
365 Surveys in combinatorics 2009, S. HUCZYNSKA, J.D. MITCHELL & C.M. RONEY-DOUGAL (eds)
366 Highly oscillatory problems, B. ENGQUIST, A. FOKAS, E. HAIRER & A. ISERLES (eds)
367 Random matrices: High dimensional phenomena, G. BLOWER
368 Geometry of Riemann surfaces, F.P. GARDINER, G. GONZÁLEZ-DIEZ & C. KOUROUNIOTIS (eds)
369 Epidemics and rumours in complex networks, M. DRAIEF & L. MASSOULIÉ
370 Theory of p-adic distributions, S. ALBEVERIO, A.YU. KHRENNIKOV & V.M. SHELKOVICH
371 Conformal fractals, F. PRZYTYCKI & M. URBANSKI
372 Moonshine: The first quarter century and beyond, J. LEPOWSKY, J. MCKAY & M.P. TUITE (eds)
373 Smoothness, regularity, and complete intersection, J. MAJADAS & A. G. RODICIO
374 Geometric analysis of hyperbolic differential equations: An introduction, S. ALINHAC
375 Triangulated categories, T. HOLM, P. JØRGENSEN & R. ROUQUIER (eds)
376 Permutation patterns, S. LINTON, N. RUŠKUC & V. VATTER (eds)
377 An introduction to Galois cohomology and its applications, G. BERHUY
378 Probability and mathematical genetics, N. H. BINGHAM & C. M. GOLDIE (eds)
379 Finite and algorithmic model theory, J. ESPARZA, C. MICHAUX & C. STEINHORN (eds)
380 Real and complex singularities, M. MANOEL, M.C. ROMERO FUSTER & C.T.C WALL (eds)
381 Symmetries and integrability of difference equations, D. LEVI, P. OLVER, Z. THOMOVA & P. WINTERNITZ (eds)
382 Forcing with random variables and proof complexity, J. KRAJÍČEK
383 Motivic integration and its interactions with model theory and non-Archimedean geometry I, R. CLUCKERS, J. NICAISE & J. SEBAG (eds)
384 Motivic integration and its interactions with model theory and non-Archimedean geometry II, R. CLUCKERS, J. NICAISE & J. SEBAG (eds)
385 Entropy of hidden Markov processes and connections to dynamical systems, B. MARCUS, K. PETERSEN & T. WEISSMAN (eds)
386 Independence-friendly logic, A.L. MANN, G. SANDU & M. SEVENSTER
387 Groups St Andrews 2009 in Bath I, C.M. CAMPBELL *et al* (eds)
388 Groups St Andrews 2009 in Bath II, C.M. CAMPBELL *et al* (eds)

London Mathematical Society Lecture Notes Series: 384

Motivic Integration and its Interactions with Model Theory and Non-Archimedean Geometry

Volume II

Edited by

RAF CLUCKERS

Université de Lille 1, France

JOHANNES NICAISE

Katholieke Universiteit Leuven, Belgium

JULIEN SEBAG

Université de Rennes 1, France

CAMBRIDGE
UNIVERSITY PRESS

University Printing House, Cambridge CB2 8BS, United Kingdom

One Liberty Plaza, 20th Floor, New York, NY 10006, USA

477 Williamstown Road, Port Melbourne, VIC 3207, Australia

314-321, 3rd Floor, Plot 3, Splendor Forum, Jasola District Centre, New Delhi - 110025, India

103 Penang Road, #05-06/07, Visioncrest Commercial, Singapore 238467

Cambridge University Press is part of the University of Cambridge.

It furthers the University's mission by disseminating knowledge in the pursuit of education, learning and research at the highest international levels of excellence.

www.cambridge.org
Information on this title: www.cambridge.org/9781107648814

First published 2011

A catalogue record for this publication is available from the British Library

Library of Congress Cataloging in Publication data
Motivic integration and its interactions with model theory and non-Archimedean geometry / edited by Raf Cluckers, Johannes Nicaise, Julien Sebag.
p. cm. – (London Mathematical Society lecture note series ; 383)
ISBN 978-0-521-14976-1 (pbk.)
1. Model theory. 2. Valued fields. 3. Analytic spaces. 4. Geometry, Algebraic.
I. Cluckers, Raf. II. Nicaise, Johannes. III. Sebag, Julien.
QA9.7.M68 2011
511.3′4 – dc23 2011021254

ISBN 978-1-107-64881-4 Paperback

Table of Contents for Volume I

Table of Contents for Volume II

Contributors

Manuel Blickle, *Johannes Gutenberg-Universität Mainz, Institut für Mathematik, 55099 Mainz, Germany;* email*: manuel.blickle@gmail.com*

Siegfried Bosch, *Mathematisches Institüt, Fachbereich Mathematik und Informatik der Universität Münster, Einsteinstrasse 62, 48149 Münster, Germany;* email*: bosch@math.uni-muenster.de*

Antoine Chambert–Loir, *Université Rennes 1, Unité de Formation et de Recherche Mathématiques, Institut de Recherche Mathématique de Rennes (IRMAR), 263 Avenue du Général Leclerc, CS 74205, 35042 Rennes Cedex, France;* email*: Antoine.Chambert-Loir@univ-rennes1.fr*

Zoé Chatzidakis, *Université Paris 7, Unité de Formation et de Recherche Mathématiques, Case 7012, Site Chevaleret, 75205 Paris Cedex 13, France;* email*: zoe@logique.jussieu.fr*

Raf Cluckers, *Université Lille 1, Laboratoire Painlevé, CNRS - UMR 8524, Cité Scientifique, 59655 Villeneuve d'Ascq Cedex, France and Katholieke Universiteit Leuven, Department of Mathematics, Celestijnenlaan 200B, 3001 Heverlee, Belgium;* email*: Raf.Cluckers@wis.kuleuven.be*

Françoise Delon, *Université Paris 7, Unité de Formation et de Recherche Mathématiques, Case 7012, Site Chevaleret, 75205 Paris Cedex 13, France;* email*: delon@logique.jussieu.fr*

Immanuel Halupczok, *Münster Universität, Mathematisches Institut und Institut für Mathematische Logik und Grundlagenforschung, Einsteinstrasse 62, 48149 Münster, Germany;* email*: math@karimmi.de*

Florian Ivorra, *Université Rennes 1, Unité de Formation et de Recherche Mathématiques, Institut de Recherche Mathématique de Rennes (IRMAR), 263 Avenue du Général Leclerc, CS 74205, 35042 Rennes Cedex, France;* email*: Florian.Ivorra@univ-rennes1.fr*

Fumiharu Kato, *Kyoto University, Department of Mathematics, Faculty of Science, Kyoto 606–8502, Japan;* email*: kato@kusm.kyoto-u.ac.jp*

Emmanuel Kowalski, *Eidgenössische Technische Hochschule Zürich, Departement Mathematik, HG G 64.1, Rämistrasse 101, 8092 Zürich, Switzerland;* email: *emmanuel.kowalski@math.ethz.ch*

François Loeser, *École Normale Supérieure, 4, rue d'Ulm, 75230 Paris Cedex 05, France;* email: *Francois.Loeser@ens.fr*

Johannes Nicaise, *Katholieke Universiteit Leuven, Department of Mathematics, Celestijnenlaan 200B, 3001 Heverlee, Belgium;* email: *Johannes.Nicaise@wis.kuleuven.be*

Karl Rökaeus, *Horteweg de Vries Instituut voor Wiskunde, Universiteit van Amsterdam, P.O. Box 94248, 1090 Ge AMSTERDAM, The Netherlands;* email: *S.K.F.Rokaeus@uva.nl*

Julien Sebag, *Université Rennes 1, Unité de Formation et de Recherche Mathématiques, Institut de Recherche Mathématique de Rennes (IRMAR), 263 Avenue du Général Leclerc, CS 74205, 35042 Rennes Cedex, France;* email: *Julien.Sebag@univ-rennes1.fr*

Michael Temkin, *Hebrew University of Jerusalem, The Hebrew University, The Edmond J. Safra Campus–Givat Ram, Jerusalem 91904, Israel;* email: *temkin@math.huji.ac.il*

Preface to the second volume

This second volume contains results that are related to motivic integration, model theory and non-archimedean geometry in various ways, with an emphasis on the mutual interactions between these fields.

The scope of these results is quite large, ranging from motives and resolution of singularities for formal schemes to Arakelov geometry and exponential sums. Motivic integration is not explicitly visible in all of the chapters, but it is often lurking in the background. For instance, resolution of singularities for formal schemes is an important ingredient in the study of motivic integrals on formal schemes, and the structure of trees in $\mathbb{Z}_p$ is a combinatorial analog of the structure of the truncation morphisms between Greenberg schemes of different levels.

The primary aim of the second volume is to illustrate the rich interactions between motivic integration, model theory and non-archimedean geometry, and their influence on problems arising in various branches of mathematics. We hope that these results will bring the reader both enjoyment and inspiration.

9

Heights and measures on analytic spaces. A survey of recent results, and some remarks

Antoine Chambert-Loir

The first goal of this paper was to survey my definition in [19] of measures on non-archimedean analytic spaces in the sense of Berkovich and to explain its applications in Arakelov geometry. These measures are analogous the measures on complex analytic spaces given by products of first Chern forms of hermitian line bundles.[1] In both contexts, archimedean and non-archimedean, they are related with Arakelov geometry and the local height pairings of cycles. However, while the archimedean measures lie at the ground of the definition of the archimedean local heights in Arakelov geometry, the situation is reversed in the ultrametric case: we begin with the definition of local heights given by arithmetic intersection theory and define measures in such a way that the archimedean formulae make sense and are valid. The construction is outlined in Section 1, with references concerning metrized line bundles and the archimedean setting. More applications to Arakelov geometry and equidistribution theorems are discussed in Section 3.

The relevance of Berkovich spaces in Diophantine geometry has now made been clear by many papers; besides [19] and [20] and the general equidistribution theorem of Yuan [59], I would like to mention the works [38, 39, 40, 30] who discuss the function field case of the equidistribution theorem, as well as the potential theory on non-archimedean curves developed simultaneously by Favre, Jonsson & Rivera-Letelier

[1] M. Kontsevich and Yu. Tschinkel gave me copies of unpublished notes from the years 2000–2002 where they develop similar ideas to construct canonical non-archimedean metrics on Calabi–Yau varieties; see also [45, 46].

Motivic Integration and its Interactions with Model Theory and Non-Archimedean Geometry (Volume II), ed. Raf Cluckers, Johannes Nicaise, and Julien Sebag. Published by Cambridge University Press.

[32, 33] and Baker & Rumely for the projective line [8], and in general by A. Thuillier's PhD thesis [55]. The reader will find many important results in the latter work, which unfortunately is still unpublished at the time of this writing.

Anyway, I found useful to add examples and complements to the existing (and non-) literature. This is done in Section 2. Especially, I discuss in Section 2.2 the relation between the reduction graph and the skeleton of a Berkovich curve, showing that the two constructions of measures coincide. Section 2.3 shows that the measures defined are of a local nature; more generally, we show that the measures vanish on any open subset where one of the metrized line bundles involved is trivial. This suggests a general definition of *strongly pluriharmonic* functions on Berkovich spaces, as uniform limits of logarithms of absolute values of invertible holomorphic functions. (Strongly pluriharmonic fonctions should only exhaust pluriharmonic functions when the residue field is algebraic over a finite field, but not in general.) In Section 2.4, we discuss polarized dynamical systems and explain the construction of canonical metrics and measures in that case. We also show that the canonical measure vanishes on the Berkovich equicontinuity locus. In fact, what we show is that the canonical metric is "strongly pluriharmonic" on that locus. This is the direct generalization of a theorem of [52] for the projective line (see also [8] for an exposition); this generalizes also a theorem of [44] that Green functions are locally constant on the classical equicontinuity locus. As already were their proofs, mine is a direct adaptation of the proof of the complex case [43]. In Section 2.5, following Gubler [41], we finally describe the canonical measures in the case of abelian varieties.

In Section 3, we discuss applications of the measures in Diophantine geometry over global fields. Once definitions are recalled out in Section 3.1, we briefly discuss in Section 3.2 the relation between Mahler measures (*i.e.*, integration of Green functions against measures) and heights. In Section 3.3, we survey the equidistribution theorems for Galois orbits of points of "small height", following the variational method of Szpiro–Ullmo–Zhang [54] and [59]. In fact, we describe the more general statement from [20]. Finally, Section 3.4 discusses positive lower bounds for heights on curves. This is inspired by recent papers [5, 49] but the method goes back to Mimar's unpublished thesis [48]. A recent preprint [58] of Yuan and Zhang establishes a similar result in any dimension.

Acknowledgments. — This paper grew out of the invitation of Johannes Nicaise and Julien Sebag to add a contribution to the proceedings of the conference "Motivic integration and its interactions with model theory and non-archimedean geometry" (ICMS, Edinburgh 2008); I thank them heartily for that.

I wrote this paper during a stay at the Institute for Advanced Study in Princeton whose winterish but warm atmosphere was extremly motivating. I acknowledge support of the Institut universitaire de France, as well of the National Science Foundation under agreement No. DMS-0635607.

During the writing of this paper, transatlantic e-contacts with Antoine Ducros have been immensely profitable. Besides him, I also wish to thank Matthew Baker and Amaury Thuillier for their interest and comments. I am also grateful to the referee for having pointed out many misprints as well as some serious inacurracies.

1 Metrized line bundles and measures

1.1 Continuous metrics

1.1.1 Definition. — Let X be a topological space together with a sheaf of local rings $\mathscr{O}_X$ ("analytic functions"); let also $\mathscr{C}_X$ be the sheaf of continuous functions on X. In analytic geometry, local functions have an absolute value which is a real valued continuous function, satisfying the triangle inequality. Let us thus assume that we have a morphism of sheaves $\mathscr{O}_X \to \mathscr{C}_X$, written $f \mapsto |f|$, such that $|fg| = |f|\,|g|$, $|1| = 1$, and $|f+g| \leq |f| + |g|$.

A line bundle on $(X, \mathscr{O}_X)$ is a sheaf L of $\mathscr{O}_X$-modules which is locally isomorphic to $\mathscr{O}_X$. In other words, X is covered by open sets U such that $\mathscr{O}_U \simeq L|U$; such an isomorphism is equivalent to a non-vanishing section $\varepsilon_U \in \Gamma(U, L)$, also called a local frame of L.

If s is a section of a line bundle L on an open set U, the value of s at a point $x \in U$ is only well-defined as an element of the stalk $L(x)$, which is a $\kappa(x)$-vector space of dimension 1. (Here, $\kappa(x)$ is the residue field of $\mathscr{O}_X$ at x.) Prescribing a metric on L amounts to assigning, in a coherent way, the norms of these values. Formally, a *metric* on L is the datum, for any open set $U \subset X$ and any section $s \in \Gamma(U, L)$, of a continuous function $\|s\|_U : U \to \mathbf{R}_+$, satisfying the following properties:

1. for any open set $V \subset U$, $\|s\|_V$ is the restriction to V of the function $\|s\|_U$;
2. for any function $f \in \mathscr{O}_X(U)$, $\|fs\| = |f|\,\|s\|$;
3. if s is a local frame on U, then $\|s\|$ doesn't vanish at any point of U.

One usually writes $\overline{L}$ for the pair $(L, \|\cdot\|)$ of a line bundle L and a metric on it.

Observe that the trivial line bundle $\mathscr{O}_X$ has a natural "trivial" metric, for which $\|1\| = 1$. In fact, a metric on the trivial line bundle $\mathscr{O}_X$ is equivalent to the datum of a continuous function h on X, such that $\|1\| = e^{-h}$.

1.1.2 The Abelian group of metrized line bundles. — Isomorphism of metrized line bundles are isomorphisms of line bundles which respect the metrics; they are called *isometries*. Constructions from tensor algebra extend naturally to the framework of metrized line bundles, compatibly with isometries. The tensor product of two metrized line bundles $\overline{L}$ and $\overline{M}$ has a natural metrization such that $\|s \otimes t\| = \|s\|\,\|t\|$, if s and t are local sections of L and M respectively. Similarly, the dual of a metrized line bundle has a metrization, and the obvious isomorphism $L \otimes L^\vee \simeq \mathscr{O}_X$ is an isometry. Consequently, isomorphism classes of metrized line bundles on X form an Abelian group $\overline{\operatorname{Pic}}(X)$. This group fits in an exact sequence

$$0 \to \mathscr{C}(X) \to \overline{\operatorname{Pic}}(X) \to \operatorname{Pic}(X),$$

where the first map associates to a real continuous function h on X the trivial line bundle endowed with the metric such that $\|1\| = e^{-h}$, and the second associates to a metrized line bundle the underlying line bundle. It is surjective when any line bundle has a metric (this certainly holds if X has partitions of unity).

Similarly, we can consider pull-backs of metrized line bundle. Let $\varphi \colon Y \to X$ be a morphism of locally ringed spaces such that $|\varphi^* f| = |f| \circ \varphi$ for any $f \in \mathscr{O}_X$. Let $\overline{L}$ be a metrized line bundle on X. Then, there is a canonical metric on $\varphi^* L$ such that $\|\varphi^* s\| = \|s\| \circ \varphi$ for any section $s \in \Gamma(U, L)$. This induces a morphism of Abelian groups $\varphi^* : \overline{\operatorname{Pic}}(X) \to \overline{\operatorname{Pic}}(Y)$.

1.2 The case of complex analytic spaces

1.2.1 Smooth metrics. — In complex analytic geometry, metrics are a very well established tool. Let us first consider the case of the projective space $\mathrm{X} = \mathbf{P}^n(\mathbf{C})$; a point $x \in \mathrm{X}$ is a $(n+1)$-tuple of homogeneous coordinates $[x_0 : \cdots : x_n]$, not all zero, and up to a scalar. Let $\pi \colon \mathbf{C}_*^{n+1} \to \mathrm{X}$ be the canonical projection map, where the index $*$ means that we remove the origin $(0, \ldots, 0)$. The fibers of π have a natural action of $\mathbf{C}^*$ and the line bundle $\mathscr{O}(1)$ has for sections s over an open set $\mathrm{U} \subset \mathbf{P}^n(\mathbf{C})$ the analytic functions F_s on the open set $\pi^{-1}(\mathrm{U}) \subset \mathbf{C}_*^{n+1}$ which are homogeneous of degree 1. The *Fubini-Study metric* of $\mathscr{O}(1)$ assigns to the section s the norm $\|s\|_{\mathrm{FS}}$ defined by

$$\|s\|_{\mathrm{FS}}([x_0 : \cdots : x_n]) = \frac{|F_s(x_0, \ldots, x_n)|}{\left(|x_0|^2 + \cdots + |x_n|^2\right)^{1/2}}.$$

It is more than continuous; indeed, if s is a local frame on an open set U, then $\|s\|$ is a $\mathscr{C}^\infty$-function on U; such metrics are called *smooth.*

1.2.2 Curvature. — Line bundles with smooth metrics on smooth complex analytic spaces allow to perform differential calculus. Namely, the *curvature* of a smooth metrized line bundle $\overline{L}$ is a differential form $c_1(\overline{L})$ of type $(1,1)$ on X. Its definition involves the differential operator

$$\mathrm{dd}^c = \frac{i}{\pi}\partial\overline{\partial}.$$

When an open set $\mathrm{U} \subset \mathrm{X}$ admits local coordinates $(z_1, \ldots, z_n)$, and $s \in \Gamma(\mathrm{U}, L)$ is a local frame, then

$$c_1(\overline{L})|_{\mathrm{U}} = \mathrm{dd}^c \log \|s\|^{-1} = \frac{i}{\pi} \sum_{1 \leq j,k \leq n} \frac{\partial^2}{\partial z_j \partial \overline{z}_k} \log \|s\|^{-1} \, \mathrm{d}z_j \wedge \mathrm{d}\overline{z}_k.$$

The Cauchy-Riemann equations ($\partial f / \partial \overline{z} = 0$ for any holomorphic function f of the variable z) imply that this formula does not depend on the choice of a local frame s. Consequently, these differential forms defined locally glue to a well-defined global differential form on X.

Taking the curvature form of a metrized line bundle is a linear operation: $c_1(\overline{L} \otimes \overline{M}) = c_1(\overline{L}) + c_1(\overline{M})$. It also commutes with pull-back: if $f \colon \mathrm{Y} \to \mathrm{X}$ is a morphism, then $f^* c_1(\overline{L}) = c_1(f^* \overline{L})$.

In the case of the Fubini-Study metric over the projective space $\mathbf{P}^n(\mathbf{C})$, the curvature is computed as follows. The open subset U_0 where the homogeneous coordinate x_0 is non-zero has local coordinates $z_1 = x_1/x_0, \ldots, z_n = x_n/x_0$; the homogeneous polynomial X_0 defines a non-vanishing section s_0 of $\mathscr{O}(1)$ on U_0 and

$$\log \|s_0\|_{\mathrm{FS}}^{-1} = \frac{1}{2}\log\left(1+\sum_{j=1}^{n}|z_j|^2\right).$$

Consequently, over U_0,

$$\begin{aligned}
c_1(\overline{\mathscr{O}(1)}_{\mathrm{FS}}) &= \frac{i}{\pi}\partial\bar\partial\log\|s_0\|_{\mathrm{FS}}^{-1}\\
&= \frac{i}{2\pi}\partial\left(\sum_{k=1}^{n}\frac{z_k}{1+\|z\|^2}\mathrm{d}\bar z_k\right)\\
&= \frac{i}{2\pi}\sum_{j=1}^{n}\frac{1}{1+\|z\|^2}\mathrm{d}z_j\wedge\mathrm{d}\bar z_j - \frac{i}{2\pi}\sum_{j,k=1}^{n}\frac{z_k\bar z_j}{(1+\|z\|^2)^2}\mathrm{d}z_j\wedge\mathrm{d}\bar z_k.
\end{aligned}$$

In this calculation, we have abbreviated $\|z\|^2=\sum_{j=1}^{n}|z_j|^2$.

1.2.3 Products, measures. — Taking the product of n factors equal to this differential form, we get a differential form of type (n,n) on the n-dimensional complex space X. Such a form can be integrated on X and the Wirtinger formula asserts that

$$\int_{\mathrm{X}} c_1(\overline{L})^n = \deg(L)$$

is the *degree* of L as computed by intersection theory. As an example, if $\mathrm{X}=\mathbf{P}^1(\mathbf{C})$, we have seen that

$$c_1(\overline{\mathscr{O}(1)}_{\mathrm{FS}}) = \frac{i}{2\pi(1+|z|^2)^2}\mathrm{d}z\wedge d\bar z,$$

where $z=x_1/x_0$ is the affine coordinate of $\mathrm{X}\setminus\{\infty\}$. Passing in polar coordinates $z=re^{i\theta}$, we get

$$c_1(\overline{\mathscr{O}(1)}_{\mathrm{FS}}) = \frac{1}{2\pi(1+r^2)^2}2r\mathrm{d}r\wedge\mathrm{d}\theta$$

whose integral over $\mathbf{C}$ equals

$$\int_{\mathbf{P}^1(\mathbf{C})} c_1(\overline{\mathscr{O}(1)}_{\mathrm{FS}}) = \int_0^\infty \frac{1}{2\pi(1+r^2)^2}2r\mathrm{d}r\int_0^{2\pi}\mathrm{d}\theta = \int_0^\infty\frac{1}{(1+u)^2}\mathrm{d}u = 1.$$

1.2.4 The Poincaré–Lelong equation. — An important formula is the Poincaré–Lelong equation. For any line bundle L with a smooth metric, and any section $s \in \Gamma(\mathrm{X}, L)$ which does not vanish identically on any connected component of X, it asserts the following equality of *currents*[2]:

$$\mathrm{dd}^c \log \|s\|^{-1} + \delta_{\mathrm{div}(s)} = c_1(\overline{L}),$$

where $\mathrm{dd}^c \log \|s\|^{-1}$ is the image of $\log \|s\|^{-1}$ under the differential operator dd^c, taken in the sense of distributions, and $\delta_{\mathrm{div}(s)}$ is the current of integration on the cycle $\mathrm{div}(s)$ of codimension 1.

1.2.5 Archimedean height pairing. — Metrized line bundles and their associated curvature forms are a basic tool in Arakelov geometry, invented by Arakelov in [2] and developped by Faltings [31], Deligne [25] for curves, and by Gillet-Soulé [34] in any dimension. For our concerns, they allow for a definition of height functions for algebraic cycles on algebraic varieties defined over number fields. As explained by Gubler [35, 36], they also permit to develop a theory of archimedean local heights.

For simplicity, let us assume that X is proper, smooth, and that all of its connected components have dimension n.

Let $\overline{L}_0, \dots, \overline{L}_n$ be metrized line bundles with smooth metrics. For $j \in \{0, \dots, n\}$, let s_j be a regular meromorphic section of L_j and let $\mathrm{div}(s_j)$ be its divisor. The given metric of L_j furnishes moreover a function $\log \|s_j\|^{-1}$ on X and a $(1,1)$-form $c_1(\overline{L}_j)$, related by the Poincaré–Lelong equation $\mathrm{dd}^c \log \|s_j\|^{-1} + \delta_{\mathrm{div}(s_j)} = c_1(\overline{L}_j)$. In the terminology of Arakelov geometry, $\log \|s_j\|^{-1}$ is a Green current (here, function) for the cycle $\mathrm{div}(s_j)$; we shall write $\widehat{\mathrm{div}}(s_j)$ for the pair $(\mathrm{div}(s_j), \log \|s_j\|^{-1})$.

Let $\mathrm{Z} \subset \mathrm{X}$ be a k-dimensional subvariety such that the divisors $\mathrm{div}(s_j)$, for $0 \leq j \leq k$, have no common point on Z. Then, one defines inductively the local height pairing by the formula:

$$\begin{aligned}(\widehat{\mathrm{div}}(s_0) \dots \widehat{\mathrm{div}}(s_k) | \mathrm{Z}) &= (\widehat{\mathrm{div}}(s_0) \dots \widehat{\mathrm{div}}(s_{k-1}) | \mathrm{div}(s_k|_\mathrm{Z})) \\ &\quad + \int_\mathrm{X} \log \|s_k\|^{-1} c_1(\overline{L}_0) \dots c_1(\overline{L}_{k-1}) \delta_\mathrm{Z}. \qquad (9.1)\end{aligned}$$

[2] The space of currents is the dual to the space of differential forms, with the associated grading; in the orientable case, currents can also be seen as differential forms with distribution coefficients.

The second hand of this formula requires two comments. 1) The divisor $\operatorname{div}(s_k|_{\mathrm{Z}})$ is a formal linear combination of $(k-1)$-dimensional subvarieties of X, and its local height pairing is computed by linearity from the local height pairings of its components. 2) The integral of the right hand side involves a function with singularities ($\log \|s_k\|^{-1}$) to be integrated against a distribution: in this case, this means restricting the differential form $c_1(\overline{L}_0)\dots c_1(\overline{L}_{k-1})$ to the smooth part of Z, multiplying by $\log \|s_k\|^{-1}$, and integrating the result. The basic theory of closed positive currents proves that the resulting integral converges absolutely; as in [34], one can also resort to Hironaka's resolution of singularities.

It is then a non-trivial result that the local height pairing is symmetric in the involved divisors; it is also multilinear. See [37] for more details, as well as [34] for the global case.

1.2.6 Positivity. — Consideration of the curvature allows to define positivity notions for metrized line bundles. Namely, one says that a smooth metrized line bundle $\overline{L}$ is *positive* (resp. *semi-positive*) if its curvature form is a positive (resp. a non-negative) $(1,1)$-form. This means that for any point $x \in \mathrm{X}$, the hermitian form $c_1(\overline{L})_x$ on the complex tangent space $\mathrm{T}_x\mathrm{X}$ is positive definite (resp. non-negative). As a crucial example, the line bundle $\mathscr{O}(1)$ with its Fubini-Study metric is positive. The pull-back of a positive metrized line bundle by an immersion is positive. In particular, ample line bundles can be endowed with a positive smooth metric; Kodaira's embedding theorem asserts the converse: if a line bundle possesses a positive smooth metric, then it is ample.

The pull-back of a semi-positive metrized line bundle by any morphism is still semi-positive. If $\overline{L}$ is semi-positive, then the measure $c_1(\overline{L})^n$ is a positive measure.

1.2.7 Semi-positive continuous metrics. — More generally, both the curvature and the Poincaré–Lelong equation make sense for metrized line bundles with arbitrary (continuous) metrics, except that $c_1(\overline{L})$ has to be considered as a current. The notion of semi-positivity can even be extended to this more general case, because it can be tested by duality: a current is positive if its evaluation on any nonnegative differential form is nonnegative. Alternatively, semi-positive (continuous) metrized line bundles are characterized by the fact that for any local frame s of $\overline{L}$ over an open set U, the continuous function $\log \|s\|^{-1}$ is *plurisubharmonic* on U. In turn, this means that for any morphism $\varphi\colon \overline{D} \to \mathrm{U}$, where

$\overline{D} = \overline{D}(0,1)$ is the closed unit disk in $\mathbf{C}$,

$$\log \|s\|^{-1} (\varphi(0)) \leq \frac{1}{2\pi} \int_0^{2\pi} \log \|s\|^{-1} (\varphi(e^{i\theta}))\mathrm{d}\theta.$$

Assume that $\overline{L}$ is semi-positive. Although products of currents are not defined in general (not more than products of distributions), the theory of Bedford–Taylor [10, 9] and Demailly [26, 27] defines a current $c_1(\overline{L})^n$ which then is a positive measure on X. There are two ways to define this current. The first one works locally and proceeds by induction: if $u = \log \|s\|^{-1}$, for a local non-vanishing section s of L, one defines a sequence (T_k) of closed positive currents by the formulae $T_0 = 1$, $T_1 = \mathrm{dd}^c u, \ldots, T_{k+1} = \mathrm{dd}^c(uT_k)$ and $c_1(\overline{L})^n = \mathrm{dd}^c(u)^n$ is defined to be T_n. What makes this construction work is the fact that at each step, uT_k is a well-defined current (product of a continuous function and of a positive current), and one has to prove that T_{k+1} is again a closed positive current. The other way, which shall be the one akin to a generalization in the ultrametric framework, consists in observing that if L is a line bundle with a continuous semi-positive metric $\|\cdot\|$, then there exists a sequence of smooth semi-positive metrics $\|\cdot\|_k$ on the line bundle L which converges uniformly to the initial metric: for any local section s, $\|s\|_k$ converges uniformly to $\|s\|$ on compact sets. The curvature current $c_1(\overline{L})$ is then the limit of the positive currents $c_1(\overline{L}_k)$, and the measure $c_1(\overline{L})^n$ is the limit of the measures $c_1(\overline{L}_k)^n$. (We refer to [47] for the global statement; to construct the currents, one can in fact work locally in which case a simple convolution argument establishes the claim.)

An important example of semi-positive metric which is continuous, but not smoth, is furnished by the Weil metric on the line bundle $\mathscr{O}(1)$ on $\mathbf{P}^n(\mathbf{C})$. This metric is defined as follows: if $\mathrm{U} \subset \mathbf{P}^n(\mathbf{C})$ is an open set, and s is a section of $\mathscr{O}(1)$ on U corresponding to an analytic function F_s on $\pi^{-1}(\mathrm{U}) \subset \mathbf{C}_*^{n+1}$ which is homogeneous of degree 1, then for any $(x_0, \ldots, x_n) \in \pi^{-1}(\mathrm{U})$, one has

$$\|s\|_{\mathrm{W}} = \frac{|F_s(x_0, \ldots, x_n)|}{\max(|x_0|, \ldots, |x_n|)}.$$

The associated measure $c_1(\overline{\mathscr{O}(1)_{\mathrm{W}}})^n$ on $\mathbf{P}^n(\mathbf{C})$ is as follows, cf. [62, 47]: the subset of all points $[x_0 : \cdots : x_n] \in \mathbf{P}^n(\mathbf{C})$ such that $|x_j| = |x_k|$ for all j, k is naturally identified with the polycircle $\mathbf{S}_1^n$ (map $[x_0 : \cdots : x_n]$ to $(x_1/x_0, \ldots, x_n/x_0)$); take the normalized Haar measure of this compact group and push it onto $\mathbf{P}^n(\mathbf{C})$.

1.2.8 Admissible metrics. — Let us say that a continuous metrized line bundle is *admissible* if it can be written as $\overline{L} \otimes \overline{M}^{\vee}$, where $\overline{L}$ and $\overline{M}$ are metrized line bundles whose metrics are continuous and semi-positive. Admissible metrized line bundles form a subgroup $\overline{\mathrm{Pic}}_{\mathrm{ad}}(\mathrm{X})$ of $\overline{\mathrm{Pic}}(\mathrm{X})$ which maps surjectively onto $\mathrm{Pic}(\mathrm{X})$ if X is projective.

The curvature current $c_1(\overline{L})$ of an admissible metrized line bundle $\overline{L}$ is a differential form of type $(1,1)$ whose coefficients are signed measures. Its nth product $c_1(\overline{L})^n$ is well-defined as a signed measure on X.

1.2.9 Local height pairing (admissible case). — The good analytic properties of semi-positive metrics allow to extend the definition of the local height pairing to the case of admissible line bundles. Indeed, when one approximates uniformly a semi-positive line bundle by a sequence of smooth semi-positive line bundles, one can prove that the corresponding sequence of local height pairings converges, the limit being independent on the chosen approximation.

The proof is inspired by Zhang's proof of the global case in [63] and goes by induction. Let us consider, for each j, two smooth semi-positive metrics on the line bundle L_j and assume that they differ by a factor e^{-h_j}. Then, the corresponding local height pairings differ from an expression of the form

$$\sum_{j=0}^{k} \int_{\mathrm{Z}} h_j c_1(\overline{L}_0) \dots \widehat{c_1(\overline{L}_j)} \dots c_1(\overline{L}_k),$$

where the written curvature forms are associated to the first metric for indices $< j$, and to the second for indices $> j$. This differential forms are positive by assumption, so that the integral is bounded in absolute value by

$$\sum_{j=0}^{k} \|h_j\|_{\infty} \int_{\mathrm{Z}} c_1(\overline{L}_0) \dots \widehat{c_1(\overline{L}_j)} \dots c_1(\overline{L}_k)$$
$$= \sum_{j=0}^{K} \|h_j\|_{\infty} (c_1(L_0) \dots \widehat{c_1(L_j)} \dots c_1(L_k)|\mathrm{Z}),$$

where the last expression is essentially a degree. (In these formulae, the factor with a hat is removed.) This inequality means that on the restriction to the space of smooth semi-positive metrics, with the topology of uniform convergence, the local height pairing is uniformly continuous.

Therefore, it first extends by continuity on the space of continuous semi-positive metrics, and then by multilinearity to the space of admissible metrics.

1.3 The case of non-archimedean analytic spaces

Let K be a complete ultrametric field. We are principally interested in finite extensions of $\mathbf{Q}_p$, but the case of local fields of positive characteristic (finite extensions of $k((T))$, for a finite field k) have proved being equally useful, as are non-local fields like the field $\mathbf{C}((T))$ of Laurent power series with *complex* coefficients. For simplicity, we will assume that K is the field of fractions of a complete discrete valuation ring K°, let π be a generator of the maximal ideal of K° and let $\tilde{K} = K^\circ/(\pi)$ be the residue field.

1.3.1 Continuous metrics. — Let X be a K-analytic space in the sense of Berkovich [12]. For simplicity, we will assume that X is the analytic space associated to a *proper* scheme over K. In that context, the general definition of continuous metrized line bundles given above makes sense.

Let us detail the example of the line bundle $\mathscr{O}(1)$ on the projective space P^n_K. A point $x \in \mathrm{P}^n_K$ possesses a complete residue field $\mathscr{H}(x)$ which is a complete extension of K and homogeneous coordinates $[x_0 : \cdots : x_n]$ in the field $\mathscr{H}(x)$. As in complex geometry, the projective space P^n_K is obtained by glueing $n+1$ copies $\mathrm{U}_0, \ldots, \mathrm{U}_n$ of the affine space A^n_K, where U_i corresponds to those points x such that $x_i \neq 0$. Recall also that A^n_K is the space of multiplicative semi-norms on the K-algebra $K[T_1, \ldots, T_n]$ which induce the given absolute value on K, together with the coarsest topology such that for any semi-norm $x \in \mathrm{A}^n_K$, the map $K[T_1, \ldots, T_n] \to \mathbf{R}$ defined by $f \mapsto x(f)$ is continuous. The kernel of a semi-norm x is a prime ideal $\mathfrak{p}_x$ of $K[T_1, \ldots, T_n]$ and x induces a norm on the quotient ring $K[T_1, \ldots, T_n]/\mathfrak{p}_x$, hence on its field of fractions $K(x)$. The completion of $K(x)$ with respect to this norm is denoted $\mathscr{H}(x)$ and is called the *complete residue field* of x. The images in $\mathscr{H}(x)$ of the intederminates T_i are denoted $T_i(x)$, more generally, the image in $\mathscr{H}(x)$ of any polynomial $f \in K[T_1, \ldots, T_n]$ is denoted $f(x)$; one has $x(f) = |f(x)|$.

Let f be a rational function on P^n_K, that is an element of $K(T_1, \ldots, T_n)$. It defines an actual function on the open set U of $\mathbf{P}^n_K$ where its denominator does not vanish; its value at a point $x \in U$ is an element of $\mathscr{H}(x)$.

More generally, Berkovich defines an analytic function on an open set U of P^n_K as a function f on U such that $f(x) \in \mathscr{H}(x)$ for any $x \in \mathrm{U}$, and such that any point $x \in \mathrm{U}$ possesses a neighbourhood $\mathrm{V} \subset \mathrm{U}$ such that $f|_{\mathrm{V}}$ is a uniform limit of rational functions without poles on V.

The line bundle $\mathscr{O}(1)$ can also be defined in a similar way to the classical case; by a similar GAGA theorem, its global sections are exactly the same as in algebraic geometry and are described by homogeneous polynomials of degree 1 with coefficients in K. If P is such a polynomial and s_P the corresponding section, then

$$\|s_P\|(x) = \frac{|P(x_0, \dots, x_n)|}{\max(|x_0|, \dots, |x_n|)}$$

where $[x_0 : \dots : x_n]$ is a system of homogeneous coordinates in $\mathscr{H}(x)$ for the point x. The function $\|s_P\|$ is continuous on P^n_K, by the very definition of the topology on P^n_K. Using the fact that $\mathscr{O}(1)$ is generated by its global sections, one deduces the existence of a continuous metric on $\mathscr{O}(1)$ satisfying the previous formula.

1.3.2 Smooth metrics. — Following [63], we now want to explain the analogues of smooth, and, later, of semi-positive metrics.

Smooth metrics come from algebraic geometry over K°, and, more generally, over the ring of integers of finite extensions of K. Let namely $\mathfrak{X}$ be a formal proper K°-scheme whose generic fibre in the sense of analytic geometry is X.[3] Let also $\mathfrak{L}$ be a line bundle on $\mathfrak{X}$ which is model of some power L^e, where $e \geq 1$. From this datum $(\mathfrak{X}, \mathfrak{L}, e)$, we can define a metric on L as follows. Let $\mathfrak{U}$ be a formal open subset of $\mathfrak{X}$ over which $\mathfrak{L}$ admits a local frame $\varepsilon_{\mathfrak{U}}$; over its generic fibre $\mathrm{U} = \mathfrak{U}_K$, for any section s of L, one can write canonically $s^e = f\varepsilon_{\mathfrak{U}}$, where $f \in \mathscr{O}_{\mathrm{X}}(\mathrm{U})$. We decree that $\|s\| = |f|^{1/e}$. In other words, the norm of a local frame on the formal model is assigned to be identically one. This makes sense because if $\eta_{\mathfrak{U}}$ is another local frame of $\mathfrak{L}$ on $\mathfrak{U}$, there exists an invertible formal function $f \in \mathscr{O}_{\mathfrak{X}}(\mathfrak{U})^*$ such that $\eta_{\mathfrak{U}} = f\varepsilon_{\mathfrak{U}}$ and the absolute value $|f|$ of the associated analytic function on U is identically equal to 1. Considering a finite cover of $\mathfrak{X}$ by formal open subsets, their generic fibers form a finite cover of X by *closed* subsets and this is enough to glue the local definitions to a continuous metric on L.

Metrics on L given by this construction, for some model $(\mathfrak{X}, \mathfrak{L}, e)$ of some power L^e of L will be said to be *smooth*.

[3] The reader might want to assume that X is the analytic space associated to a projective K-scheme X and that $\mathfrak{X}$ is a projective K°-scheme whose generic fibre equals X. This doesn't make too much a difference for our concerns.

1.3.3 Green functions; smooth functions. — Let $\overline{L}$ be a metrized line bundle and let s be a regular meromorphic section of L. Its divisor $\operatorname{div}(s)$ is a Cartier divisor in X. The function $\log \|s\|^{-1}$ is defined on the open set $\mathrm{X} \setminus |\operatorname{div}(s)|$; by analogy to the complex case, we call it a *Green function* for the divisor $\operatorname{div}(s)$. When the metric on $\overline{L}$ is smooth, the Green function is said to be smooth. The same remark applies for the other qualificatives semi-positive, or admissible, that wil be introduced later.

Let us take for L the trivial line bundle, with its canonical trivialization $s = 1$, and let us endow it with a smooth metric. By definition, we call $\log \|s\|^{-1}$ a smooth function. More generally, we define the space $\mathscr{C}^\infty(\mathrm{X})$ of (real valued) smooth functions to be the real vector space spanned by these elementary smooth functions. Observe that this definition reverses what happens in complex geometry where smooth metrics on the trivial line bundle are *defined* from the knowledge of smooth functions.

1.3.4 Example: projective space. — Let us consider the smooth metric on $\mathscr{O}(1)$ associated to the model $(\mathbf{P}^n_{K^\circ}, \mathscr{O}(1), 1)$ of $(\mathrm{P}^n_K, \mathscr{O}(1))$. Let $\mathfrak{U}_i$ be the formal open subset of $\mathbf{P}^n_{K^\circ}$ defined by the non-vanishing of the homogeneous coordinate x_i. Over, $\mathfrak{U}_i$, $\mathscr{O}(1)$ has a global non-vanishing section, namely the one associated to the homogeneous polynomial X_i. The generic fiber U_i of $\mathfrak{U}_i$ in the sense of algebraic geometry is an affine space, with coordinates $z_j = x_j/x_i$, for $0 \leq j \leq n$, and $j \neq i$. However, its generic fiber U_i in the sense of rigid geometry is the n-dimensional polydisk in this affine space defined by the inequalities $|z_j| \leq 1$. We thus observe that for any $x \in (\mathfrak{U}_i)_K$,

$$\begin{aligned} \|X_i\|(x) &= 1 \\ &= \frac{1}{\max(|z_0|, \dots, |z_{i-1}|, 1, |z_{i+1}|, \dots, |z_n|)} \\ &= \frac{|x_i|}{\max(|x_0|, \dots, |x_i|)} \\ &= \|X_i\|_{\mathrm{W}}(x). \end{aligned}$$

In other words, the Weil metric on $\mathscr{O}(1)$ is a smooth metric.

1.3.5 The Abelian group of smooth line bundles. — Let us show that *any line bundle has a smooth metric.* There is a general theory, due to Raynaud, that shows how to define formal models from rigid analytic objects. In the present case, X being projective, we may assume that L is ample and consider a closed embedding of X in a projective space P^n_K given by some power L^e. Let $\mathfrak{X}$ be the Zariski closure of X in $\mathrm{P}^n_{K^\circ}$; in concrete terms, if $I \subset K[X_0, \dots, X_n]$ is the homogeneous ideal of $i(X)$, $I \cap K^\circ[X_0, \dots, X_n]$ is the homogeneous ideal of $\mathfrak{X}$. Let then $\mathfrak{L}$ be the restriction to $\mathfrak{X}$ of the line bundle $\mathscr{O}(1)$. The triple $(\mathfrak{X}, \mathfrak{L}, e)$ is a model of L and induces a smooth metric on L.

Different models can give rise to the same metric. If $\varphi \colon \mathfrak{X}' \to \mathfrak{X}$ is a morphism of models, and $\mathfrak{L}' = \varphi^* \mathfrak{L}$, then $(\mathfrak{X}', \mathfrak{L}', e)$ defines the same smooth metric on L. Moreover, if two models $(\mathfrak{X}_i, \mathfrak{L}_i, e_i)$, for $i \in \{1, 2\}$, define the same metric, there exists a third model $(\mathfrak{X}, \mathfrak{L})$, with two morphisms $\varphi_i \colon \mathfrak{X} \to \mathfrak{X}_i$ such that the pull-backs $\varphi_i^* \mathfrak{L}_i^{e_1 e_2 / e_i}$ coincide with $\mathfrak{L}$. More precisely, if two models $\mathfrak{L}$ and $\mathfrak{L}'$ of some power L^e on a *normal* model $\mathfrak{X}$ define the same metric, then they are isomorphic. (See, *e.g.*, Lemma 2.2 of [20]; this may be false for non-normal models; it suffices that $\mathfrak{X}$ be integrally closed in its generic fiber.)

As a consequence, the set $\overline{\mathrm{Pic}}_{\mathrm{sm}}(\mathrm{X})$ of smooth metrized line bundles is a subgroup of the group $\overline{\mathrm{Pic}}(\mathrm{X})$. The group $\overline{\mathrm{Pic}}_{\mathrm{sm}}(\mathrm{X})$ fits within an exact sequence

$$0 \to \mathscr{C}^\infty(\mathrm{X}) \to \overline{\mathrm{Pic}}_{\mathrm{sm}}(\mathrm{X}) \to \mathrm{Pic}(\mathrm{X}) \to 0,$$

the last map is surjective because every line bundle admits a model. If $f \colon \mathrm{Y} \to \mathrm{X}$ is a morphism, then $f^*(\overline{\mathrm{Pic}}_{\mathrm{sm}}(\mathrm{X})) \subset \overline{\mathrm{Pic}}_{\mathrm{sm}}(\mathrm{Y})$.

1.3.6 Semi-positive metrics. — A smooth metric is said to be *ample* if it is defined by a model $(\mathfrak{X}, \mathfrak{L}, e)$ such that the restriction $\mathfrak{L}_{\tilde{K}}$ of $\mathfrak{L}$ to the closed fiber $\mathfrak{X}_{\tilde{K}}$ is ample. The Weil metric on the line bundle $\mathscr{O}(1)$ on the projective space is ample. The proof given above of the existence of smooth metrics shows, more precisely, that ample line bundles admit ample metrics, and that *the pull-back of a smooth ample metric by an immersion is a smooth ample metric.*

A smooth metric is said to be *semi-positive* if it can be defined on a model $(\mathfrak{X}, \mathfrak{L}, e)$ such that the restriction $\mathfrak{L}_{\tilde{K}}$ of $\mathfrak{L}$ to the closed fiber $\mathfrak{X}_{\tilde{K}}$ is numerically effective: for any projective curve $C \subset \mathfrak{X}_{\tilde{K}}$, the degree of the restriction to C of $\mathfrak{L}_{\tilde{K}}$ is non-negative. Ample metrics are semi-positive.

The pull-back of a smooth semi-positive metric by any morphism is semi-positive.

1.3.7 Continuous semi-positive metrics. — Let us say that a continuous metric on a line bundle L is semi-positive if it is the uniform limit of a sequence of smooth semi-positive metrics on the same line bundle L. As in the complex case, we then say that a metrized line bundle is *admissible* if it can be written as $\overline{L} \otimes \overline{M}^{\vee}$, for two line bundles L and M with continuous semi-positive metrics.

Let L be a metrized line bundle, and let $\|\cdot\|_1$ and $\|\cdot\|_2$ be two continuous metrics on L. It follows from the definition that the metrics $\|\cdot\|_{\min} = \min(\|\cdot\|_1, \|\cdot\|_2)$ and $\|\cdot\|_{\max} = \max(\|\cdot\|_1, \|\cdot\|_2)$ are continuous metrics.

Moreover, *these metrics $\|\cdot\|_{\min}$ and $\|\cdot\|_{\max}$ are smooth if the initial metrics are smooth.* Indeed, there exists a model $\mathfrak{X}$, as well as two line bundles $\mathfrak{L}_1$ and $\mathfrak{L}_2$ extending the same power L^e of L and defining the metrics $\|\cdot\|_1$ and $\|\cdot\|_2$ respectively. We may assume that $\mathfrak{L}_1$ and $\mathfrak{L}_2$ have regular global sections s_1 and s_2 on $\mathfrak{X}$ which coincide on X, with divisors $\mathfrak{D}_1$ and $\mathfrak{D}_2$ respectively. (The general case follows, by twisting $\mathfrak{L}_1$ and $\mathfrak{L}_2$ by a sufficiently ample line bundle on $\mathfrak{X}$.) The blow-up $\pi\colon \mathfrak{X}' \to \mathfrak{X}$ of the ideal $\mathfrak{I}_{\mathfrak{D}_1} + \mathfrak{I}_{\mathfrak{D}_2}$ carries an invertible ideal sheaf $\mathfrak{I}_{\mathfrak{E}} = \pi^*(\mathfrak{I}_{\mathfrak{D}_1} + \mathfrak{I}_{\mathfrak{D}_2})$, with corresponding Cartier divisor $\mathfrak{E}$. Since $\mathfrak{D}_1$ and $\mathfrak{D}_2$ coincide on the generic fiber, $\mathfrak{I}_{\mathfrak{D}_1} + \mathfrak{I}_{\mathfrak{D}_2}$ is already invertible there and π is an isomorphism on the generic fiber.

The divisors $\pi^*\mathfrak{D}_1$ and $\pi^*\mathfrak{D}_2$ decompose canonically as sums

$$\pi^*\mathfrak{D}_1 = \mathfrak{D}_1' + \mathfrak{E}, \quad \pi^*\mathfrak{D}_2 = \mathfrak{D}_2' + \mathfrak{E}.$$

Let us pose

$$\mathfrak{D}' = \mathfrak{D}_1' + \mathfrak{D}_2' + \mathfrak{E} = \mathfrak{D}_1' + \pi^*\mathfrak{D}_2 = \pi^*\mathfrak{D}_1 + \mathfrak{D}_2'.$$

An explicit computation on the blow-up shows that $(\mathfrak{X}', \mathfrak{D}', e)$ and $(\mathfrak{X}', \mathfrak{E}, e)$ are models of $\|\cdot\|_{\min}$ and $\|\cdot\|_{\max}$ respectively. In particular, these metrics are smooth.

Assume that the initial metrics are semi-positive, and that some positive power of L is effective. Then, the metric $\|\cdot\|_{\min}$ is semi-positive too. By approximation, it suffices to treat the case where the initial metrics are smooth and semi-positive. Then, the previous construction applies. Keeping the introduced notation, let us show that the restriction to the special fiber of the divisor $(D')_{\tilde{K}}$ is numerically effective. Let $C \subset \mathfrak{X}'_{\tilde{K}}$ be an integral curve and let us prove that $C\cdot(D')_{\tilde{K}}$ is nonnegative. If C is not contained in $\mathfrak{D}_1'$, then $C \cdot (\mathfrak{D}_1')_{\tilde{K}} \geq 0$, and $C \cdot (\pi^*\mathfrak{D}_2)_{\tilde{K}} = \pi_*C \cdot \mathfrak{D}_2 \geq 0$ since $(\mathfrak{D}_2)_{\tilde{K}}$ is numerically effective; consequently, $C \cdot (D')_{\tilde{K}} \geq 0$. Similarly, $C \cdot (D')_{\tilde{K}} \geq 0$ when C is not contained in $\mathfrak{D}_2'$. Since $\mathfrak{D}_1' \cap \mathfrak{D}_2' = \emptyset$, this shows that $C \cdot \mathfrak{D}'_{\tilde{K}} \geq 0$ in any case, hence $(\mathfrak{D}')_{\tilde{K}}$ is numerically effective.

This last result is the analogue in the ultrametric case to the fact that the maximum of two continuous plurisubharmonic functions is continuous plurisubharmonic. However, observe that in the complex case, the maximum or the minimum of smooth functions are not smooth in general.

1.3.8 Measures (smooth metrics). — In the non-archimedean case, there isn't yet a purely analytic incarnation of the curvature form (or current) $c_1(\overline{L})$ of a metrized line bundle $\overline{L}$, although the non-archimedean Arakelov geometry of [14] should certainly be pushed forward in that direction. However, as I discovered in [19], one can define an analogue of the measure $c_1(\overline{L})^n$ when the space X has dimension n.

The idea consists in observing the local height pairing (defined by arithmetic intersection theory) and defining the measures so that a formula analogous to the complex one holds.

Let us therefore consider smooth metrized line bundles $\overline{L}_j$ (for $0 \leq j \leq n$) as well as regular meromorphic sections s_j which have no common zero on X. There exists a proper model $\mathfrak{X}$ of X over K° and, for each j, a line bundle $\mathfrak{L}_j$ on $\mathfrak{X}$ which extends some power $L_j^{e_j}$ of L_j and which defines its metric.

Let $\mathrm{Z} \subset \mathrm{X}$ be an algebraic k-dimensional subvariety and let $\mathfrak{Z}$ be its Zariski closure in $\mathfrak{X}$; this is a $(k+1)$-dimensional subscheme of $\mathfrak{X}$. Let's replace it by its normalization or, more precisely, by its integral closure in its generic fiber. The local height pairing is then given by intersection theory, as

$$(\widehat{\mathrm{div}}(s_0)\dots\widehat{\mathrm{div}}(s_k)|\mathrm{Z}) = (c_1(\mathrm{div}(s_0|_{\mathfrak{Z}}))\dots c_1(\mathrm{div}(s_k|_{\mathfrak{Z}}))|\mathfrak{Z}) \log|\pi|^{-1},$$

where $\mathrm{div}(s_j|_{\mathfrak{Z}})$ means the divisor of s_j, viewed as a regular meromorphic section of $\mathfrak{L}_j$ over $\mathfrak{Z}$. The right hand side means taking the intersection of the indicated Cartier divisors on $\mathfrak{Z}$, which is a well-defined class of a 0-cycle supported by the special fiber of $\mathfrak{Z}$; then take its degree and multiply it by $\log|\pi|^{-1}$. (Recall that π is a fixed uniformizing element of K; is absolute value does not depend on the actual choice.)

When one views $s_k|_{\mathrm{Z}}$ as a regular meromorphic section of $\mathfrak{L}_k$ on $\mathfrak{Z}$ its divisor has two parts: the first one, say H, is "horizontal" and is the Zariski closure of the divisor $\mathrm{div}(s_k|_Z)$; the second one, say V, is vertical, *i.e.*, lies in the special fiber of $\mathfrak{Z}$ over the residue field of K°.

This decomposes the local height pairing as a sum

$$\begin{aligned}
&(\widehat{\operatorname{div}}(s_0)\dots\widehat{\operatorname{div}}(s_k)|\mathrm{Z})\\
&\qquad = (c_1(\operatorname{div}(s_0|_{\mathfrak{Z}}))\dots c_1(\operatorname{div}(s_k|_{\mathfrak{Z}}))|\mathfrak{Z})\log|\pi|^{-1}\\
&\qquad = (c_1(\operatorname{div}(s_0|_{\mathfrak{Z}}))\dots c_1(\operatorname{div}(s_{k-1}|_{\mathfrak{Z}}))|\operatorname{div}(s_k|_{\mathfrak{Z}}))\log|\pi|^{-1}\\
&\qquad = (c_1(\operatorname{div}(s_0|_{\mathfrak{Z}}))\dots c_1(\operatorname{div}(s_{k-1}|_{\mathfrak{Z}}))|H)\log|\pi|^{-1}\\
&\qquad\quad + (c_1(\operatorname{div}(s_0|_{\mathfrak{Z}}))\dots c_1(\operatorname{div}(s_{k-1}|_{\mathfrak{Z}}))|V)\log|\pi|^{-1}.
\end{aligned}$$

The first term is the local height pairing of $\operatorname{div}(s_k|_{\mathrm{Z}})$. Let us investigate the second one.

Let (V_i) be the family of irreducible components of this special fiber; for each i, let m_i be its multiplicity in the fiber. Then, the vertical component V of $\operatorname{div}(s_k|_{\mathfrak{Z}})$ decomposes as

$$V = \sum_i c_i m_i V_i,$$

where c_i is nothing but the order of vanishing of s_k along the special fiber at the generic point of V_i. Then,

$$\begin{aligned}
&(c_1(\operatorname{div}(s_0|_{\mathfrak{Z}}))\dots c_1(\operatorname{div}(s_{k-1}|_{\mathfrak{Z}}))|V)\\
&\qquad = \sum_i c_i m_i (c_1(\operatorname{div}(s_0|_{\mathfrak{Z}}))\dots c_1(\operatorname{div}(s_{k-1}|_{\mathfrak{Z}}))|V_i).
\end{aligned}$$

Since V_i lies within the special fiber of $\mathfrak{X}$,

$$(c_1(\operatorname{div}(s_0|_{\mathfrak{Z}}))\dots c_1(\operatorname{div}(s_{k-1}|_{\mathfrak{Z}}))|V_i) = (c_1(\mathfrak{L}_0)\dots c_1(\mathfrak{L}_{k-1})|V_i),$$

the multidegree of the vertical component V_i with respect to the restriction on the special fiber of the line bundles $\mathfrak{L}_0,\dots,\mathfrak{L}_{k-1}$.

One remarkable aspect of Berkovich's theory is the existence, for each i, of a unique point v_i in Z which specializes to the generic point of V_i. (Here, we use that $\mathfrak{Z}$ is integrally closed in its generic fibre.) Then,

$$\log\|s_k\|^{-1}(v_i) = c_i \log|\pi|^{-1}.$$

Finally,

$$\begin{aligned}
&(c_1(\operatorname{div}(s_0|_{\mathfrak{Z}}))\dots c_1(\operatorname{div}(s_{k-1}|_{\mathfrak{Z}}))|V)\log|\pi|^{-1}\\
&\qquad = \sum_i \log\|s_k\|^{-1}(v_i)(c_1(\mathfrak{L}_0)\dots c_1(\mathfrak{L}_{k-1})|V_i).
\end{aligned}$$

Let us sum up this calculation: we have introduced points $v_i \in \mathrm{Z}$ and decomposed the local height pairing as a sum:

$$(\widehat{\operatorname{div}}(s_0)\dots\widehat{\operatorname{div}}(s_k)|\mathrm{Z}) = (\widehat{\operatorname{div}}(s_0)\dots\widehat{\operatorname{div}}(s_{k-1})|\operatorname{div}(s_k|_{\mathrm{Z}})) \\ + \sum_i \log \|s_k\|^{-1}(v_i) m_i(c_1(\mathfrak{L}_0)\dots c_1(\mathfrak{L}_{k-1})|V_i).$$

It now remains to define

$$c_1(\overline{L}_0)\dots c_1(\overline{L}_{k-1})\delta_{\mathrm{Z}} = \sum_i m_i(c_1(\mathfrak{L}_0)\dots c_1(\mathfrak{L}_{k-1})|V_i)\delta_{v_i}, \quad (9.1)$$

where δ_{v_i} is the Dirac measure at the point $v_i \in \mathrm{Z}$. This is a *measure* on X, whose support is contained in Z, and whose total mass equals

$$\sum_i (c_1(\mathfrak{L}_0)\dots c_1(\mathfrak{L}_{k-1})|V_i) = (c_1(\mathfrak{L}_0)\dots c_1(\mathfrak{L}_{k-1})|V) \\ = (c_1(L_0)\dots c_1(L_{k-1})|\mathrm{Z}).$$

One can also check that it does not depend on the choice of the section s_k.

With this definition, the local height pairing obeys an induction formula totally analogous to the one satisfied in the complex case:

$$(\widehat{\operatorname{div}}(s_0)\dots\widehat{\operatorname{div}}(s_k)|\mathrm{Z}) = (\widehat{\operatorname{div}}(s_0)\dots\widehat{\operatorname{div}}(s_{k-1})|\operatorname{div}(s_k|_{\mathrm{Z}})) \\ + \int_{\mathrm{X}} \log \|s_k\|^{-1} c_1(\overline{L}_0)\dots c_1(\overline{L}_{k-1})\delta_{\mathrm{Z}}. \quad (9.2)$$

1.3.9 Local height pairing (admissible metrics). — With the notation of the previous paragraph, observe that the measures we have defined are positive when the smooth metrized line bundles are semi-positive. Indeed, this means that the line bundles $\mathfrak{L}_j$ are numerically effective hence, as a consequence of the criterion Nakai–Moishezon, any subvariety of the special fiber has a nonnegative multidegree.

With basically the same argment that the one we sketched in the complex case, we conclude that the local height pairing extends by continuity when semi-positive metrized line bundles are approximated by smooth semi-positive metrized line bundles. By linearity, this extends the local height pairing to admissible metrized line bundles.

1.3.10 Measures (admissible metrics). — Let us now return to semi-positive metrized line bundles $\overline{L}_0, \dots, \overline{L}_{k-1}$, approximated by smooth semi-positive metrized line bundles $\overline{L}_j^{(m)}$. I claim that for any

k-dimensional variety $\mathrm{Z} \subset \mathrm{X}$, the measures $c_1(\overline{L}_0^{(m)}) \dots c_1(\overline{L}_{k-1}^{(m)})\delta_{\mathrm{Z}}$ converge to a measure on X.

To prove the claim, we may assume that $\overline{L}_0, \dots, \overline{L}_{k-1}$ have sections $s_0, \dots, s_{k-1}$ without common zeroes on Z. Let also consider a smooth function φ on X; let $\overline{L}_k$ be the trivial line bundle with the section $s_k = 1$, metrized in such a way that $\|s_k\| = e^{-\varphi}$. Then, one has

$$\int_{\mathrm{X}} \varphi c_1(\overline{L}_0^{(m)}) \dots c_1(\overline{L}_{k-1}^{(m)})\delta_{\mathrm{Z}} = (\widehat{\operatorname{div}}(s_0)^{(m)} \dots \widehat{\operatorname{div}}(s_{k-1})^{(m)}\, \widehat{\operatorname{div}}(s_k)|\mathrm{Z});$$

writing $\overline{L}_k$ has the quotient of two ample metrized line bundles, we deduce from the existence of the local height pairing for admissible metrics that these integrals converge when $m \to \infty$. Consequently, the sequence of measures $(c_1(\overline{L}_0^{(m)}) \dots c_1(\overline{L}_{k-1}^{(m)})\delta_{\mathrm{Z}})_m$ converges to a positive linear form on the space of smooth functions. By a theorem of Gubler ([36], Theorem 7.12), which builds on the Stone-Weierstraß theorem and the compactness of the Berkovich space X, the space of smooth functions is dense in the space of continuous complex functions on X. A positivity argument, analogous to the proof that positive distributions are measures, then implies that our linear form is actually a positive measure which deserves the notation

$$c_1(\overline{L}_0) \dots c_1(\overline{L}_{k-1})\delta_{\mathrm{Z}}.$$

We then extend this definition by linearity to the case of arbitrary admissible line bundles. The total mass of this measure is again the multidegree of Z with respect to the line bundles L_j (for $0 \leq j \leq k-1$).

1.3.11 Integrating Green functions. — The definition of the convergence of a sequence of measures is convergence of all integrals against a given *continuous* compactly supported function. In applications, however, it can be desirable to integrate against more general functions. The inductive formula (9.1) for the local height pairing in the complex case, is such an example, as is the interpretation of Mahler measures of polynomials as (the archimedean component of) heights. However, its analogue (Equation 9.2) a priori holds only when $\log\|s_0\|^{-1}$ is continuous, that is when the section s_0 has no zeroes nor poles.

The fact that it still holds in the archimedean case is a theorem of Maillot [47] building on the theory of Bedford–Taylor. We proved in [20, Th. 4.1] that this relation holds in the ultrametric case too. The proof (valid both in the ultrametric and archimedean cases) works by induction, and ultimately relies on an approximation lemma according to which any semi-positive Green function g for a divisor D is an increasing limit of smooth functions (g_n) such that, for any n, $g - g_n$ is a semi-positive Green function for D. In fact, it suffices to pose $g_n = \min(g, n\log|\pi|^{-1})$; then, $g - g_n = \max(0, g - n\log|\pi|^{-1})$ is the maximum of two semi-positive Green functions, hence is semi-positive. (In the archimedean case, one needs to further regularize g_n; see [20] for details.)

The symmetry of the local height pairing then implies the following analogue of the Poincaré–Lelong formula. When $\overline{L}$ is the trivial line bundle, with the metric defined by an admissible function φ, the factor $c_1(\overline{L})$ will be written $\mathrm{dd}^c\varphi$, by analogy to the complex case.

Proposition 1.3.12 *Let φ be a smooth function on X and let $\overline{L}_1, \dots, \overline{L}_k$ be admissible metrized line bundles; let Z be a k-dimensional subvariety of X and let s be an invertible meromorphic sections of $\overline{L}_1$. Then,*

$$\int_{\mathrm{X}} \varphi c_1(\overline{L}_1)\dots c_1(\overline{L}_k)\delta_{\mathrm{Z}}$$
$$= \int_{\mathrm{X}} \varphi c_1(\overline{L}_2)\dots c_1(\overline{L}_k)\delta_{\mathrm{div}(s|\mathrm{Z})} + \int_{\mathrm{X}} \log\|s\|^{-1}\,\mathrm{dd}^c\varphi c_1(\overline{L}_2)\dots c_1(\overline{L}_k)\delta_{\mathrm{Z}}.$$

Proof Let $\overline{L}_0$ be the trivial line bundle with global section $s_0 = 1$ and metric defined by $\varphi = \log\|s_0\|^{-1}$. Let $s_1 = s$ and, for $2 \le j \le k$, let s_j be an invertible meromorphic section of L_j. Since $\mathrm{div}(s_0|_{\mathrm{Z}}) = 0$,

$$(\widehat{\mathrm{div}}(s_0)\dots\widehat{\mathrm{div}}(s_k)|\mathrm{Z}) = \int_{\mathrm{X}} \varphi c_1(\overline{L}_1)\dots c_1(\overline{L}_k)\delta_{\mathrm{Z}}$$

and

$$(\widehat{\mathrm{div}}(s_0)\,\widehat{\mathrm{div}}(s_2)\dots\widehat{\mathrm{div}}(s_k)|\,\mathrm{div}(s|_{\mathrm{Z}})) = \int_{\mathrm{X}} \varphi c_1(\overline{L}_2)\dots c_1(\overline{L}_k)\delta_{\mathrm{div}(s|\mathrm{Z})}.$$

One the other hand, the symmetry of the local height pairing implies that

$$\begin{aligned}(\widehat{\mathrm{div}}(s_0)\dots\widehat{\mathrm{div}}(s_k)|\mathrm{Z}) &= (\widehat{\mathrm{div}}(s_1)\,\widehat{\mathrm{div}}(s_0)\dots\widehat{\mathrm{div}}(s_k)|\mathrm{Z})\\ &= (\widehat{\mathrm{div}}(s_0)\,\widehat{\mathrm{div}}(s_2)\dots\widehat{\mathrm{div}}(s_k)|\,\mathrm{div}(s|_{\mathrm{Z}}))\\ &\quad + \int_{\mathrm{X}} \log\|s\|^{-1} c_1(\overline{L}_0)c_1(\overline{L}_2)\dots c_1(\overline{L}_k)\delta_{\mathrm{Z}}.\end{aligned}$$

Combining these equations, we obtain the claim. □

2 Examples

In this section, we give some examples of metrics and measures. Without mention of the contrary, we stick to the non-archimedean case; basic notation concerning K, K°, etc., is as in Section 1.3.

2.1 The projective space

Let X be the projective space P^n_K and let $\mathscr{O}(1)$ be the tautological line bundle on X, together with its Weil metric. Let us describe the associated measure, taking the opportunity to add details concerning Berkovich spaces.

As we remarked above, the Weil metric is induced by the tautological line bundle on the projective scheme $\mathfrak{X} = \mathbf{P}^n_{K^\circ}$ and is smooth. The special fiber of $\mathfrak{X}$ is the projective space $\mathbf{P}^n_{\tilde{K}}$ over the residue field of K°; it is in particular irreducible. Moreover, the degree of the tautological line bundle is equal to 1. The measure $c_1(\overline{\mathscr{O}(1)})^n$ is therefore equal to the Dirac mass at the unique point of X which reduces to the generic point of the special fiber. It remains to describe this point more precisely.

The scheme $\mathbf{P}^n_{K^\circ}$ is the union of $(n+1)$ affine open subsets $\mathfrak{U}_0, \dots, \mathfrak{U}_n$ defined by the non-vanishing of the homogeneous coordinates $x_0, \dots, x_n$. Their generic fibers in the sense of analytic geometry are $n+1$ affinoid subsets $\mathrm{U}_0, \dots, \mathrm{U}_n$, which cover P^n_K. In fact, U_i corresponds to the set of points $[x_0 : \dots : x_n]$ of P^n_K such that $|x_i| = \max(|x_0|, \dots, |x_n|)$.

To fix ideas, let us consider $i = 0$. Then, $\mathfrak{U}_0 = \operatorname{Spec}(K^\circ[T_1, \dots, T_n])$ is the affine space over K° with coordinates $T_j = x_j/x_0$. The natural topology on the algebra $K^\circ[T_1, \dots, T_n]$, and on its tensor product with K, $K[T_1, \dots, T_n]$, is induced by the *Gauß norm*

$$\|f\| = \max_{\mathbf{a} \in \mathbf{N}^n} |f_{\mathbf{a}}|, \qquad f = \sum f_{\mathbf{a}} T_1^{a_1} \dots T_n^{a_n}.$$

The completion of $K[T_1, \dots, T_n]$ for this norm is the Tate algebra and is denoted $K\langle T_1, \dots, T_n\rangle$. It consists of all power series $f = \sum f_{\mathbf{a}} T_1^{a_1} \dots T_n^{a_n}$ with coefficients in K such that $|f_{\mathbf{a}}| \to 0$ when $|\mathbf{a}| = a_1 + \dots + a_n \to \infty$; it is endowed with the natural extension of the Gauß norm, and is complete. By definition, the generic fiber U_0 of $\mathfrak{U}_0$ in the sense of analytic geometry is the Berkovich spectrum of the Tate algebra, that is the set of all multiplicative semi-norms on it which are continuous with respect to the topology defined by the Gauß norm. Since the theorem of Gauß asserts that this norm is multiplicative, it defines a point $\gamma \in \mathrm{U}_0$, which we like to call the Gauß point.

The reduction map $\mathrm{U}_0 \to \mathfrak{U}_0 \otimes \tilde{K}$ is defined as follows. Let $x \in \mathrm{U}_0$, let $\mathfrak{p}_x \subset K\langle T_1, \ldots, T_d\rangle$ be the kernel of the semi-norm x, which is also the kernel of the canonical morphism $\theta_x \colon K\langle T_1, \ldots, T_d\rangle \to \mathscr{H}(x)$. The images $T_j(x)$ of the indeterminates T_j are elements of absolute value ≤ 1 of the complete ultrametric field $\mathscr{H}(x)$; they belong to its valuation ring $\mathscr{H}(x)^\circ$. Letting $\widetilde{\mathscr{H}(x)}$ to be the residue field, there exists a unique morphism $\theta_{\overline{x}} \colon \tilde{K}[T_1, \ldots, T_d] \to \widetilde{\mathscr{H}(x)}$ such that $\theta_{\overline{x}}(T_i)$ is the image in $\widetilde{\mathscr{H}(x)}$ of $T_i(x)$. The kernel of this morphism is a prime ideal of the ring $\tilde{K}[T_1, \ldots, T_d]$ and defines a point $\overline{x}$ in the scheme $\mathfrak{U}_0 \otimes \tilde{K}$.

Let us now compute the reduction of the Gauß point γ. By definition, the field $\mathscr{H}(\gamma)$ is the completion of the Tate algebra $K\langle T_1, \ldots, T_d\rangle$ for the Gauß norm. I claim that morphism $\theta_{\overline{\gamma}}$ is injective, in other words, that the images of $T_1(\gamma), \ldots, T_d(\gamma)$ in the residue field $\widetilde{\mathscr{H}(\gamma)}$ are algebraically independent. Let $P \in K^\circ[T_1, \ldots, T_d]$ be any polynomial whose reduction $\overline{P}$ belongs to the kernel of $\theta_{\overline{\gamma}}$; this means $|P|_\gamma < 1$; in other words, the Gauß norm of P is < 1 and each coefficient of P has absolute value < 1. Consequently, $\overline{P} = 0$ and $\theta_{\overline{\gamma}}$ is injective, as claimed. This shows that $\overline{\gamma}$ is the generic point of the scheme $\mathfrak{U}_0 \otimes \tilde{K}$.

We thus have proved the following proposition.

Proposition 2.1.1 *The measure $c_1(\overline{\mathscr{O}(1)}_{\mathrm{W}})^n$ on P^n_K is the Dirac measure at the Gauß point γ.*

2.2 Semi-stable curves and reduction graphs

In this section, we assume that X is the analytic space associated to a projective curve over a field K which is complete for a discrete valuation. The semi-stable reduction theorem of Deligne–Mumford asserts that, up to replacing the base field K by a finite extension, the curve X has a projective model $\mathfrak{X}$ over K° which is regular (as a 2-dimensional scheme) and whose special fiber is reduced, with at most double points for singularities. We may also assume that the irreducible components are geometrically irreducible. We do not require, however, that $\mathfrak{X}$ is the minimal semi-stable model.

2.2.1 The reduction graph of the special fiber. — In that situation, the *reduction graph* $R(\mathfrak{X})$ is a metrized graph defined as follows. It has for vertices the irreducible components of the special fiber, with as many edges of length $\log|\pi|^{-1}$ between two vertices as the number of intersection points of the corresponding components. In a neighborhood of a double point, $\mathfrak{X}$ looks like (*i.e.*, has an étale map to) the scheme with equation $xy = \pi$ in the affine plane $\mathbf{A}^2_{K^\circ}$.

If one replaces the field K by a finite extension K', the base change $\mathfrak{X}\otimes_{K^\circ}(K')^\circ$ may no longer be regular. Indeed, $\mathfrak{X}\otimes_{K^\circ}(K')^\circ$ is étale locally isomorphic to $xy = (\pi')^e$, where π' is a uniformizing element of K', and e is the ramification index. When $e > 1$, the origin is a singular point of that scheme and one needs to blow it up repeatedly in order to obtain a regular scheme, which is a semi-stable model of $\mathrm{X}_{K'}$ over $(K')^\circ$. The two initial components are replaced by a chain of $e+1$ components, the $e-1$ intermediate ones being projective lines. In other words, $e-1$ vertices have been added, regularly spaced along each edge. One concludes that the reduction graph has not changed, as a topological space. Its metric has not changed either, since the e edges that partition an original edge (of length $\log|\pi|^{-1}$) have length $\log|\pi'|^{-1} = \frac{1}{e}\log|\pi|^{-1}$.

We say that a function on $R(\mathfrak{X})$ is piecewise linear if, up to passing to a finite extension (which replaces each edge by e edges of length equal to $1/e$th of the initial one), it is linear on each edge.

2.2.2 Drawing the reduction graph on the Berkovich space. — Let us analyse the situation from the Berkovich viewpoint. As we have seen, the generic points of the special fiber are the reductions of canonical points of X: the vertices of the graph $R(\mathfrak{X})$ naturally live in X. The same holds for the edges, but is a bit more subtle. As we have seen, blowing-up intersection points of components in the special fiber gives rise to new components, hence to new points of X. Would we enlarge the ground field and blow-up indefinitely, the constellation of points in X that we draw converges to a graph which is isomorphic to $R(\mathfrak{X})$.

According to Berkovich [13], a far more precise result holds. Let us consider a neighborhood $\mathfrak{U}$ of a singular point of the special fiber, pretending it is isomorphic to the locus defined by the equation $xy - \pi$ in $\mathbf{A}^2$; so $\mathfrak{U} = \operatorname{Spec}(K^\circ[x,y]/(xy-\pi))$. Its generic fibre is the affinoid space U defined by the equality $|xy| = |\pi|$ in the unit polydisk $\mathrm{B}^2 = \mathscr{M}(K\langle x,y\rangle)$. The affinoid algebra of U is the quotient

$$K\langle x,y\rangle/(xy-\pi)$$

whose elements f are (non-uniquely) represented by a series

$$\sum_{m,n=0}^{\infty} a_{m,n}x^m y^n,$$

with $a_{m,n} \to 0$ when $m+n \to \infty$. However, observing that x is invertible in this algebra, with inverse $\pi^{-1}y$, so that $y = \pi x^{-1}$, we can replace each product $x^m y^n$ by $\pi^n x^{m-n}$, leading to an expression of the form

$$f = \sum_{n \in \mathbf{Z}} a_n x^n,$$

where $|a_n| \to 0$ when $n \to +\infty$ and $|a_n| \, \pi^{-n} \to 0$ when $n \to -\infty$. Such an expression is now unique, and is called the Laurent expansion of f.

It leads to a natural family $(\gamma_r)_{r \in [0, \log|\pi|^{-1}]}$ of multiplicative semi-norms on the algebra $\mathscr{O}(\mathrm{U})$, parametrized by the unit interval in $\mathbf{R}$. Namely, for each real number $r \in [0, \log |\pi|^{-1}]$, we can set

$$\gamma_r(f) = \max_{n \in \mathbf{Z}} |a_n| \, e^{-rn}, \qquad f = \sum_{n \in \mathbf{Z}} a_n x^n \in \mathscr{O}(\mathrm{U}).$$

Obviously, γ_r is a norm on $\mathscr{O}(\mathrm{U})$ which extends the absolute value of K; its multiplicativity is proved analogously to that of the Gauß norm. It is easy to check that the map $[0, \log |\pi|^{-1}] \to \mathrm{U}$ defined by $r \mapsto \gamma_r$ is continuous (this amounts to the fact that the maps $r \mapsto \gamma_r(f)$ are continuous), hence defines an parametrized path in the topological space U.

Let $S(\mathfrak{U})$ be its image (with the induced distance); Berkovich calls it the skeleton of the formal scheme obtained by completing $\mathfrak{U}$ along its special fibre. A point u in U has two coordinates $(x(u), y(u))$ in the completed residue field $\mathscr{H}(u)$ which are elements of absolute value ≤ 1 satisfying $x(u)y(u) = \pi$. In particular,

$$r(u) = \log |x(u)|^{-1} \in [0, \log |\pi|^{-1}].$$

The map $\rho \colon u \mapsto \gamma_{r(u)}$ is a continuous function from U to $S(\mathfrak{U})$.

Let us compute the image of γ_r by this map. By definition of γ_r, one has

$$|x(\gamma_r)| = \gamma_r(x) = e^{-r},$$

hence $r(\gamma_r) = r$ and $\rho(\gamma_r) = \gamma_r$. In other words, the map ρ is a retraction of U onto the skeleton $S(\mathfrak{U})$.

The special fiber of $\mathfrak{U}$ is defined by the equation $xy = 0$ in $\mathbf{A}^2_{\widetilde{K}}$, hence has two components. One can check that the point γ_0 reduces to the generic point of the component with equation $y = 0$, while $\gamma_{\log|\pi|^{-1}}$ reduces to the generic point of the component with equation $x = 0$.

These constructions have to be done around each singular point of the special fiber of $\mathfrak{X}$, locally for the étale topology of $\mathfrak{X}$. Berkovich proves that they can be glued, so that the graph $R(\mathfrak{X})$ is again canonically interpreted as an actual metrized graph drawn on the analytic space X; we write $\iota\colon R(\mathfrak{X}) \hookrightarrow \mathrm{X}$ for the canonical embedding. The map ι admits a continuous retraction $\rho\colon \mathrm{X} \to R(\mathfrak{X})$.

Although we will not use this fact, we must mention that the retraction ρ is a deformation retraction.

2.2.3 Metrized line bundles and the reduction graph. — A construction of S. Zhang [61], building on prior results of Chinburg–Rumely [22], furnishes continuous metrics on divisors from continuous functions on the reduction graph $R(\mathfrak{X})$. It works as follows. First of all, if $P \in \mathrm{X}(K)$ is a rational point, there is a unique morphism $\varepsilon_P\colon \operatorname{Spec} K^\circ \to \mathfrak{X}$ which extends the point P viewed as a morphism from $\operatorname{Spec} K$ to X. The image of this section is a divisor D_P on $\mathfrak{X}$ and the line bundle $\mathscr{O}(D_P)$ on $\mathfrak{X}$ defines a smooth metric on $\mathscr{O}(P)$; we write $\overline{\mathscr{O}(P)}_{\mathfrak{X}}$ for the corresponding metrized line bundle. We also define μ_P as the Dirac measure at the vertex of the graph corresponding to the (unique) irreducible component of the special fiber by which D_P passes through. The construction and the notation is extended by additivity for divisors which are sums of rational points. More generally, if P is only a closed point of X, we do this construction after the finite extension $K(P)/K$, so that P becomes a sum of rational points, using for model the minimal resolution of $\mathfrak{X} \otimes K(P)^\circ$ described earlier.

If f is any continuous function on $R(\mathfrak{X})$ and D a divisor on X, the metrized line bundle $\mathscr{O}(D+f)_{\mathfrak{X}}$ is deduced from $\overline{\mathscr{O}(D)}_{\mathfrak{X}}$ by multiplying the metric by e^{-f}. When f is piecewise linear, this metrized line bundle is smooth. To prove that, we may extend the scalars and assume that D is a sum of rational points $\sum n_j P_j$ and that f is linear on each edge corresponding to an intersection point of components of the special fiber. Letting (V_i) being the family of these components, and writing v_i for the vertex of $R(\mathfrak{X})$ corresponding to V_i, the divisor

$$\sum_j n_j D_{P_j} + \sum_i f(v_i) V_i \tag{9.1}$$

defines the metrized line bundle $\mathscr{O}(D+f)_{\mathfrak{X}}$.

In this context, Zhang has defined a curvature operator, which associates to a metrized line bundle a distribution on the graph $R(\mathfrak{X})$, defined in such a way that

- for any divisor D on X, $\operatorname{curv}(\overline{\mathscr{O}(D)_{\mathfrak{X}}}) = \mu_D$;
- for any continuous function f, $\operatorname{curv}(\overline{O}(f)_{\mathfrak{X}}) = -\Delta f$, where Δ is the Laplacian operator of the graph $R(\mathfrak{X})$,

and depending linearly on the metrized line bundle. The following lemma compares this construction with the general one on Berkovich spaces.

Lemma 2.2.4 *Let $\overline{L} = \overline{\mathscr{O}(D+f)}_{\mathfrak{X}}$ be a metrized line bundle on X associated to a divisor D on X and a continuous function f on the graph $R(\mathfrak{X})$. If it is semi-positive, resp. admissible in the sense of [61], then it is semi-positive, resp. admissible in the sense of this article, and one has*

$$c_1(\overline{L}) = \iota_* \operatorname{curv}(\overline{\mathscr{O}(D+f)}) \log |\pi|^{-1}.$$

In other words, the measure $c_1(\overline{L})$ is supported by the graph $R(\mathfrak{X})$ where it coincides essentially with Zhang's curvature.

Proof We first assume that f is linear on each edge of $R(\mathscr{X})$ and that D is a sum of rational points of X. Then, $\overline{L}$ corresponds to the line bundle $\mathfrak{L}$ on the model $\mathfrak{X}$ given by Equation 9.1. By definition, the measure $c_1(\mathfrak{L})$ is computed as follows. It is a sum, for all components V_i of the special fiber, of $\deg(\mathfrak{L}|V_i) \log |\pi|^{-1}$ times the Dirac measure at the corresponding point v_i of $R(\mathfrak{X})$. In particular, it is supported by $R(\mathfrak{X})$. Then,

$$\deg(\mathfrak{L}|V_i) = \sum_j n_j \begin{Bmatrix} 1 & \text{if } D_{P_j} \text{ passes through } V_i; \\ 0 & \text{otherwise} \end{Bmatrix} + \sum_j f(V_j)(V_i, V_j),$$

where (V_i, V_j) is the intersection number of the divisors V_i and V_j. That D_{P_j} passes through V_i means exactly that $\rho(P_j) = v_i$. Moreover, if $j \neq i$, then $(V_i, V_j) = m_{i,j}$ is just the number of intersection points of V_i and V_j, while

$$(V_i, V_i) = (V_i, \sum_j V_j) - \sum_{j \neq i} (V_i, V_j) = -\sum_{j \neq i} (V_i, V_j),$$

since the whole special fiber is numerically equivalent to zero. Consequently,

$$\sum_j f(v_j)(V_i, V_j) = \sum_{j \neq i} m_{i,j} (f(V_j) - f(V_i)).$$

Observe that this is the sum, over all edges from V_i, of the derivative of f along this edge. Comparing with the definitions given by Zhang in [61], one finds, for any function g on $R(\mathfrak{X})$

$$\begin{aligned}\sum_i \deg(\mathfrak{L}|V_i) g(v_i) &= \sum_j n_j g(\rho(P_j)) + \sum_i \langle \delta f(v_i), g\rangle \\ &= \int_{R(\mathfrak{X})} g\,(\mu_D + \delta f) \\ &= \int_{R(\mathfrak{X})} g \operatorname{curv}(\overline{\mathscr{O}(D+f)}).\end{aligned}$$

This proves the claimed formula when f is linear on each edge of $\mathfrak{X}$ and D is a sum of rational points.

By working over an appropriate finite extension of K, it extends to the case where f is only piecewise linear, D being any divisor on X.

Zhang defines $\overline{\mathscr{O}(D+f)}_{\mathfrak{X}}$ to be semi-positive if f is a uniform limit of piecewise linear functions f_n such that $\operatorname{curv}(\overline{\mathscr{O}(D+f_n)_{\mathfrak{X}}}) \geq 0$. The metrized line bundle $\overline{L}$ is then the limit of the metrized line bundles $\overline{L}_n$ corresponding to models $\mathfrak{L}_n$ (on appropriate models $\mathfrak{X}_n$ of X after some extension of scalars) of $\mathscr{O}(D)$. By the previous computation, these metrics are smooth and $c_1(\overline{L}_n) \geq 0$. Reversing the computation, this means that $\mathfrak{L}_n$ is numerically effective on $\mathfrak{X}_n$, hence $\overline{L}$ is semi-positive. By definition of the measure $c_1(\overline{L})$, one has

$$\begin{aligned}c_1(\overline{L}) &= \lim_n c_1(\overline{L}_n) = \lim_n \iota_* \operatorname{curv}(\overline{\mathscr{O}(D+f_n)}) \\ &= \iota_* \lim_n \operatorname{curv}(\overline{\mathscr{O}(D+f_n)}) = \iota_* \operatorname{curv}(\overline{\mathscr{O}(D+f)}).\end{aligned}$$

The case of an admissible metrized line bundle follows by linearity. □

2.3 Local character of the measures

The definition of the measures associated to metrized line bundles is global in nature. Still, the main result of this section implies that they are local.

Definition 2.3.1 Let X be an analytic space. A function on X is said to be *strongly pluriharmonic* if it is locally a uniform limit of functions of the form $a \log |u|$, where $a \in \mathbf{R}$ and u is holomorphic and nonvanishing.

There is a general theory of harmonic functions on curves due to Thuillier [55] (see also [33, 8] on the projective line; note that the definition of a strongly harmonic function of the latter reference is different from the one adopted here). Strongly pluriharmonic functions are harmonic in their sense. Indeed, logarithms of absolute values of invertible holomorphic functions are harmonic, and harmonic functions are preserved by uniform limits (Prop. 2.3.20 and 3.1.2 of [55]). In fact, when the residue field of K is algebraic over a finite field, any harmonic function is locally of the form $a \log |u|$, where $a \in \mathbf{R}$ and u is an invertible holomorphic function (*loc.cit.*, Theorem 2.3.21).

This is not necessarily the case for more general fields K: there are harmonic functions over analytic curves which are not locally equal to the logarithm of the absolute value of an invertible function; examples require to consider curves of genus ≥ 1. In a conversation with A. Ducros, we devised the following example of a one-dimensional affinoid space. Let $\mathfrak{E}$ be an elliptic scheme over K°, let o be the origin in $\mathfrak{E}_{\tilde{K}}$ and let p be a *non-torsion* rational point in $\mathfrak{E}_{\tilde{K}}$; let $\mathfrak{X}$ be the blow-up of $\mathfrak{E}$ at the point p. Let then $\mathfrak{U}$ be its open subset obtained by removing the point o as well as a smooth point in the exceptional divisor of the blow-up; its generic fiber U is the desired affinoid space — it is the complementary subset in the elliptic curve E_K to two small disjoint disks. One can prove that the space of harmonic functions on U is 2-dimensional, and that all holomorphic invertible functions on U have constant absolute value.

Definition 2.3.2 Let $\overline{L}$ be a metrized line bundle on an analytic space X and let U be an open subset of X. One says that $\overline{L}$ is strongly pluriharmonic on U if for any local frame s of L defined on an open subset $\mathrm{V} \subset \mathrm{U}$, $\log \|s\|^{-1}$ is strongly pluriharmonic on V.

A metrized line bundle is strongly pluriharmonic on U if it admits, in a neighbourhood of any point of U, a local frame whose norm is identically equal to 1. (For the converse to hold, one would need to introduce a notion of local frame for real line bundles.)

Proposition 2.3.3 *Let* X *a the analytic space associated to a proper K-scheme. Let $\overline{L}_1, \overline{L}_2, \dots, \overline{L}_k$ be admissible metrized line bundles on* X. *Let* Z *be a k-dimensional Zariski closed subset of* X. *Assume that $\overline{L}_1$ is strongly pluriharmonic on* U. *Then, the support of the measure $c_1(\overline{L}_1) \dots c_1(\overline{L}_k)\delta_{\mathrm{Z}}$ is disjoint from* U.

Proof One has to show that for any continuous function φ with compact support contained in U

$$\int_{\mathrm{X}} \varphi\, c_1(\overline{L}_1) \dots c_1(\overline{L}_k)\delta_{\mathrm{Z}} = 0.$$

By Gubler's theorem, the space of smooth functions is dense in the space of continuous functions on X. Using the fact that the maximum and the minimum of smooth functions are still smooth, one proves that the space of smooth functions with compact support contained in U is dense in the space of continuous functions with compact support contained in U, for the topology of uniform convergence. We thus may assume that φ is smooth, with compact support contained in U. Finally, we may also assume that the metric on the line bundles $\overline{L}_2, \dots, \overline{L}_k$ are smooth.

We may argue locally and assume that L_1 has a meromorphic section s whose divisor $\operatorname{div}(s)$ is disjoint from U. Up to shrinking U again, we may assume that there exists a sequence (u_n) of rational functions without zeroes nor poles on U such that $\log \|s\| = \lim \log |u_n|^{1/n}$.

According to Prop. 1.3.12, one has

$$\int_{\mathrm{X}} \varphi\, c_1(\overline{L}_1) \dots c_1(\overline{L}_k) \delta_{\mathrm{Z}}$$
$$= \int_{\mathrm{X}} \varphi c_1(\overline{L}_2) \dots c_1(\overline{L}_k) \delta_{\operatorname{div}(s|_{\mathrm{Z}})} + \int_{\mathrm{X}} \log \|s\|^{-1} \mathrm{dd}^c \varphi c_1(\overline{L}_2) \dots c_1(\overline{L}_k) \delta_{\mathrm{Z}}.$$

The first term vanishes because $\operatorname{div}(s|_{\mathrm{Z}})$ and the support of φ are disjoint. The second is the limit of

$$\int_{\mathrm{X}} \log |u_n|^{-1/n} \mathrm{dd}^c \varphi c_1(\overline{L})^{k-1} \delta_{\mathrm{Z}}.$$

Using the fact that $\operatorname{div}(u_n) \cap \mathrm{U}$ is empty and applying the same computation, the term of index n equals

$$\frac{1}{n} \int_{\mathrm{X}} \varphi\, c_1(\overline{M}_n) c_1(\overline{L}_2) \dots c_1(\overline{L}_k) \delta_{\mathrm{Z}}.$$

where $\overline{M}_n$ is the trivial metrized line bundle $\mathscr{O}_{\mathrm{X}}$, and its meromorphic section u_n replacing s. But this integral is zero, by the formula (9.1) which defines the measure associated to smooth metrized line bundles. □

2.4 Polarized dynamical systems

We now explain another example of metrized line bundles: the canonical metric associated to a dynamical system.

Lemma 2.4.1 *Let* X *be the analytic space associated to a proper K-scheme and let $f \colon \mathrm{X} \to \mathrm{X}$ be a finite morphism. Let L be a line bundle on* X*, d an integer such that $d \geq 2$ and an isomorphism $\varepsilon \colon f^*L \simeq L^d$.*

The line bundle L possesses a unique continuous metric such that the isomorphism ε is an isometry. If L is ample, then this metric is semi-positive.

In essence, this result, or at least its proof, goes back to Tate's construction of the "Néron–Tate" canonical height for abelian varieties. In the slightly different language of local heights and Néron functions, it has been proved by Call–Silverman [17]. In the asserted form, it is due to Zhang [63].

Proof Let us first prove uniqueness. If $\overline{L}$ and $\overline{L}'$ are two metrics on L, let φ be the continuous function such that $\|\cdot\|' = e^{-\varphi}\|\cdot\|$. Assuming that ε is an isometry for these two metrics, one obtains the following equation

$$\varphi(f(x)) = d\varphi(x),$$

for any $x \in \mathrm{X}$. Since X is compact, φ is bounded and this equation implies that $\|\varphi\|_\infty \leq \frac{1}{d}\|\varphi\|_\infty$. Since $d \geq 2$, one concludes that $\varphi \equiv 0$.

For the existence, one begins with any continuous metric $\overline{L}_0$ on L. Let us then consider the sequence of metrics $(\overline{L}_n)$ on L induced by the pull-backs on $L^d = \varepsilon f^*L$, $L^{d^2} = (\varepsilon f^*)^2 L$, etc., hence on L. Since $d \geq 2$, a similar contraction argument as the one used for uniqueness shows that this is a Cauchy sequence of metrics on L; consequently, it converges to a continuous metric on L. If $\overline{L}_0$ is chosen to be semi-positive, which we may if L is ample, then all of the metrized line bundles $\overline{L}_n$ are semi-positive, hence the canonical metric is semi-positive.

Concretely, in the non-archimedean case, one begins with a model $(\mathfrak{X}_0, \mathfrak{L}_0, e)$ such that $\mathfrak{L}_0$ is numerically effective. Then one considers the map $f\colon X \to \mathfrak{X}_0$ and the normalization $\mathfrak{X}_1$ of $\mathfrak{X}_0$ in X; this is a projective model $\mathfrak{X}_1$, equiped with a finite morphism $f_1\colon \mathfrak{X}_1 \to \mathfrak{X}_0$ extending f. Moreover, $\mathfrak{L}_1 = f_1^*\mathfrak{L}_0$ is a model of f^*L^e which is identified with L^{ed} via the fixed isomorphism ε. Iterating this construction defines a sequence $(\mathfrak{X}_n, \mathfrak{L}_n, ed^n)$ of models of (X, L), with finite morphisms $f_n\colon \mathfrak{X}_n \to \mathfrak{X}_{n-1}$ such that $f_n^*\mathfrak{L}_{n-1} = \mathfrak{L}_n$. The metric on L defined by any of these models is semi-positive, hence so is their uniform limit. □

2.4.2 The canonical measure. — The measure $c_1(\overline{L})^n$ on X defined by the metrized line bundle $\overline{L}$ is a very important invariant of the dynamical system. It satisfies the functional equations

$$f^*c_1(\overline{L})^n = d^n c_1(\overline{L})^n \quad \text{and} \quad f_*c_1(\overline{L})^n = c_1(\overline{L})^n.$$

The first follows by a general functorial property proved in [19]; it implies the second. The support of the canonical measure is therefore totally invariant under f.

2.4.3 The Fatou set. — Generalizing results of Kawaguchi–Silverman in [44] and Baker–Rumely [8], we want to show here that the canonical measure vanishes on any open set U of X where the sequence $(f^n|_{\mathrm{U}})$ of iterates of f is equicontinuous.

Let U be an open set in X and $\mathscr{F}$ be a family of continuous maps from U to X. One says that this family is equicontinuous if for any $x \in \mathrm{U}$ and any finite open covering (V_j) of X, there exists a neighbourhood U_x of x in U such that for any $\varphi \in \mathscr{F}$, there exists an index j such that $\varphi(\mathrm{U}_x) \subset \mathrm{V}_j$. (This definition is adapted from Definition 10.63 in [8]; it is the definition of equicontinuity associated to the canonical uniform structure of the *compact* space X.)

We define the equicontinuous locus of f as the largest open subset E_f of X over which the sequence of iterates of f is equicontinuous.

Proposition 2.4.4 *If L is ample, then the metric $\overline{L}$ is strongly pluri-harmonic on E_f.*[4]

Proof The proof is inspired from the above-mentioned sources, which in turns is an adaptation of the complex case [43] (see also [56]).

We may replace L by a positive power of itself and assume that it is very ample, induced by a closed embedding of X in P^n, and that the natural map $\Gamma(\mathbf{P}^n, \mathscr{O}(d)) \to \Gamma(\mathrm{X}, \mathscr{O}(d))$ is surjective. Then, there are homogeneous polynomials $(F_0, \ldots, F_n)$, of degree d, with coefficients in K, and without common zeroes on X, such that $f([x_0 : \cdots : x_n]) = [F_0(x) : \cdots : F_n(x)]$ for any $x = [x_0 : \cdots : x_n] \in \mathrm{P}^n$. One considers the polynomial map $F\colon \mathrm{A}^{n+1} \to \mathrm{A}^{n+1}$; it lifts a rational map on P^n which extends the morphism f.

For $(x_0, \ldots, x_n) \in \mathrm{A}^{n+1}$, define $\|x\| = \max(|x_0|, \ldots, |x_n|)$. The Weil metric on $\mathscr{O}(1)$ is given by

$$\log \|s_P(x)\|^{-1} = \log |P(x)|^{-1} + \deg(P) \log \|x\|,$$

where P is an homogeneous polynomial, s_P the corresponding global section of $\mathscr{O}(\deg(P))$, and x is a point of A^{n+1} such that $P(x) \neq 0$. The restriction to X of this metric is a semi-positive metric $\|\cdot\|_0$ on L. The construction of the canonical metric on L introduces a sequence of

[4] The ampleness assumption should not be necessary for the result to hold.

semi-positive metrics $\|\cdot\|_n$ on L; these metrics are given by the following explicit formula

$$\log \|s_P(x)\|_k^{-1} = \log |P(x)|^{-1} + \deg(P) d^{-k} \log \left\| F^{(k)}(x) \right\|,$$

where $F^{(k)} : \mathrm{A}^{n+1} \to \mathrm{A}^{n+1}$ is the kth iterate of F.

The convergence of this sequence is therefore equivalent to the convergence of the sequence $(d^{-k} \log \|F^{(k)}\|)_k$ towards a continuous fonction on the preimage of X under the projection map $\mathrm{A}^{n+1} \setminus \{0\} \to \mathrm{P}^n$. The limit is usually called the homogeneous Green function.

For $0 \leq i \leq n$, let V_i be the open set of points $x = [x_0 : \cdots : x_n] \in \mathrm{P}^n$ such that $|x_i| > \frac{1}{2} \|x\|$. They form an open covering of P^n; their intersections with X form an open covering of X.

Fix $x \in \mathrm{E}_f$ and let U be an open neighbourhood of x such that for any positive integer k, there exists $i \in \{0, \dots, n\}$ such that $f^k(\mathrm{U}) \subset \mathrm{V}_i$. For any i, let N_i be the set of integers k such that $f^k(\mathrm{U}) \subset \mathrm{V}_i$. Let us consider any index i such that N_i is infinite; to fix ideas, let us assume that $i = 0$. The canonical norm of a section s_P at a point $y \in \mathrm{U}$ is given by

$$\begin{aligned}
\log \|s_P(y)\|^{-1} &= \log |P(y)|^{-1} + \deg(P) \lim_{\substack{k \to \infty \\ k \in N_0}} d^{-k} \log \left\| F^{(k)}(y) \right\| \\
&= \log |P(y)|^{-1} \\
&\quad + \deg(P) \lim_{\substack{k \to \infty \\ k \in N_0}} d^{-k} \Bigg(\log \left| F_0^{(k)}(y) \right| \\
&\qquad\qquad + \log \max_{0 \leq i \leq m} \left| F_i^{(k)}(y) / F_0^{(k)}(y) \right| \Bigg).
\end{aligned}$$

Observe that $[F_0^{(k)}(y) : \cdots : F_m^{(k)}(y)]$ are the homogeneous coordinates of the point $f^k(y)$. Since $y \in \mathrm{U}$ and $f^k(\mathrm{U}) \subset \mathrm{V}_0$, one has $\left| F_i^{(k)}(y) \right| \leq 2 \left| F_0^{(k)}(y) \right|$, so that the last term is bounded by $d^{-k} \log 2$ and uniformly converges to 0 on U. Finally, uniformly on U,

$$\log \|s_P(y)\|^{-1} = \log |P(y)|^{-1} + \deg(P) \lim_{\substack{k \to \infty \\ k \in N_0}} d^{-k} \log \left| F_0^{(k)}(y) \right|.$$

This shows that $\log \|s_P\|^{-1}$ is strongly harmonic on U, as claimed. □

Corollary 2.4.5 *The canonical measure $c_1(\overline{L})^n$ vanishes on E_f.*

Proof It suffices to apply Prop. 2.3.3. □

2.4.6 Remarks. — 1) The particular case $X = P^n$ generalizes Theorem 6 in [44] according to which canonical metrics are locally constant on the classical Fatou set (meaning that the norm of a non-vanishing local section is locally constant). Indeed, the restriction to the set of smooth rigid points of a strongly harmonic function is locally constant. This follows from the fact that any such point has an affinoid neighbourhood U which is a polydisk, so that the absolute value of any invertible function on U, hence any harmonic function on U is constant.

2) In the case $X = P^1$, Fatou and Julia sets in the Berkovich framework have been studied by Rivera-Letelier [52] and Benedetto [11]; see also [8] for a detailed exposition of the theory and further references. An example of Rivera-Letelier on the projective line (Example 10.70 of [8]) shows that the equicontinuity locus E_f may be smaller than the complement of the support of the measure $c_1(\overline{L})$.

Anyway, this proposition suggests the interest of a general study of Fatou sets and of pluripotential theory on Berkovich spaces. For example, is there an interesting theory of pseudoconvexity for Berkovich spaces? Is it related to Stein spaces? By analogy to the complex case (see [56]), are Berkovich Fatou components pseudoconvex? Stein?

2.5 Abelian varieties

Let us assume throughout this section that X is an Abelian variety. For any integer m, let $[m]$ be the multiplication-by-m endomorphism of X.

2.5.1 Canonical metrics. — Let L be a line bundle on X. Let 0 be the neutral element of X and let us fix a trivialization L_0 of L at 0.

The line bundle $L^{\otimes 2}$ is canonically decomposed as the tensor product of an even and an odd line bundle:

$$L^{\otimes 2} = (L \otimes [-1]^* L) \otimes (L \otimes [-1]^* L^{-1}).$$

By the theorem of the cube, an even line bundle L satisfies $[m]^*L \simeq L^{\otimes m^2}$, while for an odd line bundle L, one has $[m]^*L \simeq L^{\otimes m}$; moreover, there are in each case a unique isomorphism compatible with the trivialization at the origin. By Lemma 2.4.1, an even (resp. an odd) line bundle possesses a canonical continuous metric making this isomorphism an isometry. This furnishes a *canonical metric* on $L^{\otimes 2}$, hence on L. According to this lemma, this metric is semi-positive if L is ample and even. Using a Lemma of Künnemann, ([18], Lemme 2.3), one proves that this also holds if L is algebraically equivalent to 0. In any case, the canonical metrics are admissible.

2.5.2 The case of good reduction. — When the variety X has good reduction, the canonical metrics and the associated measures are fairly easy to describe. Indeed, let $\mathfrak{X}$ be the Néron model of X over K°, an Abelian scheme. For any line bundle L on X there is a unique line bundle $\mathfrak{L}$ on $\mathfrak{X}$ which extends L and which admits a trivialization at the 0 section extending the given one over K. By the theorem of the cube for the Abelian scheme $\mathfrak{X}$, the isomorphism $[m]^*L \simeq L^{\otimes m^a}$ (with $a = 1$ or 2, according to whether L is odd or even) extends uniquely to an isomorphism $[m]^*\mathfrak{L} \simeq \mathfrak{L}^{\otimes m^a}$. This implies that the canonical metrics are smooth, induced by these models.

The description of the canonical measures on $\mathfrak{X}$ follows at once. Let ξ be the point of X whose reduction is the generic point of the special fiber of $\mathfrak{X}$. Then, for any family $(L_1, \dots, L_n)$ of line bundles on X, one has

$$c_1(\overline{L}_1) \dots c_1(\overline{L}_n) = \deg(c_1(L_1) \dots c_1(L_n))\delta_\xi.$$

We see in particular that they only depend on the classes of the line bundles L_j modulo numerical equivalence.

2.5.3 Gubler's description. — W. Gubler [41] has computed the canonical measures when the Abelian variety has bad reduction. We describe his result here.

Up to replacing K by a finite extension, we assume that X has split semi-stable reduction. Raynaud's uniformization involves an analytic group E which is an extension of an abelian variety with good reduction Y by a split torus $\mathrm{T} \simeq \mathbf{G}_{\mathrm{m}}^t$, where $t \in \{1, \dots, n\}$ — the so-called Raynaud extension of X. One has $t \geq 1$ since we assume *bad* reduction; moreover, $\dim \mathrm{Y} = \dim \mathrm{E} - t = n - t$. There is a morphism $p \colon \mathrm{E} \to \mathrm{X}$, whose kernel is a discrete subgroup M of $\mathrm{E}(K)$, so that the induced map $\mathrm{E}/\Lambda \to \mathrm{X}$ is an isomorphism. When $t = n$, one says that X has totally degenerate reduction, and the morphism p is the rigid analytic uniformization of the abelian variety X.

Moreover, E is constructed as a contracted product $(\mathrm{E}_1 \times \mathrm{T})/\mathrm{T}_1$ from an extension E_1 of Y by the "unit subtorus" T_1 of T (defined by the equalities $|T_j(x)| = 1$ for $j \in \{1, \dots, t\}$ and $x \in \mathrm{T}$). The natural map $\lambda_{\mathrm{T}} \colon \mathrm{T} \to \mathbf{R}^t$ defined by

$$x \mapsto (-\log|T_1(x)|, \dots, -\log|T_t(x)|)$$

is continuous and surjective; it admits a canonical section ι_{T} which maps a point $(u_1, \dots, u_t) \in \mathbf{R}^t$ to the semi-norm

$$f \mapsto \sup_{\mathbf{m} \in \mathbf{Z}^t} a_{\mathbf{m}} e^{-m_1 u_1 - \dots - m_t u_t}, \qquad \text{for } f = \sum_{\mathbf{m}} a_{\mathbf{m}} T_1^{m_1} \cdots T_t^{m_t} \in \mathscr{O}(\mathrm{T}).$$

The map λ_{T} extends uniquely to a morphism $\lambda \colon \mathrm{E} \to \mathbf{R}^t$ whose kernel contains E_1. The image $\Lambda = \lambda(M)$ is a lattice of $\mathbf{R}^t$, and the morphism p induces a continuous proper morphism $\rho \colon \mathrm{X} \to \mathbf{R}^t/\Lambda$. Composing the section ι_{T} with the projection p furnishes a section $\iota \colon \mathbf{R}^t/\Lambda \to \mathrm{X}$ of ρ. Its image is the *skeleton* of X. Gubler's theorem ([41], Cor. 7.3) is the following:

Theorem 2.5.4 *Let $L_1, \dots, L_n$ be line bundles on* X. *The canonical measure $c_1(\overline{L}_1) \dots c_1(\overline{L}_n)$ is the direct image by ι of the unique Haar measure on $\mathbf{R}^t/\Lambda$ whose total mass is* $\deg(c_1(L_1) \dots c_1(L_n))$.

3 Applications to Arakelov geometry

We now describe some applications of the previous considerations to arithmetic geometry over global fields.

3.1 Adelic metrics and heights

3.1.1 Adelic metrics. — Let F be either a number field (arithmetic case), or a finite extension of the field of rational functions over a constant field (geometric case). Let $M(F)$ be the set of normalized absolute values on F. Let X be a projective variety over F. Any $v \in M(F)$ gives rise to a complete valued field F_v, and to an analytic space X_v over F_v: if v is archimedean, $X_v = X(\overline{F_v})$, while X_v is the Berkovich analytic space attached to X_{F_v} if v is ultrametric.

If L is a line bundle on X, an *adelic metric* on L is a family $(\|\cdot\|_v)_{v \in M(F)}$ of continuous metrics on the induced line bundles over the analytic spaces X_v. We require the following supplementary compatibility assumption: there exists a model $(\mathfrak{X}, \mathscr{L}, e)$ over the ring of integers of F inducing the given metrics at almost all places v. An adelic metric is said to be semi-positive, *resp.* admissible if it is so at all places of F.

Line bundles on X endowed with an adelic metric form a group $\overline{\mathrm{Pic}}(X)$; admissible line bundles form a subgroup $\overline{\mathrm{Pic}}_{\mathrm{ad}}(X)$. If $f \colon Y \to X$ is any morphism, there is a natural morphism of groups $f^* \colon \overline{\mathrm{Pic}}(X) \to \overline{\mathrm{Pic}}(Y)$; it maps $\overline{\mathrm{Pic}}_{\mathrm{ad}}(X)$ into $\overline{\mathrm{Pic}}_{\mathrm{ad}}(Y)$.

3.1.2 Heights. — Consider line bundles $\overline{L}_0, \dots, \overline{L}_n$ with admissible adelic metrics. Let Z be a subvariety of X of dimension k and $s_0, \dots, s_k$ invertible meromorphic sections of $L_0, \dots, L_k$ whose divisors have no common intersection point on Z. For any $v \in M(F)$, we have recalled in Sections 1.2.5, 1.2.9 and 1.3.9 the definitions of the local height pairing

$$(\widehat{\operatorname{div}}(s_0) \dots \widehat{\operatorname{div}}(s_k)|Z)_v$$

where the index v indicates the corresponding place of F. The global height is the sum, over all $v \in M(F)$, of these local heights:

$$(\widehat{\operatorname{div}}(s_0) \dots \widehat{\operatorname{div}}(s_k)|Z) = \sum_{v \in M(F)} (\widehat{\operatorname{div}}(s_0) \dots \widehat{\operatorname{div}}(s_k)|Z)_v.$$

It inherits from the local heights their multilinear symmetric character.

Let us replace s_k by another invertible meromorphic section fs_k. Then,

$$\begin{aligned}(\widehat{\operatorname{div}}(s_0) \dots \widehat{\operatorname{div}}(fs_k)|Z) &= \sum_{v \in M(F)} (\widehat{\operatorname{div}}(s_0) \dots \widehat{\operatorname{div}}(fs_k)|Z)_v \\ &= \sum_{v \in M(F)} (\widehat{\operatorname{div}}(s_0) \dots \widehat{\operatorname{div}}(s_k)|Z)_v \\ &\quad + \sum_{v \in M(F)} \int_{X_v} \log |f|^{-1} c_1(\overline{L}_0) \dots c_1(\overline{L}_{k-1}) \delta_{Z_v}.\end{aligned}$$

In particular, if Z is a point $z \in X(F)$, then $\delta_{Z_v} = \delta_z$ is the Dirac mass at z and

$$(\widehat{\operatorname{div}}(s_0)|Z) = \sum_{v \in M(F)} \log \|s_0\|_v^{-1}(z).$$

Let us observe that it is independent on the choice of the chosen meromorphic section s_0, provided it is regular at z. Any other section has the form fs_0, for some invertible meromorphic function f on X. Then,

$$\begin{aligned}(\widehat{\operatorname{div}}(fs_0)|Z) &= \sum_{v \in M(F)} \log \|fs_0\|_v^{-1}(z) \\ &= \sum_{v \in M(F)} \log \|s_0\|_v^{-1}(z) + \sum_{v \in M(F)} \log |f|_v^{-1}(z) \\ &= (\widehat{\operatorname{div}}(fs_0)|Z)\end{aligned}$$

since, by the product formula, the second term vanishes.

By induction on the dimension of Z, and using the commutativity of the local height pairings, it follows that the global height only depends on the metrized line bundles, and not on the actual chosen sections $s_0, \dots, s_k$. We denote it by

$$(\widehat{c}_1(\overline{L}_0) \dots \widehat{c}_1(\overline{L}_k)|Z).$$

Again, it is multilinear symmetric in the metrized line bundles $\overline{L}_0, \dots, \overline{L}_k$. By the same argument, it only depends on their isomorphism classes in $\overline{\mathrm{Pic}}_{\mathrm{ad}}(X)$.

It satisfies a projection formula: for any morphism $f\colon Y \to X$ and any k-dimensional subvariety Z of Y,

$$(\widehat{c}_1(f^*\overline{L}_0) \dots \widehat{c}_1(f^*\overline{L}_k)|Z) = (\widehat{c}_1(\overline{L}_0) \dots \widehat{c}_1(\overline{L}_k)|f_*(Z)),$$

where the cycle $f_*(Z)$ is defined as $\deg(Z/f(Z))f(Z)$ if Z and $f(Z)$ have the same dimension, so that $f\colon Z \to f(Z)$ is generically finite, of some degree $\deg(Z/f(Z))$. If Z and $f(Z)$ don't have the same dimension, one sets $f_*(Z) = 0$.

3.1.3 Heights of points. — The height of an algebraic point is an important tool in Diophantine geometry. If $\overline{L}$ is a line bundle with an adelic metric on X, then for any point $P \in X(F)$, viewed as a closed subscheme of X, one has

$$h_{\overline{L}}(P) = (\widehat{c}_1(\overline{L})|P) = \sum_v \log \|s\|_v^{-1}(P),$$

where s is any meromorphic section on L which has neither a zero nor a pole at P. More generally, let $P \in X(\overline{F})$ be an algebraic point and let $[P]$ be the corresponding closed point of X. Then,

$$h_{\overline{L}}(P) = \frac{1}{[F(P):F]}(\widehat{c}_1(\overline{L})|[P])$$

is the height of P with respect to the metrized line bundle $\overline{L}$. In fact, restricted to points, these definitions apply to any, not necessary admissible, continuous metric on L. The resulting function is a representative of the classical height function relative to L (which is only defined up to the addition of a bounded function).

Observe also the following functorial property of the height: If $f\colon Y \to X$ is a morphism and $P \in Y(\overline{F})$, then $h_{f^*\overline{L}}(P) = h_{\overline{L}}(f(P))$. Finally, recall that if F is a global field, then the height with respect to a metrized ample line bundle $\overline{L}$ satisfies Northcott's finiteness property: for any integers d and B, there are only finitely many points $P \in X(\overline{F})$ such that $[F(P) : F] \leq d$ and $h_{\overline{L}}(P) \leq B$.

3.1.4 Zhang's inequality. — The essential minimum of the height $h_{\overline{L}}$ is defined as

$$e(\overline{L}) = \sup_{\emptyset \neq U \subset X} \inf_{P \in U(\overline{F})} h_{\overline{L}}(P),$$

where the supremum runs over non-empty open subsets of X. If L is big, then $e(\overline{L})$ is a real number. Another way to state its definition is the following: for any real number B, then the set

$$\{P \in X(\overline{F})\,;\, h_{\overline{L}}(P) \leq B\}$$

is Zariski dense if $B > e(\overline{L})$, and is not Zariski dense if $B < e(\overline{L})$.

Assume that $\overline{L}$ is an ample line bundle on X, equipped with a semi-positive adelic metric. The (geometric/arithmetic) Hilbert-Samuel theorem implies the following inequality

$$e(\overline{L}) \geq \frac{(\widehat{c}_1(\overline{L})^{n+1}|X)}{(n+1)(c_1(L)^n|X)}.$$

(See Zhang [63], as well as [40, 30] for more details in the geometric case.) When X is a curve and F is a number field, Autissier [3] proved that the inequality holds for any ample line bundle with an admissible adelic metric (see [19]); this extends to the geometric case.

3.2 Mahler measures and heights of divisors

In this section, we assume that X is a projective geometrically integral smooth curve of positive genus g over F. For any place $v \in M(F)$, let X_v be the corresponding analytic curve.

3.2.1. — Let f be an invertible meromorphic function on X. Let us view it as an invertible meromorphic section of the trivial metrized line bundle $\overline{\mathscr{O}_X}$. Let $\overline{L}$ be any line bundle on X with an admissible adelic metric. Then,

$$(\widehat{c}_1(\overline{L})\widehat{c}_1(\overline{\mathscr{O}_X})|X) = 0.$$

Moreover, according to Theorem 1.3 of [20] (see Section 1.3.11),

$$(\widehat{c}_1(\overline{L})\widehat{c}_1(\overline{\mathscr{O}_X})|X) = (\widehat{c}_1(\overline{L})|\operatorname{div}(f)) + \sum_{v \in M(F)} \int_{X_v} \log|f|_v^{-1}\, c_1(\overline{L})_v.$$

In other words, this furnishes an integral formula for the height (relative to $\overline{L}$) of any divisor which is rationally equivalent to 0:

$$(\widehat{c}_1(\overline{L})|\operatorname{div}(f)) = \sum_{v\in M(F)} \int_{X_v} \log|f|_v\, c_1(\overline{L})_v.$$

3.2.2 Néron–Tate heights. — We want to apply this formula to a specific metrized line bundle on X. The Jacobian J of X is an Abelian variety of dimension g. We also choose a divisor D of degree 1 on X and correspondingly fix an embedding ι of X into J. (For this, we may need to enlarge the ground field F.) Finally, we let Θ be the theta divisor of J, defined as the image of X^{g-1} by the map $(x_1,\dots,x_{g-1}) \mapsto \sum_{j=1}^{g-1}\iota(x_j)$.

As described above, the line bundle $\mathscr{O}_J(\Theta)$ admits a canonical metrization; this induces a metrization on its inverse image $L = \iota^*\mathscr{O}_J(\Theta)$ on X. The metrized line bundle $\overline{\mathscr{O}_J(\Theta)}$ gives rise to the (theta) Néron–Tate height on J. Consequently, decomposing $\operatorname{div}(f) = \sum n_P P$, we obtain

$$\begin{aligned}(\widehat{c}_1(\overline{L})|\operatorname{div}(f)) &= \sum n_P[F(P):F]\widehat{h}_\Theta(\iota(P)) \\ &= \sum n_P[F(P):F]\widehat{h}_\Theta([P-D]),\end{aligned}$$

where D is the fixed divisor of degree 1 on X.

3.2.3 Canonical measures. — Since L has degree g, the measure $c_1(\overline{L})_v$ on X_v has total mass g; let us define a measure of total mass 1 on X_v by

$$\mu_v = \frac{1}{g}c_1(\overline{L})_v.$$

When v is archimedean, the measure μ_v is the Arakelov measure on the Riemann surface $X_v(\mathbf{C})$. Let us recall its definition. Consider an orthonormal basis $(\omega_1,\dots,\omega_g)$ of $H^0(X,\Omega^1_X)$, *i.e.*, a basis satisfying the relations

$$\int_{X_v(\mathbf{C})} \omega_j \wedge \overline{\omega_k} = \delta_{j,k} = \begin{cases} 1 & \text{if } j=k; \\ 0 & \text{otherwise.}\end{cases}$$

Then,

$$\mu_v = \frac{1}{g}\sum_{j=1}^{g} \omega_j \wedge \overline{\omega_j}.$$

Let us now assume that v is ultrametric. By a theorem of Heinz [42], the metric on the line bundle $\overline{L}$ coincides with the canonical metric defined by Zhang [61] using the reduction graph of the minimal regular model of X; see also [6] for a related interpretation in the framework of tropical geometry. This allows in particular to compute the measure μ_v: the reader will find in [61, Lemma 3.7] a quite explicit formula for μ_v, involving the physical interpretation of the graph as an electric network. (Zhang's computation generalizes Theorem 2.11 of the prior paper [22] by Chinburg and Rumely, the normalization is slightly different; see also [7].)

3.2.4 Superelliptic curves. — The formulas of this section combine to the following: if $\operatorname{div}(f) = \sum n_P P$ is a divisor of an invertible meromorphic function on X,

$$\sum n_P \widehat{h}_\Theta([P-D]) = \sum_{v\in M(F)} \int_{X_v} \log|f(x)|_v \, \mathrm{d}\mu_v(x).$$

As pointed out by R. De Jong [24], the case of superelliptic curves is particularly interesting. Indeed, such curves are presented as a ramified μ_N-covering $x\colon X \to \mathbf{P}^1$ of the projective line, which is totally ramified over the point at infinity, given by an equation $y^N = a(x)$, where a is a polynomial of degree $m > N$, prime to N. One has $g = \frac{1}{2}(N-1)(m-1)$.

Let us take for the divisor D the single point O over the point at infinity. For each point P in $X(F)$, $x - x(P)$ is a rational function on X which has a single pole of order N at infinity, and which vanishes along the fiber $x^{-1}(x(P))$ of x. The group of automorphisms of X acts transitively on this fiber, and respects the metrics, so that all of these points have the same Néron-Tate height. This implies the following formula

$$\widehat{h}_\Theta(P-O) = \frac{1}{N} \sum_{v\in M(F)} \int_{X_v} \log|x - x(P)|_v \, \mu_v$$

of [24]. The elliptic Mahler measure, defined by [29, 28] as a Shnirelman integral is therefore a natural integral when viewed on Berkovich spaces.

3.3 An equidistribution theorem

3.3.1 Bogomolov's conjecture. — Let X be a projective smooth curve of genus $g \geq 2$ and let $\overline{L}$ be an ample line bundle on X with a canonical metric inducing the Néron–Tate height. When F is a number field, Bogomolov conjectured in [15] that $e(\overline{L}) > 0$; this conjecture has been shown by Ullmo [57]. Its generalization to a subvariety X of an Abelian variety A, L being an ample line bundle on A with a canonical metric, asserts that $e(X, \overline{L}) > 0$ when X is not the translate of an abelian subvariety by a torsion point; it has been shown by Zhang [64].

Since $h_{\overline{L}}(P) = 0$ for any algebraic point $P \in A(\overline{F})$ which is a torsion point, these theorems imply in turn a theorem of Raynaud [50, 51] (formerly, a conjecture of Manin and Mumford) that the torsion points lying in a subvariety X of an abelian variety are not Zariski dense in X, unless X is itself the translate of an abelian subvariety by a torsion point.

The analogues of Bogomolov's and Zhang's conjecture in the geometric case is still open in general; see [39, 23] and the references therein for partial results.

3.3.2. — The proofs by Ullmo and Zhang of Bogomolov's conjecture make a fundamental use of an equidistribution principle which had been discovered together with Szpiro [54]. Let us first introduce a terminology: say a sequence (or a net) of algebraic points in a variety X over a number field is *generic* if any strict subvariety of X contains at most finitely many terms of the sequence.

Let $\overline{L}$ be a line bundle on X with a semi-positive adelic metric. The idea of the equidistribution principle is to consider a generic sequence (x_j) such that $h_{\overline{L}}(x_j) \to e(\overline{L})$, *i.e.*, realizing the equality in Zhang's inequality, and to use this inequality further, as a variational principle. Let v be a place of F; for any n, let $\delta(x_j)_v$ be the probability measure on X_v which gives any conjugate of x_j the same mass, $1/[F(x_j) : F]$. The equidistribution theorem states that for a generic sequence (x_j), the sequence of measures $(\delta(x_j)_v)$ on X_v converges vaguely towards the measure $c_1(\overline{L})_v^n/(c_1(L)^n|X)$.

In these papers, the equidistribution property was only investigated at an archimedean place, but the introduction of the measures on Berkovich spaces was motivated by potential equidistribution theorems on those. In [19], I was able to prove general results on curves only. Indeed, unless X is a curve, I needed an ampleness assumption on the metrized line bundle $\overline{L}$ in order to apply Zhang's inequality to slight variations of it. This requirement has been removed by a paper of Yuan [59] who could understand *arithmetic volumes* beyond the ample case. Yuan's proof is an arithmetic analogue of an inequality of Siu [53] which Faber [30] and Gubler [40] used to prove the geometric case of the equidistribution theorem.

In [20], we considered more general variations of the metrized line bundles. The discussion in that article was restricted to the arithmetic case but the arguments extend to the geometric case.

Theorem 3.3.3 *Let X be a projective variety of dimension n over F. Let $\overline{L}$ be an ample line bundle on X with a semi-positive adelic metric such that $e(\overline{L}) = (\widehat{c}_1(\overline{L})^{n+1}|X) = 0$. Let (x_j) be a generic sequence of algebraic points in X such that $h_{\overline{L}}(x_j) \to 0$. Then, for any line bundle $\overline{M}$ on X with an admissible adelic metric,*

$$\lim_{j\to\infty} h_{\overline{M}}(x_j) = \frac{(\widehat{c}_1(\overline{L})^n \widehat{c}_1(\overline{M})|X)}{(c_1(L)^n|X)}.$$

The particular case stated above is equivalent to *loc.cit.*, Lemma 6.1, as one can see by by multiplying the metric on $\overline{L}$ by an adequate constant at some place of F. Taking for M the trivial line bundle $\mathscr{O}_X$, with an admissible metric, one recovers the equidistribution theorems of Yuan, Faber and Gubler.

3.4 Lower bounds for heights and the Hodge index theorem

In the final section, we use the Hodge index theorem in Arakelov geometry to establish positive lower bounds for heights on curves. The results are inspired by recent papers [5, 49], and the proofs are borrowed from [48]. After they were conceived, I received the preprint [58] which proves a similar result in any dimension.

3.4.1 The arithmetic Hodge index theorem. — Let X be a projective smooth curve over F, let $\overline{L}$ be a line bundle of degree 0 on X, with an admissible metric. Let $\overline{L}_0$ be the same line bundle with the canonical metric: if X has genus ≥ 1, this is the metric induced by an embedding of X into its Jacobian, if X is of genus 0, then $\overline{L}_0$ is the trivial metrized line bundle. The metrized line bundle $\overline{L}\otimes\overline{L}_0^{-1}$ is the trivial line bundle, together with an admissible metric which is given by a function f_v at the place v of F.

A formula of Faltings–Hriljac expresses $(\widehat{c}_1(\overline{L}_0)^2|X)$ as minus twice the Néron–Tate height of the point of J corresponding to L. More generally,

$$(\widehat{c}_1(\overline{L})^2|X) = -2\widehat{h}_{\mathrm{NT}}([L]) + \sum_{v\in M(F)} \mathscr{D}(f_v),$$

where for each $v \in M(F)$,

$$\mathscr{D}(f_v) = \int_{X_v} f_v \,\mathrm{dd^c}(f_v)$$

is the *Dirichlet energy* of f_v. This is a non-positive quadratic form which vanishes if and only if f_v is constant. For more details, I refer to [16] at archimedean places and [55] at ultrametric places. (When X has genus 0, $L \simeq \mathscr{O}_X$ and the term $\widehat{h}_{\mathrm{NT}}([L])$ has to be interpreted as 0.)

As a consequence, $(\widehat{c}_1(\overline{L})^2|X) \leq 0$. Let us analyse the case of equality. Since they are nonpositive, all terms in the formula above have to vanish. Consequently, $[L]$ is a torsion point in the Jacobian, and all functions f_v are constant. We will say that some power of $\overline{L}$ is constant

Proposition 3.4.2 *Let F be a number field, let X be a projective smooth curve over F. Let $\overline{L}$ and $\overline{M}$ be two admissible metrized line bundles over X. Assume that $\deg(L) = \ell$, $\deg(M) = m$ are positive. and $(\widehat{c}_1(\overline{L})^2|X) = (\widehat{c}_1(\overline{M})^2|X) = 0$. Then, the essential minimum of $\overline{L} \otimes \overline{M}$ satisfies the following inequality:*

$$e(\overline{L} \otimes \overline{M}) \geq -\frac{1}{2(\ell+m)\ell m}(\widehat{c}_1(m\overline{L} - \ell\overline{M})^2|X).$$

Moreover, the right hand side of this inequality is always nonnegative and vanishes if and only if some power of $\overline{L}^m \otimes \overline{M}^{-\ell}$ is constant.

Proof By Zhang's inequality (see [19]), one has

$$e(\overline{L} + \overline{M}) \geq \frac{1}{2(\ell+m)}(\widehat{c}_1(\overline{L} + \overline{M})^2|X).$$

Since $(\widehat{c}_1(\overline{L})^2|X) = (\widehat{c}_1(\overline{M})^2|X) = 0$ by assumption, we observe that

$$(\widehat{c}_1(\overline{L} + \overline{M})^2|X) = 2(\widehat{c}_1(\overline{L})\widehat{c}_1(\overline{M})|X) = -\frac{1}{\ell m}(\widehat{c}_1(m\overline{L} - \ell\overline{M})^2|X).$$

This shows the first claim.

Since mL and ℓM have the same degree, viz. ℓm, the rest of the proposition follows from the negativity properties of the height recalled above. □

3.4.3. — Assume that (x_n) is a generic sequence of points such that $h_{\overline{L}}(x_n)$ tends to 0. By Theorem 3.3.3, $h_{\overline{M}}(x_n)$ converges to

$$\frac{1}{\ell}(\widehat{c}_1(\overline{L})\widehat{c}_1(\overline{M})|X).$$

Except when both lower bounds are zero, this is strictly bigger than the lower bound of the proposition, which is equal to

$$\frac{1}{\ell+m}(\widehat{c}_1(\overline{L})\widehat{c}_1(\overline{M})|X).$$

In other words, the greedy obvious method to find points of small height for $\overline{L}+\overline{M}$ that first minimizes the height $h_{\overline{L}}$, only works up to the factor $(\ell+m)/\ell > 1$.

3.4.4 An example. — Let us give some explicit formulae for the lower-bound above, in some particular cases. We consider $X = \mathbf{P}^1$ over $\mathbf{Q}$ and the metrized line bundle $\overline{\mathscr{O}(1)}_{\mathrm{W}}$. Let φ and ψ be polynomials with integral coefficients, of degrees ℓ and m respectively; let us put $\overline{L} = \varphi^*\overline{\mathscr{O}(1)}_{\mathrm{W}}$, $\overline{M} = \psi^*\overline{\mathscr{O}(1)}_{\mathrm{W}}$. The line bundle $\overline{L}^m \otimes \overline{L}^{-\ell}$ is trivial and its metric is given by a family of functions (f_v). Since φ and ψ have integral coefficients, $f_v = 0$ at all finite places. Moreover, since $g_{\overline{L}}(x) = \log\max(|\varphi(x)|, 1)$ and $g_{\overline{M}}(x) = \log\max(|\psi(x)|, 1)$ are the Green functions for the divisors $\ell[\infty]$ and $m[\infty]$ respectively, one has

$$f_\infty(x) = \log\frac{\max(|\varphi(x)|^m, 1)}{\max(|\psi(x)|^\ell, 1)}.$$

Then,

$$\mathrm{dd}^c f_\infty = \frac{m}{2\pi}\mathrm{d}\,\mathrm{Arg}\,\varphi(x)\wedge\delta_{|\varphi(x)|=1} - \frac{\ell}{2\pi}\mathrm{d}\,\mathrm{Arg}\,\psi(x)\wedge\delta_{|\psi(x)|=1}.$$

From this, we deduce that

$$\mathscr{D}(f_\infty) = \frac{\ell m}{2\pi}\left(\int_{|\psi(x)|=1}\log\max(|\varphi(x)|, 1)\mathrm{d}\,\mathrm{Arg}\,\psi(x) + \int_{|\varphi(x)|=1}\log\max(|\psi(x)|, 1)\mathrm{d}\,\mathrm{Arg}\,\varphi(x)\right),$$

the two others terms vanishing. In fact, Stokes's formula implies that the two terms within the parentheses in the previous formula are equal and we have

$$\mathscr{D}(f_\infty) = \frac{\ell m}{\pi}\int_{|\varphi(x)|=1}\log\max(|\psi(x), 1|)\mathrm{d}\,\mathrm{Arg}\,\varphi(x).$$

The simplest case to study is for $\varphi(x) = x^\ell$. Then,

$$\mathscr{D}(f_\infty) = \frac{\ell m}{\pi}\int_0^{2\pi}\log\max(\left|\psi(e^{i\theta})\right|, 1)\,\mathrm{d}\theta$$

is $2\ell m$ times the logarithm of the variant $\mathrm{M}^+(\psi)$ of the Mahler measure of ψ:

$$\mathrm{M}^+(\psi) = \exp\left(\frac{1}{2\pi}\int_0^{2\pi}\log\max(\left|\psi(e^{i\theta})\right|, 1)\,\mathrm{d}\theta\right).$$

In fact, Jensen's formula implies that

$$\mathrm{M}^{+}(\psi)=\exp\left(\frac{1}{(2\pi)^2}\int_0^{2\pi}\log\left|\psi(e^{i\theta_1})-e^{i\theta_2}\right|\,\mathrm{d}\theta_1\mathrm{d}\theta_2\right)$$

is the Mahler measure $\mathrm{M}(\psi(x)-y)$ of the 2-variables polynomial $\psi(x)-y$.

Consequently, except for finitely many exceptions, any algebraic point $x\in\mathbf{P}^1(\overline{\mathbf{Q}})$ satisfies

$$\ell h(x)+h(\psi(x))\geq\frac{1}{\ell+m}\log\mathrm{M}(\psi(x)-y).$$

For $\ell=1$ and $\psi(x)=1-x$, we obtain that up to finitely many exceptions,

$$h(x)+h(1-x)\geq\frac{1}{2}\log\mathrm{M}(1-x-y)\approx 0.161538.$$

In that particular case, Zagier [60] has proved a much more precise result: except for 5 explicit points in $\mathbf{P}^1$,

$$h(x)+h(1-x)\geq\frac{1}{2}\log\left(\frac{1+\sqrt{5}}{2}\right)\approx 0.240606.$$

Observe also that if (x_j) is a sequence of points such that $h(x_j)\to 0$, Theorem 3.3.3 implies that $h(1-x)\to\log\mathrm{M}(1-x-y)\approx 0.323076$.

3.4.5 Application to dynamical systems. — Let us assume that $\overline{L}$ and $\overline{M}$ are the metrized line bundles $\overline{\mathscr{O}(1)}_{\varphi}$ and $\overline{\mathscr{O}(1)}_{\psi}$ attached to rational functions φ and ψ of degres d and e respectively, with $d\geq 2$ and $e\geq 2$. Let us write h_φ and h_ψ for the height relative to these metrized line bundles; we call them the canonical heights. The isometry $\varphi^*\overline{\mathscr{O}(1)}_{\varphi}\simeq\overline{\mathscr{O}(1)}_{\varphi}^{d}$ and the functorial properties of the height imply that for any $x\in\mathbf{P}^1(\overline{F})$, $h_\varphi(\varphi(x))=dh_\varphi(x)$ and $h_\psi(\psi(x))=eh_\psi(x)$. In particular, preperiodic points for φ (*i.e.*, points with finite forward orbit) satisfy $h_\varphi(x)=0$. Moreover,

$$\begin{aligned}d^2(\widehat{c}_1(\overline{\mathscr{O}(1)}_{\varphi})^2|\mathbf{P}^1)&=(\widehat{c}_1(\varphi^*\overline{\mathscr{O}(1)}_{\varphi})^2|\mathbf{P}^1)\\&=(\widehat{c}_1(\overline{\mathscr{O}(1)}_{\varphi})^2|\varphi_*\mathbf{P}^1)\\&=d(\widehat{c}_1(\overline{\mathscr{O}(1)}_{\varphi})^2|\mathbf{P}^1),\end{aligned}$$

hence $(\widehat{c}_1(\overline{\mathscr{O}(1)}_{\varphi})^2|\mathbf{P}^1)=0$ since $d\neq 0,1$. Similarly, preperiodic points of ψ satisfy $h_\psi(x)=0$, and $(\widehat{c}_1(\overline{\mathscr{O}(1)}_{\psi})^2|\mathbf{P}^1)=0$.

In the arithmetic case, or over function fields over a finite field, Northcott's finiteness theorem implies easily that points x such that $h_\varphi(x)=0$

are preperiodic for φ, and similarly for ψ. This is not true in general: for example, if φ is constant, all constant points have height 0 but only countably many of them are preperiodic; more generally isotrivial rational functions, *i.e.* rational functions which are constant after conjugacy by an automorphism of $\mathbf{P}^1$ will furnish counterexamples. A theorem of Baker [4] shows that the converse is true: if φ is not isoconstant, then a point of height zero is then preperiodic; the proof relies on a detailed analysis of a Green function relative to the diagonal on $\mathbf{P}^1 \times \mathbf{P}^1$. (In a more general context than the case of polarized dynamical systems, Chatzidakis and Hrushovski [21] proved that the Zariski closure of the orbit of a point of canonical height zero is isoconstant.)

Let us show how Prop. 3.4.2 implies some of the results of Baker and DeMarco [5], and of Petsche, Szpiro and Tucker [49].

Proposition 3.4.6 *In the geometric case, let us assume that ψ is non-isotrivial; if F is a function field over an infinite field, let us moreover assume that it is a polynomial. The following are then equivalent:*

1. *the heights h_φ and h_ψ coincide;*
2. *φ and ψ have infinitely many common preperiodic points;*
3. *the essential lowest bound of $h_\varphi + h_\psi$ is zero;*
4. *the equilibrium measures μ_φ and μ_ψ are equal at all places;*
5. *the metrized line bundles $\mathscr{O}(1)_\varphi$ and $\mathscr{O}(1)_\psi$ are isomorphic, up to a family of constants (c_v) such that $\prod c_v = 1$.*

Proof The arguments are more or less formal from Prop. 3.4.2; let us detail them anyway for the sake of the reader.

1)⇒2). Like any rational map, φ has infinitely many preperiodic points in $\mathbf{P}^1(\overline{F})$, and they satisfy $h_\varphi(x) = 0$. If $h_\varphi = h_\psi$, then they also satisfy $h_\psi(x) = 0$. Under the assumptions of the proposition, they are preperiodic for ψ.

2)⇒3) is obvious, for common preperiodic points of φ and ψ satisfy $h_\varphi(x) + h_\psi(x) = 0$.

3)⇒4). By Prop. 3.4.2, the line bundle $\mathscr{O}(1)_\varphi - \mathscr{O}(1)_\psi$ has the constant metric at all places. In particular, the local measures μ_φ and μ_ψ coincide at all places.

4)⇒5). Let s be a non zero global section of $\mathscr{O}(1)$. For any place v, $f_v = \log(\|s\|_{v,\varphi} / \|s\|_{v,\psi})$; one has $\mu_{v,\psi} - \mu_{v,\varphi} = \mathrm{dd}^c f_v$, hence $\mathrm{dd}^c f_v = 0$. By the maximum principle of [55], f_v is constant. Moreover,

$$0 = (\widehat{c}_1(\overline{\mathscr{O}(1)}_\psi)^2|X) = (\widehat{c}_1(\overline{\mathscr{O}(1)}_\varphi)^2|X) + \sum_v \log c_v = \sum_v \log c_v.$$

5)⇒1). This is obvious. □

3.4.7 Remarks. — 1) The restrictive hypotheses on ψ have only been used to establish the implication 1)$\Rightarrow$2).

2) Of course, many other results can be established by the same reasoning, in particular the number field case of Theorem 1.2 of [5]. Let us also recall that the support of the equilibrium measure μ_φ is the Julia set $J(\varphi)$. If one can prove that $J(\varphi) \neq J(\psi)$ at some place, then none of the assertions of Prop. 3.4.6 can possibly hold. Similarly, Theorem 1.1 of that article can be seen as the conjunction of our proposition and an independent complex analytic study of generalized Mandelbrot sets (Prop. 3.3).

3) The main result of [58] is that a variant of the implication (4)$\Rightarrow$(5) also holds in a more general setting: two semi-positive metrics on a line bundle which define the same measure at a place v differ by multiplication by a constant. The given proof works for curves.

4) We also recall that an implication similar to (1)$\Rightarrow$(5) holds for general metrized line bundles on arithmetic varieties, as proven by [1]: if $\overline{L}$ and $\overline{M}$ are line bundles with adelic metrics such that $h_{\overline{L}} = h_{\overline{M}}$, then $\overline{L} \otimes \overline{M}^{-1}$ is torsion in the Arakelov Picard group $\widehat{\mathrm{Pic}}(X)$: the heights determine the metrics.

References

[1] Agboola, Adebisi, and Pappas, G. 2000. Line bundles, rational points and ideal classes. *Math. Res. Letters*, **7**, 709–717.

[2] Arakelov, Sergei Juri. 1974. Intersection theory of divisors on an arithmetic surface. *Izv. Akad. Nauk SSSR Ser. Mat.*, **38**(6), 1167–1180.

[3] Autissier, Pascal. 2001. Points entiers sur les surfaces arithmétiques. *J. reine angew. Math.*, **531**, 201–235.

[4] Baker, Matthew. 2009. A Finiteness Theorem for Canonical Heights Attached to Rational Maps over Function Fields. *J. reine angew. Math.*, **626**, 205–233. `arXiv:math.NT/0601046`.

[5] Baker, Matthew, and DeMarco, Laura. 2009. *Preperiodic points and unlikely intersections.* To appear in Duke math. J; arXiv:0911.0918.

[6] Baker, Matthew, and Faber, Xander. 2009. Metric properties of the tropical Abel-Jacobi map. To appear in *Journal of Algebraic Combinatorics.* **33** (3), 349–381.

[7] Baker, Matthew, and Rumely, Robert. 2007. Harmonic Analysis on Metrized Graphs. *Canad. J. Math.*, **59**(2), 225–275.

[8] Baker, Matthew, and Rumely, Robert. 2010. *Potential Theory and Dynamics on the Berkovich Projective Line.* Surveys and Mathematical Monographs, vol. 159. Amer. Math. Soc.

[9] Bedford, Eric. 1993. Survey of pluri-potential theory. In: *Several complex variables.* Math. Notes, no. 38. Stockholm, 1987/1988: Princeton Univ. Press.

[10] Bedford, Eric, and Taylor, B.A. 1982. A new capacity for plurisubharmonic functions. *Acta Math.*, **149**, 1–40.

[11] Benedetto, Robert L. 1998. Fatou components in p-adic dynamics. *PhD Thesis.*, 1–98.

[12] Berkovich, Vladimir G. 1990. *Spectral theory and analytic geometry over non-Archimedean fields.* Mathematical Surveys and Monographs, vol. 33. Providence, RI: American Mathematical Society.

[13] Berkovich, Vladimir G. 1999. Smooth p-adic analytic spaces are locally contractible. *Invent. Math.*, **137**(1), 1–84.

[14] Bloch, Spencer, Gillet, Henri, and Soulé, Christophe. 1995. Non-archimedean Arakelov theory. *J. Algebraic Geometry*, **4**, 427–485.

[15] Bogomolov, Fedor Alekseivich. 1980. Points of finte order on abelian varieties. *Izv. Akad. Nauk. SSSR Ser. Mat.*, **44**(4), 782–804, 973.

[16] Bost, Jean-Benoît. 1999. Potential theory and Lefschetz theorems for arithmetic surfaces. *Ann. Sci. École Norm. Sup.*, **32**(2), 241–312.

[17] Call, Gregory, and Silverman, Joseph. 1993. Canonical heights on varieties with morphisms. *Compositio Math.*, **89**, 163–205.

[18] Chambert-Loir, Antoine. 2000. Points de petite hauteur sur les variétés semi-abéliennes. *Ann. Sci. École Norm. Sup.*, **33**(6), 789–821.

[19] Chambert-Loir, Antoine. 2006. Mesures et équidistribution sur des espaces de Berkovich. *J. reine angew. Math.*, **595**, 215–235. math.NT/0304023.

[20] Chambert-Loir, Antoine, and Thuillier, Amaury. 2009. Mesures de Mahler et équidistribution logarithmique. *Ann. Inst. Fourier (Grenoble)*, **59**(3), 977–1014.

[21] Chatzidakis, Zoé, and Hrushovski, Ehud. 2008. Difference fields and descent in algebraic dynamics. I. *J. Inst. Math. Jussieu*, **7**(4), 653–686.

[22] Chinburg, Ted, and Rumely, Robert. 1993. The capacity pairing. *J. reine angew. Math.*, **434**, 1–44.

[23] Cinkir, Zubeyir. 2009. *Zhang's Conjecture and the Effective Bogomolov Conjecture over function fields.* arXiv:0901.3945.

[24] De Jong, Robin. 2009. *Canonical height and logarithmic equidistribution on superelliptic curves.* arXiv:0911.1271.

[25] Deligne, P. 1987. Le déterminant de la cohomologie. Pages 93–177 of: *Current trends in arithmetical algebraic geometry (Arcata, Calif., 1985).* Contemp. Math., vol. 67. Providence, RI: Amer. Math. Soc.

[26] Demailly, Jean-Pierre. 1985. Mesures de Monge-Ampère et caractérisation géométrique des variétés algébriques affines. *Mém. Soc. Math. France*, 124.

[27] Demailly, Jean-Pierre. 1993. Monge-Ampère operators, Lelong numbers and Intersection theory. Pages 115–193 of: *Complex analysis and geometry.* Univ. Ser. Math. New York: Plenum.

[28] Everest, G. R. 1999. On the elliptic analogue of Jensen's formula. *J. London Math. Soc. (2)*, **59**(1), 21–36.

[29] Everest, G. R., and Fhlathúin, Bríd Ní. 1996. The elliptic Mahler measure. *Math. Proc. Cambridge Philos. Soc.*, **120**(1), 13–25.

[30] Faber, X. W. C. 2009. Equidistribution of dynamically small subvarieties over the function field of a curve. *Acta Arith.*, **137**(4), 345–389.

[31] Faltings, Gerd. 1984. Calculus on arithmetic surfaces. *Ann. of Math. (2)*, **119**(2), 387–424.

[32] Favre, Charles, and Jonsson, Mattias. 2004. *The valuative tree.* Lecture Notes in Mathematics, vol. 1853. Berlin: Springer-Verlag.

[33] Favre, Charles, and Rivera-Letelier, Juan. 2007. *Théorie ergodique des fractions rationnelles sur un corps ultramétrique.* arXiv:0709.0092.

[34] Gillet, Henri, and Soulé, Christophe. 1990. Arithmetic intersection theory. *Publ. Math. Inst. Hautes Études Sci.*, **72**, 94–174.

[35] Gubler, Walter. 1997. Heights of subvarieties over M-fields. Pages 190–227 of: Catanese, F. (ed), *Arithmetic geometry.* Symp. Math., vol. 37.

[36] Gubler, Walter. 1998. Local heights of subvarieties over non-archimedean fields. *J. reine angew. Math.*, **498**, 61–113.

[37] Gubler, Walter. 2003. Local and canonical heights of subvarieties. *Ann. Scuola Norm. Sup. Pisa*, **2**(4), 711–760.

[38] Gubler, Walter. 2007a. The Bogomolov conjecture for totally degenerate abelian varieties. *Invent. Math.*, **169**(2), 377–400.

[39] Gubler, Walter. 2007b. Tropical varieties for non-Archimedean analytic spaces. *Invent. Math.*, **169**(2), 321–376.

[40] Gubler, Walter. 2008. Equidistribution over function fields. *Manuscripta Math.*, **127**(4), 485–510.

[41] Gubler, Walter. 2010. Non-archimedean canonical measures on abelian varieties. *Compositio Math.* **146** (3), 683–730.

[42] Heinz, Niels. 2004. Admissible metrics for line bundles on curves and abelian varieties over non-Archimedean local fields. *Arch. Math. (Basel)*, **82**(2), 128–139.

[43] Hubbard, John H., and Papadopol, Peter. 1994. Superattractive fixed points in $\mathbf{C}^n$. *Indiana Univ. Math. J.*, **43**(1), 321–365.

[44] Kawaguchi, Shu, and Silverman, Joseph H. 2009. Nonarchimedean Green functions and dynamics on projective space. *Math. Z.*, **262**(1), 173–197.

[45] Kontsevich, Maxim, and Soibelman, Yan. 2001. Homological mirror symmetry and torus fibrations. Pages 203–263 of: *Symplectic geometry and mirror symmetry (Seoul, 2000).* World Sci. Publ., River Edge, NJ.

[46] Kontsevich, Maxim, and Soibelman, Yan. 2006. Affine structures and non-Archimedean analytic spaces. Pages 321–385 of: *The unity of mathematics.* Progr. Math., vol. 244. Boston, MA: Birkhäuser Boston.

[47] Maillot, Vincent. 2000. Géométrie d'Arakelov des variétés toriques et fibrés en droites intégrables. *Mém. Soc. Math. France*, 129.

[48] Mimar, Arman. 1997. *On the preperiodic points of an endomorphism of* $\mathbf{P}^1 \times \mathbf{P}^1$. PhD Thesis, Columbia University.

[49] Petsche, Clayton, Szpiro, Lucien, and Tucker, Thomas J. 2009. *A dynamical pairing between two rational maps.* arXiv:0911.1875.

[50] Raynaud, Michel. 1983a. Courbes sur une variété abélienne et points de torsion. *Invent. Math.*, **71**(1), 207–233.

[51] Raynaud, Michel. 1983b. Sous-variétés d'une variété abélienne et points de torsion. Pages 327–352 of: Artin, Michael, and Tate, John (eds), *Arithmetic and Geometry. Papers dedicated to I.R. Shafarevich*. Progr. Math., no. 35. Birkhäuser.

[52] Rivera-Letelier, Juan. 2003. Dynamique des fonctions rationnelles sur des corps locaux. *Astérisque*, xv, 147–230. Geometric methods in dynamics. II.

[53] Siu, Yum Tong. 1993. An effective Matsusaka big theorem. *Ann. Inst. Fourier (Grenoble)*, **43**(5), 1387–1405.

[54] Szpiro, Lucien, Ullmo, Emmanuel, and Zhang, Shou-Wu. 1997. Équidistribution des petits points. *Invent. Math.*, **127**, 337–348.

[55] Thuillier, Amaury. 2005. *Théorie du potentiel sur les courbes en géométrie non archimédienne. Applications à la théorie d'Arakelov*. Ph.D. thesis, Université de Rennes 1.

[56] Ueda, Tetsuo. 1994. Fatou sets in complex dynamics on projective spaces. *J. Math. Soc. Japan*, **46**(3), 545–555.

[57] Ullmo, Emmanuel. 1998. Positivité et discrétion des points algébriques des courbes. *Ann. of Math.*, **147**(1), 167–179.

[58] Yuan, Xiniy, and Zhang, Shou-Wu. 2009. *Calabi theorem and algebraic dynamics*. `http://www.math.columbia.edu/~szhang/papers/calabi%20and%20semigroup.pdf`.

[59] Yuan, Xinyi. 2008. Big line bundles on arithmetic varieties. *Invent. Math.*, **173**, 603–649. `arXiv:math.NT/0612424`.

[60] Zagier, Don. 1993. Algebraic numbers close to both 0 and 1. *Math. Comp.*, **61**(203), 485–491.

[61] Zhang, Shou-Wu. 1993. Admissible pairing on a curve. *Invent. Math.*, **112**(1), 171–193.

[62] Zhang, Shou-Wu. 1995a. Positive line bundles on arithmetic varieties. *J. Amer. Math. Soc.*, **8**, 187–221.

[63] Zhang, Shou-Wu. 1995b. Small points and adelic metrics. *J. Algebraic Geometry*, **4**, 281–300.

[64] Zhang, Shou-Wu. 1998. Equidistribution of small points on abelian varieties. *Ann. of Math.*, **147**(1), 159–165.

10

C-minimal structures without density assumption

Françoise Delon

Partially financed by Marie Curie European Network MODNET (MRTN–CT–2004–512234) and the project ANR–06–BLAN–0183.

A *C-relation* is the ternary relation induced by a meet-semi-lattice tree on the set of its branches[1], namely: $C(x; y, z)$ if the intersection of x and z is strictly contained in the intersection of y and z. An ultrametric distance, in particular a valuation on a field, defines a C-relation: $C(x; y, z)$ iff $d(x, y) < d(y, z)$. A C-structure is a set equipped with a C-relation and possibly additional structure. Automorphism groups were the original motivation, in the middle of the 80's, of model theorists who considered C-structures, as C-relations provide natural examples of 2-homogeneous structures. Initially there was no connection with stability theory, another area from model theory, which can be considered to have started with an article of Michael Morley published in 1965, and has developed vigorously and continuously since then.

At the same time (around 1985), it was noticed by several people, including van den Dries, Knight, Pillay, Steinhorn, that some techniques from stability theory could be carried out in structures which are definitely not stable, namely certain linearly ordered (infinite) structures, that they called o-minimal ("o" for "order"). About ten more years were needed before Haskell, Macpherson and Steinhorn introduced dense C-minimal structures and described a common framework which comprised both strongly minimal structures (the simplest among stable structures), o-minimal structures and C-minimal structures. These are all structures which are characterized by the fact that definable sets

[1] whence the "C" which refers to "chains"

Motivic Integration and its Interactions with Model Theory and Non-Archimedean Geometry (Volume II), ed. Raf Cluckers, Johannes Nicaise, and Julien Sebag.
Published by Cambridge University Press.

in one-space (in one variable) are very simple. In a strongly minimal structure, only finite and cofinite sets are definable, *i.e.* Boolean combinations[2] of singletons; in an o-minimal structure, by definition linearly ordered, only Boolean combinations of rational intervals (: convex subsets, with endpoints in the structure or $\pm\infty$). In a C-minimal structure, by definition equipped with a C-relation, only Boolean combinations of "cones" or "thick cones" (the generalisation of "open" and "closed" balls from ultrametric spaces) are allowed. In this way a notion of "relative minimality" emerges: whatever the complexity of the structure is in higher dimension, what can be said about a single variable, can be said using no quantifiers, only equality in strongly minimal structures, the order relation in o-minimal structures, and the C-relation in C-minimal structures.

In this contribution we review what is known about C-minimal structures, but we work in a more general framework than the one that had been considered previously: we allow isolated points in the topology defined by the C-relation. In this way, strong minimality and o-minimality become particular instances of C-minimality. This goes in the direction of the most recent developments of pure model theory, the generalization of methods from stability theory to the unstable context of the so-called dependent theories ("NIP"). This also allows us to understand in a much more precise way the analogies and differences beetween o-minimality and C-minimality as well as the connections to stability.

The paper is organized as follows. In the first section we define C-relations, C-sets, their canonical tree and the topology with which they are equipped. In section 2 we introduce C-minimal structures and study their basic properties. We also present some examples and links from C-minimality to strong minimality, o-minimality, stability and dependence. In section 3 we describe unary definable subsets of C-minimal structures. In section 4 we consider C-minimal structures that are dense or discrete for the topology induced by the relation C. We present some of the results of Haskell and Macpherson about dense C-minimal structures and explain how they can be applied in general C-minimal structures. We show that a C-minimal structure can be definably cut into a definable dense part and a discrete one. Finally in Section 5 we deal with the exchange property. We present briefly Zilber's trichotomy in strongly minimal and o-minimal structures, and state first results in

[2] always finite for us

C-minimal structures by Maalouf. We also mention criteria by Haskell and Macpherson for the exchange property, and give a consequence of exchange on the partition dense/discrete.

Everything about the non-dense case is new. Consequently many of our examples are new, too. We had to adapt and slightly modify the construction of the canonical tree, originally due to Adeleke and Neumann. In Section 3, our proof of the main result is new, as well as some consequences. Section 4 is almost completely new. This is the reason why, in these places, we give a detailed presentation including full proofs.

Plan:

1. C-sets and their Adeleke-Neumann representation
2. C-minimality
3. Unary definable sets
4. Dense and discrete C-minimal structures – The partition dense/discrete
5. Geometric dense C-minimal structures and the question of the trichotomy

Conventions

1. We often distinguish between a structure $\mathbb{M}$ and its underlying set, then denoted M, for example for a field, $\mathbb{F} = \langle F; +, \cdot \rangle$.
2. In model theoretic context, the notation "$x \in M$" means generally that x is a finite tuple from M, not necessarily a singleton.
3. Equality is, by convention, always an authorized symbol in the formulas we are considering. Consequently we never mention it in the language of a structure.

1 C-sets and their Adeleke-Neumann representation

Definition 1.1 A *C-relation* is a ternary relation, usually written C, satisfying the four universal axioms:

1. $C(x; y, z) \to C(x; z, y)$
2. $C(x; y, z) \to \neg C(y; x, z)$
3. $C(x; y, z) \to [C(x; y, w) \vee C(w; y, z)]$
4. $x \neq y \to C(x; y, y)$.

C is called *dense* if it additionally satisfies:

5. $x \neq y \to \exists z \neq y \quad C(x; y, z)$
6. $\exists x \; \exists y \quad x \neq y$.

A *C-set* is a set equipped with a C-relation.

Examples

- The C-relation induced by a ultrametric distance: $C(x;y,z) \leftrightarrow d(y,z) < d(y,x) = d(x,z)$.
- The trivial C-relation: $C(x;y,z) \leftrightarrow y = z \neq x$ (a particular case of the previous example).
- C-relations induced by linear orders: $C(x;y,z) \leftrightarrow [(x < y,z) \vee y = z \neq x]$.
- In a tree (: order T in which, for any $x \in T$, the set $\{y \in T; y < x\}$ is linearly ordered), sets of branches (: maximal chains) equipped with the *canonical* C-relation: $C(\alpha;\beta,\gamma)$ iff $\alpha \cap \beta = \alpha \cap \gamma \subset \beta \cap \gamma$ ("$\subset$" means here strict inclusion). This example is canonical as is shown in Proposition 1.5.

Be aware that, as defined in [MacSte] or [HMac1], a C-relation is always dense (it is defined there respectively by axioms $1+2+3+5$ or $1+2+3+5+6$) and that our definition of density does not follow [AN]. There are many motivations for our choices. The first one is presented in Fact 1.2: we take as axioms defining C-relations the universal part of axioms previously considered. Secondly, part of the results proved in [HMac1] or [MacSte] does not use density. Thirdly there are contexts where density is not a relevant assumption, such as when considering pure C-sets or valued groups. The analogy to o-minimality is also a strong motivation, see Section 4. Finally strongly minimal and o-minimal structures become C-minimal structures under this definition, see Section 2.

Fact 1.2 *Axioms (1) to (4) axiomatize the universal part of the theory defined by the conjunction (1) & (2) & (3) & (5) [& (6)].*

Proof. 1. (2) & (3) & (5) $\Rightarrow$ (4):
(5): $x \neq y \rightarrow \exists z \neq y \quad C(x;y,z)$
(3) with $w = y$: $C(x;y,y) \vee C(y;y,z)$
(2) with $x = y$: $C(y;y,z) \rightarrow \neg C(y;y,z)$, hence: $\neg C(y;y,z)$
therefore $C(x;y,y)$.
2. Given a C-set M consider $E := 2^\omega$ equipped with the relation C induced by the ultrametric distance ranging in ω inverse (namely, if $\delta = (\delta_i)_{i\in\omega}$ and $\delta' = (\delta'_i)_{i\in\omega}$ are elements from E, $d(\delta,\delta')$ is by definition the first i where δ_i and δ'_i are different if there is such an i, and 0 otherwise), and the product $M \times E$ with the relation $C((m,e)_1;(m,e)_2,(m,e)_3)$ iff $C(m_1;m_2,m_3)_M \ \vee \ [(m_1 = m_2 = m_3) \wedge C(e_1;e_2,e_3)_E]$.

In this way $M \times E$ is dense and, for any $e \in E$, the map $m \mapsto (m, e)$ defines a C-isomorphism from M onto $M \times \{e\} \subseteq M \times E$; thus any universal consequence of axioms (1) to (6) is true in M.
This kind of construction will be used several times in sections 2, 4 and 5. □

Remarks. 1. The trivial C-relation is not dense, nor is the C-relation induced by a linear order.
2. The relation C' defined from a linear order by setting $C'(x; y, z)$ iff $x < y, z$ satisfies axioms (1) to (3), but not (4).

Definition 1.3 Call a tree *good* if it has the following properties:
- it is a *meet* semi-lattice (: any two elements x and y have an infimum, or meet, $x \wedge y$, which means: $x \wedge y \leq x, y$ and $(z \leq x, y) \to z \leq x \wedge y)$,
- there are maximal elements, or *leaves*, everywhere (: $\forall x \ \exists y \ (y \geq x \wedge \neg \exists z > y))$,
- any element is either a leaf or a *branching point* (: of the form $x \wedge y$ for some distinct x and y).

Theorem 1.4 *C-sets and good trees are bi-interpretable classes.*

Proof. One direction is easy: In T a good tree, maximal elements may be identified with branches via the map $\alpha \mapsto \{\beta \in T; \beta \leq \alpha\}$. Thus, if $Br_l(T)$ denotes the set of branches of T having a leaf, the structure $(T, <, Br_l(T), \in)$ is definable in $(T, <)$. The converse is a consequence of the following result.

Proposition 1.5 *Given a C-set M, there is a unique good tree such that M is isomorphic to the set of branches with leaf, equipped with the canonical C-relation. This tree is called* the canonical tree of M.

Proof. The existence is almost the representation theorem of Adeleke and Neumann ([AN], 12.4). Let us describe their construction. Given a C-set (M, C), define on M^2 the binary relations $\preccurlyeq$ and R by

$$(\alpha, \beta) \preccurlyeq (\gamma, \delta) :\Leftrightarrow \neg C(\gamma; \alpha, \beta) \wedge \neg C(\delta; \alpha, \beta)$$

$$(\alpha, \beta) R (\gamma, \delta) :\Leftrightarrow \neg C(\alpha; \gamma, \delta) \wedge \neg C(\beta; \gamma, \delta) \wedge \neg C(\gamma; \alpha, \beta) \wedge \neg C(\delta; \alpha, \beta).$$

Adeleke and Neumann work in fact with the set of pairs of distinct elements of M, instead of M^2 as we do, and reverse the order. About the choice of the order: we just prefer our trees with leaves up and roots down! About the diagonal: their construction is convenient for a dense C-relation but not in general (see the footnote eight lines below), hence our slight modification, which yields furthermore the definability of (M, C) in its "canonical tree". Our modification does not affect proofs we quote from [AN] since those are formal deductions from the axioms defining C-relations. We sketch below the demonstration.

Fact 1.6 *([AN], 12.1) The relation $\preccurlyeq$ is a pre-order and R is the equivalence relation that it naturally determines.* $\dashv$

Define $T(M)$ or simply $T := M^2/R$ and, for $\alpha \in M$, $\Gamma_\alpha := \{(\alpha, \beta) \mod R; \beta \in M\}$[3]; $\preccurlyeq$ induces therefore an order on T.

Fact 1.7 *T with the above order is a good tree and the Γ_α are maximal chains in T.*

Proof. That the $\Gamma_\alpha \setminus \{\alpha\}$ are chains is proved in [AN], 10.2; now, for any $\alpha, \beta, \gamma \in M$, $(\alpha, \alpha) \succcurlyeq (\alpha, \beta)$ holds, as does $(\alpha, \alpha) \not\preccurlyeq (\beta, \gamma)$ if $\beta \neq \alpha$ or $\gamma \neq \alpha$; thus, for any $\alpha \in M$, (α, α) is a maximal element of Γ_α and the chain Γ_α is maximal. That T is a meet-semi-lattice tree is proved in [AN], 12.3 (indeed: $[(\alpha, \beta) \mod R] \wedge [(\alpha, \delta) \mod R]$ is equal to $(x, y) \mod R$, For some x, y in $\{\alpha, \beta, \gamma, \delta\}$ that depend on the way $\alpha, \beta, \gamma, \delta$ are linked by C). By definition of R, any element from T is a branching point or a leaf, so T is good. $\dashv$

The map: $M \to T$, $\alpha \mapsto (\alpha, \alpha) \mod R$, is the representation we were looking for, identifying $\langle M, C \rangle$ with $Br_l(T)$ equipped with the canonical C-relation. Now, in any good tree Θ such that $Br_l(\Theta)$ and M are isomorphic C-sets, the branch representing some $\alpha \in M$ must be $\{\alpha \wedge \beta; \beta \in M\}$. The construction of $T(M)$ is then canonical. □

Proposition 1.8 *Let F be the set of leaves of $T(M)$. Then $\langle M, C \rangle$ and $\langle T(M), \wedge, F \rangle$ are first-order bi-interpretable, quantifier free and without parameters, and M and $F(T(M))$ are definably isomorphic. Therefore an embedding $M \subseteq N$ induces an embedding $T(M) \subseteq T(N)$ and the groups of automorphisms of $\langle M, C \rangle$ and of $\langle T(M), \wedge \rangle$ are canonically isomorphic.* □

Examples of canonical trees.

- Let M be a set equipped with the trivial C-relation. The canonical tree $T(M)$ has a root r; its other points are in one-to-one correspondance with M, see below left.

[3] In the way the construction is carried out in [AN], proof of 12.4, it is not true in general that the Γ_α are maximal chains. Think for example of the C-set $\{a, b, c\}$ with $C(a; b, c)$ (thus $\neg C(c; b, a) \wedge \neg C(b; c, a)$), for which $(a, c) R (a, b) \prec (c, b)$, therefore $\Gamma_a = \{(a, c)/R\}$ is strictly included in $\Gamma_b = \Gamma_c = \{(a, c)/R, (b, c)/R\}$. It is true if C is dense.

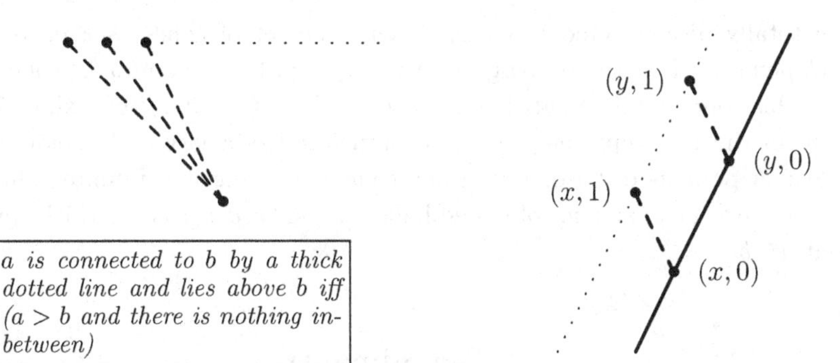

a is connected to b by a thick dotted line and lies above b iff (a > b and there is nothing in-between)

- If L is a linearly ordered set without maximal element, the canonical tree of L equipped with the corresponding C-relation is a "comb": $T(L) = L \times \{0,1\}$ with $L \times \{0\}$ ordered as L, and for $x < y$ in L, $(x,1) \wedge (y,1) = (x,0)$, see above right.

We will therefore, when working with a C-set M, make constant use of its canonical tree, which we will denote $T(M)$ or just T. We extend the meet of the tree, $\wedge : T \times T \to T$, to $M \times M \to T$ and $M \times T \to T$ in the obvious way. When we want to distinguish between an element x in M and the corresponding branch in the canonical tree, we denote $br\, x$ this last one; for α a branch in T, $\alpha^- := \{c \in \alpha; c \text{ is not the leaf of } \alpha\}$.

Examples of complete axiomatization

1. Two sets equipped with the trivial C-relation are elementarily equivalent iff they have the same cardinality, seen as an element of $\mathbb{N} \cup \{\infty\}$.
2. Two sets equipped with C-relations induced by linear orders are elementary equivalent iff the two orders are. Indeed order and C-relation are bi-interpretable without parameters, see the proof of Proposition 2.6.
3. Fact 1.8 implies that a complete axiomatization of a C-structure is provided by a complete axiomatization of its canonical tree. As an example, *covering* (see 4 and 5 lines below) sets of branches of 2-homogeneous meet-semi-lattice trees are axiomatized by the system of axioms describing the number ($\in \mathbb{N} \cup \{\infty\}$) of branches at each node, the (linear) order type of each branch, which must be dense with maximal element and without first element, and finally: $\forall c,\ \exists d \in F,\ c \leq d$ (which means that the set of branches with leaf *covers* the tree).

Topology

In a C-set, *cones* are the analogues of "open" balls in ultrametric spaces: $\{x; C(a;x,b)\}$ is the *cone of* b *at* $a \wedge b$ $(a \neq b)$. Note that two cones either are disjoint, or one is contained in the other. Hence, a C-relation defines a totally disconnected topology, having the set of cones as a basis of clopen sets. It is the topology we will work with[4]. A point x is isolated in this topology iff it satisfies $\exists y \neq x \ \forall z \neq x \ \neg C(y; x, z)$ (cf. axiom 5), iff, in the tree representation, the branch x has a last node below its leaf. A point is **not** isolated iff any cone containing it is infinite. Thick cones are the analogues of closed balls: $\{x; \neg C(x; a, b)\}$ is the thick cone at $a \wedge b$.

2 C-minimality

Minimality properties of a theory introduced in this section express that the class of definable subsets of any model is not "too complicated" as a Boolean algebra. Historically, *strongly minimal theories* were considered first, they are the simplest *stable theories.* Then various generalisations appeared...

2.1 Various notions of minimality

Definition 2.1 A C-structure is a C-set possibly equipped with additional structure.
A C-structure $\mathbb{M}$ is called C-minimal iff any definable subset of M is definable by a quantifier free formula in the pure language $\{C\}$, and this holds also in any $\mathbb{M}' \equiv \mathbb{M}$.

We emphasize that the definable sets considered here are subsets of the structure itself and not of a Cartesian power. The stability under elementary equivalence makes C-minimality a property of the complete theory of $\mathbb{M}$ rather than of the structure $\mathbb{M}$ itself. It has also the following consequence.

Fact 2.2 *Let $\mathbb{M}$ be a C-minimal structure. Then, given a formula without parameters $\varphi(x, y)$, where x is a single variable and y a tuple of variables, there is an integer n such that, for any parameter $c \in M$, $\varphi(x, c)$ is equivalent in $\mathbb{M}$ to a Boolean combination of cones or thick cones of complexity at most n (by which I mean of the form $\bigcup_{i \leq n} \bigcap_{j \leq n} D_{i,j}$ with the $D_{i,j}$ (possibly thick) cones or complements of such).*

Proof. Indeed, that $\varphi(x, c)$ is equivalent to a Boolean combination of cones or thick cones of complexity at most n can be expressed as

[4] It is not the topology chosen in [AN] (they coincide when every cone is infinite). If C is induced by a linear order, it is in general not the order topology.

follows:

$$\exists (a_{i,j}, b_{i,j})_{i,j \leq n} \bigvee_{\substack{\varepsilon \in \{1,-1\}^{n^2} \\ \delta \in \{1,-1\}^{n^2}}} \forall x \, [\, \varphi(x,c) \longleftrightarrow \bigvee_i \bigwedge_j x \in D^{\varepsilon(i,j)}_{\delta(i,j)}(a_{i,j}, b_{i,j}) \,],$$

where $D^1_1(a,b)$ is the cone of a at $a \wedge b$, $D^1_{-1}(a,b)$ is the thick cone at $a \wedge b$, and $D^{-1}_{\varepsilon_0}(a,b)$ is the complement of $D^1_{\varepsilon_0}(a,b)$. The fact follows by compactness. □

A typical example of C-minimal structures is that of algebraically closed valued fields with the C-relation induced by the valuation.

The notion of C-minimal structures has been modelled on strong minimal and o-minimal structures, which we present now.

Definition 2.3 A structure $\mathbb{M}$ is *strongly minimal* iff any definable subset of M is finite or cofinite, and this holds also in any $\mathbb{M}' \equiv \mathbb{M}$.

Examples of strongly minimal structure are sets with no structure (another way to say that the language is empty), torsion free divisible groups, and algebraically closed fields.

In the same way, in any linearly ordered structure $(M, <)$, "rational intervals" *i.e.* intervals with endpoints in $M \cup \{-\infty\} \cup \{\infty\}$, are definable.

Definition 2.4 A linearly ordered structure $\mathbb{M}$ is o-*minimal* iff any definable subset of M is a finite boolean combination of rational intervals, or equivalently, a finite disjoint union of rational intervals. (Then this holds in any $\mathbb{M}' \equiv \mathbb{M}$[5].)

Dense linear orders, divisible ordered groups and real closed ordered fields are o-minimal.

These three notions of minimality may be understood as *relative* minimalities. Consider two languages $\mathscr{L}_0 \subseteq \mathscr{L}$, an $\mathscr{L}$-structure $\mathbb{M}$ and $\mathbb{M}_0 := \mathbb{M} \restriction \mathscr{L}_0$. We say that $\mathbb{M}$ is *minimal relative to the language* $\mathscr{L}_0$, if any subset of M definable in $\mathbb{M}$ is quantifier free definable in the language $\mathscr{L}_0$ (see [MacSte]). Thus, strongly minimal structures are structures minimal relative to the empty language, o-minimal structures minimal relative to the order, and C-minimal structures minimal relative to the C-relation. C-minimal structures are related to strongly minimal or o-minimal structures by the following:

[5] Result far from being trivial, see [PiSte2].

Theorem 2.5 [[HMac1](2.7)] *Let $\mathbb{M}$ be a C-minimal structure and T its canonical tree. Then:*
1. Any element, say x, from M is o-minimal as a branch of T: any $\mathbb{M}$-definable subset of $br\, x$ is a finite union of rational intervals.
2. The set of cones at any node c of T is either finite or strongly minimal: for any $\mathbb{N} \succeq \mathbb{M}$, any $\mathbb{N}$-definable set of cones at c is finite or cofinite (here, the set of all cones at c is considered as a definable subset of $\mathbb{M}^{eq}$[6]).

As an example, if M is an algebraically closed non trivially valued field with C the relation induced by the valuation, then any branch of $T(M)$, equipped with the structure induced by $\mathbb{M}$, is isomorphic to the valuation group as an ordered group, and each node is isomorphic to the residue field[7]. Of course the converse of proposition 2.5 is not true in general (a valued field with an algebraically closed residue field and a divisible valuation group is in general not algebraically closed, hence not C-minimal, see proposition 2.8), but is true when C is "degenerate":

Proposition 2.6 *Let $\mathbb{M}$ be a C-structure. If the C-relation is trivial, then $\mathbb{M}$ is C-minimal if and only if it is strongly minimal. If the C-relation comes from a linear order, then $\mathbb{M}$ is C-minimal if and only if it is o-minimal.*

Proof. The trivial C-relation and equality are quantifier free interdefinable without parameters. The C-relation associated to a linear order is quantifier free definable from the order without parameters. Conversely: $x > y$ iff $y \neq m$ (m a possible maximal element) and $C(y; x, z)$ for some $z \neq y$ (if $y \neq m$, there is some $z > y$, and then $C(y; x, z) \leftrightarrow y < x$), which shows that C and $<$ are interdefinable with quantifiers (and without parameters), but also that the unary relation $x > a$ is definable from C with parameters without quantifiers. Therefore quantifier-free formulas with parameters in one variable are interdefinable. □

So C-minimality is a natural common framework for strong minimality, o-minimality and dense C-minimality. We will see in Subsection 2.3 why such a notion is interesting.
Other analogies and/or differences between these notions of minimality will be presented in Section 5.

[6] see the definition in [C] 2.34
[7] See [HHrMac] for more informations about this "metastable" structure.

Let us mention other notions of minimality well-adapted to valued fields:
- V-minimality for expansions of algebraically closed valued fields (see [HrK])
- P-minimality for expansions of p-adic and p-adically closed fields (see [HMac2])
- b-minimality for expansions of Henselian valued fields (see [ClLoe1] and [ClLoe2])
- weak o-minimality, which includes real closed fields equipped with a convex valuation.

See [Mac] for a presentation of these notions and of some analogies between them. See also [Adl] about the connection with other notions of minimality which arise naturally from a model theoretic point of view.

2.2 Examples of C-minimal structures

1. Strongly minimal structures with the trivial C-relation; **o-minimal structures** with the C-relation induced by the order.

2. Pure C-sets corresponding to good trees with elementarily equivalent o-minimal branches and the same number N of cones at each node ($N \in \mathbb{N} \cup \{\infty\}$).

3. As already mentioned, **algebraically closed valued fields** with the C-relation induced by the valuation are C-minimal. This result has a converse:

Definition 2.7 C is compatible with a binary function $*$ iff
$C(x; y, z) \to C(t * x; t * y, t * z) \wedge C(x * t; y * t, z * t)$.
A *C-field* is a field equipped with a C-relation compatible with the two laws.

Proposition 2.8 ([HMac1]) *Algebraically closed valued fields are C-minimal C-fields. Conversely any C-field carries a valuation satisfying $C(x; y, z) \leftrightarrow [v(x-y) = v(x-z) < v(y-z)]$ and any C-minimal C-field is algebraically closed.*

4. Let us call a group equipped with a C-relation compatible with the group law, a *C-group*. On a **C-minimal C-group**, C need not stem from a valuation, think of the multiplicative group of an (algebraically closed) valued field, where $v(x, y) \geq \min\{v(x), v(y)\}$ is not true in general. Structural properties of general C-minimal groups are proved in

[MacSte]: the form of orbits of nodes from the canonical tree under the action of the group is an important issue. Note:

Proposition 2.9 ([Si1] and [Si2]) *Dense C-minimal valued groups are nilpotent-by-finite; they are in general not Abelian-by-finite.*

Abelian C-minimal valued groups have been completely classified. Fundamental "building blocks" are: finite groups, the additive valued groups of $\mathbb{Z}_p$ and $\mathbb{Q}_p$, $(\mathbb{Q}, +, w)$ where $w = \max\{v_p, 0\}$, p-valued p-Prüfer groups, and strongly minimal groups with the trivial C-relation [DSi].
C-minimal vector spaces have been classified [Maa3], and also some C-minimal modules (Gonenc Onay).

5. The valued field $(\mathbb{C}_p, v_p)$, enriched with all strictly convergent analytic functions is C-minimal ([LR]).

6. So C-minimal structures form a broad class containing many classical algebraic structures. Furthermore C-minimality allows **a certain combinatorial freedom**. For example, for $\mathbb{M}$ a C-minimal purely relational C-structure and G a cone in M, then G and often $M \setminus G$ are C-minimal (see Definition 3.11 and Lemma 3.12). Some kind of inverse construction is described in [DMo], gluing together two C-minimal structures into a third one. We present now two other constructions.

Connection of finitely many $\mathscr{L}$-structures
Let the $\mathscr{L}_i$, for $i = 1, \ldots, r$, be languages containing C and with no other symbol common to any two of them, and $\mathbb{M}_i$ an $\mathscr{L}_i$-structure. Suppose $r \geq 2$. Let $\mathscr{L} := \bigcup_i \mathscr{L}_i \cup \{R_1, \ldots, R_r\}$, where the R_i are unary predicates, and M be the disjoint union of the M_i. We interpret in M each R_i as M_i, each $\mathscr{L}_i \setminus \{C\}$ on R_i in the sense of $\mathbb{M}_i$, and, for $x, y, z \in M$, then $\mathbb{M} \models C(x; y, z)$ iff, for some index i, either $[x, y, z \in M_i$ and $\mathbb{M}_i \models C(x; y, z)]$ or $y, z \in M_i \not\ni x$. In this way M is an $\mathscr{L}$-structure and $T(M)$ is the disjoint union of the $T(M_i)$ and a root.
We call $\mathbb{M}$ the *connection* of the $\mathbb{M}_i$ and it is denoted $\mathbb{M}_1 \sqcup \cdots \sqcup \mathbb{M}_r$.[8]
Any $\mathscr{L}$-formula in the variables $x_1, \ldots, x_n$ is equivalent in $\mathbb{M}$ to a Boolean combination of formulas of the form $\bigwedge\!\!\!\bigwedge_{0 \leq k \leq s-1} (x_{j_k} \in M_i) \wedge \psi(x_{j_0}, \ldots, x_{j_{s-1}})$, for some integer $s \leq n$, $j \in n^s$, $i \in \{1, \ldots, r\}$ and some $\mathscr{L}_i$-formula ψ.

[8] As the $\mathbb{M}_i$ are disjoint inside $\mathbb{M}_1$ and represented by predicates from $\mathscr{L}$, the construction can be carried out even if $\mathscr{L}_i \setminus \{C\}$ and $\mathscr{L}_j \setminus \{C\}$ are not disjoint for some $i \neq j$.

In M each M_i is the cone (at the root) of any of its elements. Consequently, **$\mathbb{M}$ is C-minimal iff each $\mathbb{M}_i$ is.**

Extension of a C-structure by a pure C-set

Let $\mathscr{L}$ be a relational language containing C, $\mathbb{M}$ an $\mathscr{L}$-structure, and E a pure C-set. Consider the structure $\mathbb{N} =: E \rtimes \mathbb{M}$ in the language $\mathscr{L} \cup \{\sim\}$, defined as follows:

- the underlying set is $N := E \times M$, with $\pi : E \rtimes M \to M$ the canonical projection and $x \sim y :\Leftrightarrow \pi(x) = \pi(y)$;
- if $x \sim y \sim z$, then $\mathbb{N} \models C(x; y, z) :\Leftrightarrow E = (x \mod \sim) \models C(x; y, z)$;
- if $\neg(x \sim y \sim z)$, then $\mathbb{N} \models C(x; y, z) :\Leftrightarrow \mathbb{M} \models C(\pi(x); \pi(y), \pi(z))$;
- for $R \in \mathscr{L}$, R n-ary and $R \neq C$, then $\mathbb{N} \models R(x_1, \ldots, x_n) :\Leftrightarrow \mathbb{M} \models R(\pi(x_1), \ldots, \pi(x_n))$.

Thus the canonical tree of $E \rtimes \mathbb{M}$ consists of the tree of M in which each leaf is replaced by a copy of the tree of E.

Proposition 2.10 *Suppose that E is homogenous*[9] *and such that $T(E)$ has a root. Then $E \rtimes \mathbb{M}$ is C-minimal iff E and $\mathbb{M}$ are.*

We will give the proof in section 4, before applying this proposition.

2.3 Link with stability and dependence

Stability theory is a body of results and techniques, highly powerful by reason of its generality. Originally largely work of Saharon Shelah, it has more recently been completed by the developments of *geometric stability*, mainly due to Zilber and Hrushovski. We give a flavour of these latter developments in Section 5. For about ten years, characteristic properties of stable theories have been cut up into two groups, giving rise to two "opposite" generalizations of stability. Theories satisfying the first group are called *simple* ([KiPi]) and those satisfying the second group *dependent*, or still *without the independence property*, "NIP" (see for example [HrPi2]). C-minimal theories are dependent and mainly unstable, so they offer a natural place in which to analyze the evolution from stable to dependent[10]. We will now be a bit more precise. All the general matter about stability, ω-stability and dependence presented here can be found in [S] or [Pi], most of it in [Mar].

Instability and independence of a complete theory express or/and measure some kind of complexity of uniformly definable subsets in models.

[9] By *homogeneous* we mean that all elements have the same type over the empty set.

[10] See [Adl] for other considerations of this kind.

Definition 2.11 Let T be a complete theory. A formula $\varphi(x,y)$[11] has the independence property in T if for any integer n,

$$T \vdash \exists (x_i)_{i<n},\ \exists (y_j)_{j\subseteq\{0,\dots,n-1\}},\ \bigwedge_{\substack{i<n\\ i\in j}} \varphi(x_i,y_j) \wedge \bigwedge_{\substack{i<n\\ i\notin j}} \neg\varphi(x_i,y_j).$$

T has the independence property if some formula has the independence property. In the opposite case, T is called dependent or NIP.

Fact 2.12 *If T has the independence property, then some formula $\varphi(x,y)$ with x a singleton has the independence property.*

We can take a characterisation of unstable theories as a definition:

Theorem 2.13 *T is unstable iff it has the independence property or, for some integer n and any model $\mathbb{M}$ of T, there is a preorder on M^n which is definable in T and has chains of any finite length.*

Among stable theories, ω-stable theories behave particularly well. In models of these theories, any definable set D has an ordinal rank, its *Morley rank*, $MR(D)$.

Definition 2.14 Let $\mathbb{M}$ be an $\aleph_0$-saturated model and D a definable subset in some M^n. We define inductively the relation "$MR(D) \geq \alpha$" as follows:

$MR(D) \geq 0$ iff $D \neq \emptyset$;

if α is a limit ordinal, then $MR(D) \geq \alpha$ iff $MR(D) \geq \beta$ for any ordinal $\beta < \alpha$;

for any ordinal α, $MR(D) \geq \alpha + 1$ iff there exist pairwise disjoint definable subsets $(D_i; i \in \omega)$ of M^n such that $D_i \subseteq D$ and $MR(D_i) \geq \alpha$ for any $i \in \omega$.

Then we define:

$MR(\emptyset) = -1$,

$MR(D) = \alpha$ iff $[MR(D) \geq \alpha$ and $MR(D) \not\geq \alpha + 1]$,

$MR(D) = \infty$ iff $[MR(D) \geq \alpha$ for any ordinal $\alpha]$.

A complete theory is ω-stable if, in any $\aleph_0$-saturated model, any definable set has an ordinal Morley rank. A structure is ω-stable if its complete theory is.

Note that $MR(D) = 0$ iff D is finite, and that strongly minimal theories have Morley rank 1 (a non empty definable subset of an $\aleph_0$-saturated model has $MR = 1$ iff it is infinite iff cofinite).

[11] x and y are here finite tuples.

Fact 2.15 *ω-stable theories are stable.* □

Proposition 2.16 *A C-minimal structure is dependent.*
It is stable iff there is a finite bound on the length of branches of its canonical tree. Such a C-structure is ω-stable of finite Morley rank.

Proof. For dependence, see [MacSte], Lemma 3.3. In the stable case, the only thing to prove is that a C-structure $\mathbb{M}$ with a finite bound on the length of its branches has finite Morley rank; it is clear by induction that, given some node $c = a \wedge b$, the definable set $\{x \in M; \neg C(x; a, b)\}$ has MR at most the depth of c (: maximal length of branches above c). □
More generally, for any C-structure $\mathbb{M}$ and any definable subset D of M, D is MR-ranked iff its tree $\{x \wedge y \in T(M); x, y \in D\}$ has finite depth.

3 Unary definable sets

A systematic investigation of the definable subsets (1 variable) of algebraically closed valued fields was carried out by Jan E. Holly in [Ho]. Later on, a general study was done by Deirdre Haskell and Dugald Macpherson in [HMac1] for definable subsets of M^n, $\mathbb{M}$ an arbitrary dense C-minimal structure. We should also mention [HrK], where the structures under consideration are expansions of algebraically closed valued fields, but where some of the results are valid in full generality. The decomposition considered here applies only to unary definable sets (1 variable). The formal results to be proved are the same as in [HMac1], namely that every definable set is a finite disjoint union of "Swiss cheeses", this partition can be refined to a partition into "basic sets", and finally by putting together some of these basic sets we get a partition into pieces definable over the same set of parameters as the set we began with. But the decomposition we provide here is explicite and it is uniform in the definable subset we want to partition. We shall freely use part of the terminology introduced in [Ho] and [HMac1].

Let $\mathbb{M}$ be a C-structure, and let T be its canonical tree.

Definition 3.1 For a and b distinct elements from M, $\Gamma(a \wedge b, b) := \{x; C(a; x, b)\}$ is the *cone of b at $a \wedge b$*. We also use the notation $\Gamma(C, d)$ when $C, d \in T$ with $C < d$, or $C \in T, d$ a branch of T and $C \in d$. If $T(M)$ has no root we also consider M as a cone, namely the cone of any of its elements at "$-\infty$".
For a and $b \in M$, $\underline{\Gamma}(a \wedge b) := \{x; \neg C(x; a, b)\}$ is the thick cone at $a \wedge b$; if $a \neq b$ this equals the union of all cones at $a \wedge b$.

In both cases, $a \wedge b$ ($\in T$) is the basis of the cone.
A Swiss cheese (a generalization of [Ho], Definition 3.5) is a (definable) subset of M of the following type: $F = G \setminus \bigcup_{i=1}^{s} G_i$, where G is a possibly thick cone or M, and the G_i possibly thick cones. A basis for G is called a basis for F.

Note that a definable set may be both a cone and a thick cone: for $c, d \in T$, $c < d$ without any point from T lying between c and d, the thick cone at d is the cone of d at c. An *explicit* cone (or thick cone) will correspond to an explicit formula $C(a; x, b)$ (or $\neg C(x; a, b)$) where a and b are parameters. The basis is in this case unambiguously $a \wedge b$.

Lemma 3.2 *1. Given two cones, each one possibly thick, either they are disjoint, or one contains the other.*
2. A Swiss cheese which is not a singleton is open in M.

Proof. 1. Clear.
2. Let $E := G \setminus \bigcup_{i=1}^{s} G_i$, G of basis c, each G_i of basis c_i, and let d be the leaf of x. If E contains x and is not the singleton $\{x\}$ then c and the $c_i \wedge d$ are strictly smaller than d, thus E contains the cone of x at $Max\{c, c_i \wedge d; i = 1, \ldots, s\}$. □

Proposition 3.3 *Let $\mathbb{M}$ be a C-minimal structure. Then every definable subset of M is a finite disjoint union of Swiss cheeses.*

Proof. By definition of C-minimality the set has a defining formula quantifier free in the pure language $\{C\}$. We take this formula in disjunctive form, and furthermore with pairwise contradictory terms (using

$$[(\bigwedge_i A_i) \vee (\bigwedge_j B_j)] \longleftrightarrow [(\bigwedge_i A_i) \vee \bigvee_i (\neg A_i \wedge \bigwedge_j B_j)]).$$

Now, the sets defined by an atomic formula are M, $\emptyset$, singletons, cones and complements of thick cones. Thus, by Lemma 3.2 (1), each term of the disjunction defines a Swiss cheese or the empty set. □

Corollary 3.4 *Let $\mathbb{M}$ be a C-minimal structure, $D \subseteq M$ a definable subset and $x \in D$. If x is not isolated in D then there is a cone G with $x \in G \subseteq D$. If D is infinite, then it contains a cone.*

Proof. Let E be a Swiss cheese from the decomposition of D with $x \in E$. By Lemma 3.2.(2) either E is a singleton or (x is isolated in E iff in M iff in D). If D is infinite, then it contains an infinite Swiss cheese, which must be open in M. □

Corollary 3.5 *Two disjoint open definable subsets have disjoint closures.*

Proof. Let D and E be disjoint open definable subsets of M and assume that there is $x \in \overline{D} \cap \overline{E}$. Since $D \cap E$ is empty, wlog $x \notin D$, then x is not isolated in M; but it is isolated in $\neg D$, thus there is some cone Γ_D containing x and such that $\Gamma_D \setminus \{x\} \subseteq D$. Now either x is not in E and in the same way there is some cone Γ_E containing x and such that $\Gamma_E \setminus \{x\} \subseteq E$, which yields a contradiction since any cone containing x is infinite, or $x \in E$ and E contains a cone containing x, contradiction. □

Given a definable set, we will now produce an explicit partition into Swiss cheeses of a specific form.

Definition 3.6 An almost thick cone at $n \in T$ is the thick cone at n with, in case n is not a leaf of T, finitely many (possibly no) cones at n removed; n is a basis of this almost thick cone.
The interval of bases c and d ($c, d \in T, c < d$) is the (definable) set $]c, d[:= \Gamma(c, d) \setminus \underline{\Gamma}(d)$, the cone of d at c with the thick cone at d removed; if T has no root, then $]-\infty, d[$ denotes $M \setminus \underline{\Gamma}(d)$.
A basic set is a (definable) subset of M of one of the following types: cone, almost thick cone, interval, or, if T has no root, M or M with a thick cone removed.

Proposition 3.7 *Let $\mathbb{M}$ be a C-minimal structure and $A \subseteq M$. Then every A-definable subset of M is a finite disjoint union of $(acl^{eq}A)$-definable basic sets.*

Proof. Let D be a definable set. By the definition of C-minimal, D is a Boolean combination of explicit, possibly thick, cones. Let C be the set of bases of these cones. Consider

$$T_0 := \{d \in T;\ \text{for some } c \in C, c \geq d\}$$

$$T_1 := \{c \in T;\ \exists \alpha \in D,\ c \in \alpha \ \text{ and } \ \exists \alpha \in M \setminus D,\ c \in \alpha\}.$$

Note: $T_1 \subseteq T_0$. Since C is finite, T_0 and hence T_1 has finitely many branches. We have the equivalence $T_1 = \emptyset$ iff ($D = M$ or $\emptyset$), then there is nothing to prove when $T_1 = \emptyset$. So suppose from now on $T_1 \neq \emptyset$ and define:

$U :=$ {suprema of branches from T_1}
$B :=$ {branching points of T_1}

$S := \{c \in T_1 \setminus (U \cup B)$; the thick cone at c without the cone of <u>the</u> branch of T_1 intersects non trivially both D and $\neg D\}$

$V := \big\{$infima $\in T_1 \setminus (U \cup B \cup S)$ of intervals on branches of T_1 which are maximal for being contained in $\{c \in T_1 \setminus (U \cup B \cup S)$; the thick cone at c without the cone of <u>the</u> branch of T_1 is contained entirely either in D or in $\neg D\}\big\}$

$F := U \cup B \cup S \cup V$.

Note that F is finite: U and B are finite because T_1 has finitely many branches, S is finite since it is contained in C, and V is finite by o-minimality of all branches of T_1. Furthermore T_1, U, B, S, V and F are definable from D (independently from the explicit definition of D that has been chosen); thus, if D is definable with parameters from A, then so are T_1, U, B, S and V, and $F \subseteq acl^{eq} A$. Because F is finite, it has a root $r(F)$; for each branch α of T_1 and $c \in T_1$ such that $c \in \alpha$, c has a successor in F on α, it is denoted $c^+(\alpha)$. For $c \in S \cup V$, c^+ denotes $c^+(\alpha)$ for α the unique branch of T_1 at c, and $\Gamma(c, T_1)$ the cone at c of any $d \in T_1, d > c$.

Lemma 3.8 *Suppose $D \neq M, \emptyset$. Then there is a partition of D into the finite union of the following sets:*

1. $\bigcup_{c \in U} [\underline{\Gamma}(c) \cap D]$
2. $\bigcup_{c \in B} [(\underline{\Gamma}(c) \setminus \bigcup_{\alpha \text{ branch of } T_1 \text{ at } c} \Gamma(c, \alpha)) \cap D]$
3. $\bigcup_{c \in B} \bigcup_{\alpha \text{ branch of } T_1 \text{ at } c} [(\Gamma(c, \alpha) \setminus \underline{\Gamma}(c^+(\alpha))) \cap D]$
4. $\bigcup_{c \in S} [(\underline{\Gamma}(c) \setminus \Gamma(c, T_1)) \cap D]$
5. $\bigcup_{c \in S} [(\Gamma(c, T_1) \setminus \underline{\Gamma}(c^+)) \cap D]$
6. $\bigcup_{c \in V} [(\underline{\Gamma}(c) \setminus \Gamma(c, T_1)) \cap D]$
7. $\bigcup_{c \in V} [(\Gamma(c, T_1) \setminus \underline{\Gamma}(c^+)) \cap D]$
8. *... if T has no root...* $[(M \setminus \underline{\Gamma}(r(F))) \cap D]$.

In 1, 2, 4, 6 each term between square brackets is a (possibly empty) union of cones of basis c; it is in particular an almost thick cone or a finite union of cones. In 3, 5, 7 each interval $\Gamma(c, \alpha) \setminus \underline{\Gamma}(c^+(\alpha)$ or $\Gamma(c, T_1) \setminus \underline{\Gamma}(c^+)$ is contained entirely either in D or in $\neg D$ as is the term in 8.

Proof. The partition (with some terms possibly empty) is clear. Now, if $c \in T_1$, a cone at c which is not the cone of some branch of T_1 (if $c \in B$ such a cone may not exist) is contained entirely either in D or in $\neg D$.

This proves that each term between square brackets in 1, 2, 4, 6 is a union of cones of basis c and the conclusion follows by definability. For c and d from T_1 such that $c < d$ and there is no point $f \in F$ with $c \leq f < d$, then $\underline{\Gamma}(c) \setminus \underline{\Gamma}(d)$ is contained entirely either in D or in $\neg D$. This implies that, for $c, d \in T_1$ such that $c < d$, even if $c \in F$, still $]c, d[$ is contained entirely either in D or in $\neg D$ (indeed $\Gamma(c, d) = \bigcup\{\underline{\Gamma}(e); e \in T,\ c < e \leq d\}$) and proves the last assertion. ⊣

The previous lemma yields a finite partition of D into basic sets which are $(acl^{eq} A)$-definable. □

In the partition given by Proposition 3.7 we now put A-conjugate basic sets together, which produces a partition into A-definable subsets, not only $(acl^{eq} A)$-definable. Sets U, B, S, V and F introduced in the proof are A-definable and therefore partitioned into orbits under the action of A-automorphisms of $\mathbb{M}$. These orbits are finite antichains of the canonical tree T.

Definition 3.9 For antichains U and V of T, we define:
- the order $U < V\ :\iff \forall x \in U,\ \exists y \in V,\ x < y$ and $\forall y \in V,\ \exists x \in U,\ x < y$ (given y this x is unique);
- for antichains $U < V$ in T, the subset $]U, V[$ of M, definable if U and V are finite, consisting of the union of cones of elements from V at nodes from U, with the thick cones at nodes from V removed (*i.e.* $]U, V[= \bigcup\{]u, v[; u \in U, v \in V\ \&\ u < v\}$). If T has no root we extend this notation to $]-\infty, U[$ of M, which will denote the complement of the union of thick cones at all $u \in U$.

Proposition 3.10 *An A-definable subset of M is a finite disjoint union of A-definable subsets of following type:*
- *$]U, V[$ for A-definable finite antichains U and V such that $U < V$ and: $\forall v, w \in V, \forall u \in U, (v \neq w\ \&\ w > u\ \&\ v > u) \Rightarrow u = v \wedge w$,*
- *M or $]-\infty, \{b\}[$ for an A-definable $b \in T$,*
- *union of A-conjugates of some $(A \cup \{b\})$-definable union of cones at b, for $b \in T \cap acl^{eq} A$.*

Proof. It is easy to make the partition obtained in Proposition 3.7 stable under A-automorphisms: in 1, 2, 4 and 6 we choose to present each piece between brackets as a union of cones each time it is possible, *i.e.* this piece is union of *finitely* many cones, and as a unique Swiss cheese otherwise. Then we take the union of all conjugates of any given piece. □

Example. Let $\mathbb{M}$ be a C-minimal structure. Suppose there is $a \in T(M)$ such that: there are exactly two branches at a; a has a successor, b, on one of them and has no successor on the other one; a has no predecessor. Let G be the cone of b at a and H the other cone. Then G is also the thick cone at b; a does not belong to $T(M \setminus G)$ and defines there a proper cut, therefore H is not quantifier-free definable with C only in $M \setminus G$.

Definition 3.11 A definable subset E of M is called C-minimal if, for any $\mathbb{M}$-definable $F \subseteq M$, $E \cap F$ is a Boolean combination of cones and thick cones of E.

Thus in the above example, $M \setminus G$ is not C-minimal. This is the only obstruction, as we see now.

Lemma 3.12 *Let G be a cone (respectively a thick cone) in M. Then $M \setminus G$ is C-minimal as a definable subset of M except if the basis of G is as a (respectively b) in the above example.*

Proof. Let G be a cone at a in M. If there are at least three cones at a then $T(M \setminus G) = T(M) \setminus T(G)$. If there are exactly two cones then $T(M \setminus G) = T(M) \setminus T(G) \setminus \{a\}$. In the two cases, a cone, possibly thick, in M at a node from $T(M \setminus G)$ is still a cone of same kind in $M \setminus G$. If there are exactly two cones at a, let $H \neq G$ be the other cone. If a has a predecessor a^- in $T(M)$ then H is a cone at a^-; if a has, in $T(M)$, a successor a^+ on the branch contained in H, then H is the thick cone at a^+. Thus the trace on $M \setminus G$ of a subset of M which is quantifier free definable using only C, is still quantifier free definable using only C, in $M \setminus G$. Same analysis for G a thick cone. □

Proposition 3.13 *Any A-definable subset of $\mathbb{M}$ is a finite disjoint union of A-definable subsets which are C-minimal as definable subsets of M.*

Proof. In the partition given by Proposition 3.10 only sets $]U, V[$ are not *a priori* C-minimal. We may suppose V contains no proper A-definable subset and, by Lemma 3.12, is such that any $v \in V$ has in $T(M)$ a predecessor v^-, v^- itself without predecessor. Set $W := \{v^-; v \in V\}$ and $D = \bigcup_{w \in W} [\underline{\Gamma}(w) \setminus \bigcup_{v \in V, v^- = w} \Gamma(w, v)]$. Then $]U, V[$ is the disjoint union of $]U, W[$ and D, and $]U, W[$ and D are both C-minimal. □

Corollary 3.14 *Let $D \subseteq M$ be definable with parameters from A and $x \in D$. If x is not algebraic over A, then there is an infinite $(acl^{eq} A)$-definable basic set F such that $x \in F \subseteq D$. In particular, if x is not isolated in M but is isolated in D, then it is algebraic over A.*

Proof. Follows from Lemma 3.2 (2) and Proposition 3.7. □

Proposition 3.15 *Let $A \subseteq M$ be algebraically closed and $x \in M \setminus A$. Then one of the 3 following happens:*
1. every ($acl^{eq}A$)-definable set containing x contains (M or) an infinite cone which has a basis in $acl^{eq}A$ and contains x;
2. (x is isolated in M and) there exists some node c on the branch x, at finite distance from the leaf of x, $c \in acl^{eq}A$, where the cone $G := \Gamma(c, x)$ of x is finite, there are infinitely many cones having same type as G over $A \cup \{c\}$, and such that any ($acl^{eq}A$)-definable set which contains x contains an almost thick cone of basis c, ($acl^{eq}A$)-definable and containing x.
3. There exists some node t on the branch x, $t \notin acl^{eq}A$, such that every $acl^{eq}(A)$-definable set containing x contains an ($acl^{eq}A$)-definable interval of the form $]c, d[$ or $]-\infty, d[$ with $c < t = (x \wedge d) < d$.

Proof. If $\neg 1$ let D_0 be some infinite $acl^{eq}(A)$-definable basic set which contains x and does not contain any infinite cone with a basis in $acl^{eq}A$ and containing x. Assume first that D_0 is an almost thick cone, of basis c; by the proof of proposition 3.7, c may be taken algebraic over A, hence the cone G of x at c must be finite, so that G and x are interalgebraic over $\emptyset$; since x is not algebraic over A and c is, infinitely many cones at c have same type as G over $A \cup \{c\}$. In other words G realizes the generic type over $A \cup \{c\}$ of the infinite strongly minimal structure induced by $\mathbb{M}$ at the node c. If now D_0 is an interval, say $]c, d[$, let $t := x \wedge d$. If $t \in acl^{eq}A$, again, the cone of x at t must be finite, there must be infinitely many cones at t and we are in case 2. For two such intervals D_0 and D_0' with the corresponding t and t', $D_0 \cap D_0'$ contains a subset of same type containing x only if $t = t'$. □
As a consequence:

Proposition 3.16 *Let $\kappa \geq |\mathscr{L}|$ be a cardinal, $\mathbb{M}$ a κ^+-saturated C-minimal $\mathscr{L}$-structure, $A \subseteq M$ an algebraically closed set of parameters with $|A| \leq \kappa$, and $x \in M \setminus A$. Define*

$$U_1 := \{u \in br\, x; \exists v \in (br\, x \cap acl^{eq}A), v \geq u\},$$

$$U_2 := \{u \in T; u \in acl^{eq}A \text{ and } u > u_1\}$$

$$D_1 := \bigcap_{u \in U_1} \Gamma(u, x), \quad D_2 := \bigcap_{u \in U_2} \underline{\Gamma}(u), \textit{ and}$$

$$D := \{y \in M; x \textit{ and } y \textit{ have same type over } A \cup U_1\}.$$

Then we are in one of the following situations:
1. either the cone of x at any basis in $acl^{eq}A$ is infinite and $U_2 = \emptyset$, then $D_1 \subseteq D$,
2. or the cone of x at any basis in $acl^{eq}A$ is infinite and $U_2 \neq \emptyset$, then $D_1 \setminus D_2$ is contained in D; in this case, for any $y \in D_1 \setminus D_2$, $x \wedge y$ does not belong to $acl^{eq}A$.
3. Or x is contained in some finite cone G with basis $c \in acl^{eq}A$, such that D contains all cones at c except at most κ of them.

Proof. Everything follows from Proposition 3.15 except the last assertion, a priori we only know that all cones except at most κ of them have same type as G. Now these cones must be homogeneous (see proposition 2.10, footnote 9, for the definition): a definable subset of a C-minimal structure and its complement can not both non trivially intersect infinitely many cones at a same node (see [HMac1] Fact 1 on page 129, which does not require C dense). □

4 Dense and discrete C-minimal structures The partition dense/discrete

As already mentioned, at first only the dense C-minimal structures had been considered. Group actions were an important motivation and people were interested in structures with transitivity properties (see for example the introduction of [MacSte]) and density is a natural condition in this context. Dense structures behave well and it is hence natural to assume density when one considers questions that are still completely open, like the trichotomy conjecture for example.

There are other contexts where density is not a relevant hypothesis. An example is when valued groups are under consideration, where the quotient of a dense group by a convex subgroup need not be dense, and density, although making some things easier, does not play any essential role. There is also the example given by o-minimality, which does not assume the density of the order. We have taken the analogy between C-minimality and o-minimality as a thread and an inspiration, the partition, up to finitely points, of a C-minimal structure into a dense part and a discrete one, that we present below, is an illustration of this process.

Finally general C-minimal structures provide also an interesting framework in the sense of pure model theory, as we have already explained.

4.1 Dense C-minimal structures

Let $\mathbb{M}$ be a dense C-minimal structure. In [HMac1] Deirdre Haskell and Dugald Macpherson have proven a cellular decomposition for definable subsets of any M^n. This yields in particular a notion of dimension. Cells are defined by induction on n. We will not define general cells here, only say that a cell in M is a disjoint union $D = D_1 \cup \cdots \cup D_k$, where the D_i are points, cones, almost thick cones or intervals and the automorphism group of D (with respect to the structure $\mathbb{M}$) acts transitively on the set of D_i.

Theorem 4.1 ([HMac1], Theorem 4.1) *Let $\mathbb{M}$ be a dense C-minimal structure and $D_1, \ldots, D_r$ definable subsets of M^n. Then there is a partition of M^n into finitely many cells such that each D_i is a union of some of the cells.*

Definition 4.2 If $X \subseteq M^n$ the topological dimension of X is

$$\text{topdim}\,(X) := \max\{m \in \mathbb{N} \text{ there is a projection } \pi : M^n \to M^m \text{ such that } \pi(X) \text{ has interior in } M^n\}.$$

Theorem 4.3 ([HMac1], Theorem 4.4) *Let $D_1, \ldots, D_r$ be definable subsets of M^n. Then* $\text{topdim}\,(\bigcup_{i=1}^r D_i) = \max\{\text{topdim}\, D_i; 1 \leq i \leq r\}$. *In particular if D is a definable subset of M^n without interior then D is not dense in any open subset of M^n.*

We describe now more precisely the part of this work which concerns definable unary functions. Let us first state the simplest case.

Definition 4.4 Let M be a dense C-structure. A partial function $f : M \to M$ is a local isomorphism if for any $x \in dom(f)$, there is a cone V of x in $dom(f)$ such that, for any $\alpha, \beta, \gamma \in V, C(\alpha; \beta, \gamma) \Leftrightarrow C(f(\alpha); f(\beta), f(\gamma))$.

Proposition 4.5 *Let $\mathbb{M}$ be a dense C-minimal structure and $f : M \to M$ a partial definable function. Then $dom(f)$ can be written as a disjoint union $F \cup I \cup K$, such that F is finite, I and K are definable, f is a local isomorphism and continuous on I and f is locally constant on K.*

Proof. This is part of Proposition 3.2 from [HMac1]. □

General statements of this kind consider functions ranging not only in M but in certain "imaginaries" (for this term, see [C] Subsection 2.7), and we need additional definitions.

An antichain of $T(M) =: T$ inherits the C-relation: $C(a;b,c)$ iff $C(\alpha;\beta,\gamma)$ for any branches $\alpha \ni a$, $\beta \ni b$, $\gamma \ni c$ of T. So we may define a *partial isomorphism* $f : M \to T$ as a partial one-to-one map which ranges into an antichain in T and such that: $\forall \alpha,\beta,\gamma \in dom(f), C(\alpha;\beta,\gamma) \Leftrightarrow C(f(\alpha);f(\beta),f(\gamma))$. We extend this notion to functions ranging not only into M (or T) but more generally into $\binom{M}{n} :=$ {subsets of M of size n} (or $\binom{T}{n}$), for some $n \in \mathbb{N}$. Such a map is a *local strong multi-isomorphism*[12] if

- its domain is dense,
- if f ranges into $\binom{T}{n}$, then $\bigcup\{f(x); x \in dom(f)\}$ is an antichain in T,
- for any $\alpha \in dom(f)$, there exist a cone Γ such that $\alpha \in \Gamma \subseteq dom(f)$, and cones $\Gamma_1, \ldots, \Gamma_n$ in M if f ranges into $\binom{M}{n}$ (cones in $im(f)$ if $im(f) \subseteq \binom{T}{n}$) such that, for each i, $1 \leq i \leq n$, and each $x \in \Gamma$, $\Gamma_i \cap f(x)$ is a singleton and the function $\Gamma \to \Gamma_i$ induced by f is a C-isomorphism between Γ and Γ_i;

$f : M \to \binom{T}{n}$ is a *local multi-isomorphism* if there are $s \leq n$, a local strong multi-isomorphism $\hat{f} : M \to \binom{T}{s}$ with $dom(\hat{f}) = dom(f)$, and a definable antichain A of T such that each $a \in A$ lies above some element from $\bigcup\{f(x); x \in dom(f)\}$ and, for all $x \in dom(f)$, $f(x) = \{y \in A; y \geq b$[13] for some $b \in \hat{f}(x)\}$.
A partial function $f : M^m \to \binom{Z}{n}$, where Z is either M or T, is *reducible to the family* $\{f_1, \ldots, f_k\}$ if, for each i, $1 \leq i \leq k$, $f_i : M^m \to \binom{Z}{s_i}$ is a definable partial function, $1 \leq s_i \leq n$, $dom(f_i) = dom(f)$, $\Sigma s_i = n$, and for every $\alpha \in dom(f)$, $f(\alpha) = \bigcup_{1 \leq i \leq k} f_i(\alpha)$.

We can now state the next proposition, which follows from Proposition 3.11 of [HMac1], and that we will use to describe possible interactions between the dense and the discrete parts of a C-minimal structure.

Proposition 4.6 *Let $\mathbb{M}$ be a dense C-minimal structure and $f : M \to \binom{Z}{n}$ a definable partial function, where Z is either M or T. Then the domain of f can be written as a disjoint union of a finite set F and finitely many definable open sets K_i such that, on each K_i, f is continuous and reducible to a family of functions, each of which is a local multi-isomorphism or locally constant.*

Note that, for $n = 1$ and f ranging into M, this statement is Proposition 4.5.

[12] We are not following word for word [HMac1], where the definition seems incomplete.
[13] This appears incorrectly as "$y > b$" in [HMac1].

4.2 Discrete C-minimal structures

Proposition 4.7 *Let D be a definable subset of M. Then, given $N \in \mathbb{N}$, there is a finite $F \subseteq D$ such that, for any $x \in D \setminus F$ and any possibly thick cone G containing x, if $|G| \geq N$ then $|G \cap D| \geq N$.*

Proof. By Proposition 3.7, we need only consider the case where D is a cone, an almost thick cone or an interval. It is trivial for a cone since in this case either D and G are disjoint, or one of them is contained in the other. Let D be an almost thick cone, d a basis of D and G a possibly thick cone of basis c. If $0 < |D \cap G|$ finite $< |G|$ then, either G is a cone and $c < d$, or G is a thick cone and $c \leq d$. In both cases $D = D \cap G$ is finite. If now D is the interval $]d, e[:= \{x \in M; d < (x \wedge e) < e\}$ and G a cone, again if $0 < |D \cap G|$ finite $< |G|$ then $G = \Gamma(c, e)$ for some $c < e$ at finite distance from e. Given N there is a first c such that $d < c < e$ and $|]c, e[| < N$, say c_0. Then $]c_0, e[$ must contain any point from D which is in some cone of size $\geq N$ intersecting D in a set of size $< N$. When G is a thick cone we argue in the same way with $[c, e[$ in place of $]c, e[$. □

Corollary 4.8 *Let $x \in M \supseteq A$. Assume $\mathbb{M}$ is (infinite and) $|A|^+$-saturated. Then x is algebraic over A iff there exists some A-definable $D \subseteq M$ and a possibly thick cone G in M such that $x \in G \cap D$ and $0 < |D \cap G| < |G|$ (cardinals ranging in $\omega \cup \{\infty\}$).*

Proof. The previous proposition proves $\Leftarrow$.
Conversely any $x \in M$ algebraic over A is contained in some finite A-definable $D \subseteq M$. Either there is some cone G of cardinality $> |D|$ containing x, then G must decrease when intersecting D, or $T(M)$ has a root r, at finite distance from x. In this case $\underline{\Gamma}(r) = M$. □

This last corollary can be seen as a generalisation of the property: " In a dense C-minimal structure, isolated elements from D are algebraic on parameters defining D" (Corollary 3.14).

4.3 The partition dense/discrete of C-minimal structures

Proposition 4.9 *Any C-minimal structure can be written as a disjoint union $F \cup De \cup Di$, where F is finite, De is dense, Di is discrete. If furthermore De and Di are maximal open, then they are unique and $\emptyset$-definable.*

Proof. De and Di must be defined as follows:
$De := \{x \in M$; there is a cone containing x, any point of which is a limit (in M)$\}$,
$Di := \{$isolated points of $M\}$,

which are clearly $\emptyset$-definable. If $M \setminus (De \cup Di)$ is infinite, it contains an infinite swiss cheese E, any point of which is a limit, therefore E contains a cone of limit points, contradiction with the definition of De. □

By Proposition 3.13 this partition can be refined to a finite partition into C-minimal $\emptyset$-definable pieces. Let D be such a piece, $\emptyset$-definable in the C-minimal structure $\mathbb{M}$ and C-minimal as such. Make D into a structure $\mathbb{D}$ by putting into the language of $\mathbb{D}$ an n-ary predicate for each $\mathbb{M}$-definable subset of D^n and each n. Then the structure $\mathbb{D}$ is C-minimal. Indeed, by C-minimality of $\mathbb{M}$, any $\mathbb{D}$-definable subset of D is quantifier-free definable with C only, and by Fact 2.2 this remains true for any $\mathbb{D}' \equiv \mathbb{D}$. Therefore the study of general C-minimal structures reduces to the study of discrete ones, dense ones and possible interactions between them.

Remarks. When comparing with the o-minimal case:
1. In o-minimal structures, the same definitions of De and Di lead to a similar partition. But, in o-minimal structures, dense and discrete parts are *orthogonal*, in the sense that any definable map from one part into the other one has a finite range ([PiSte2] Proposition 2.3), and a discrete structure is necessarily *trivial* ([PiSte1]), as we will define in Section 5. It is definitely not the case in C-minimal structures. We will say more about this point in Subsections 4.4 and 5.2.
2. Points from F are algebraic over the empty set. They are exactly the limits of isolated points. No $x \in F$ is a proper limit of limit points (use 3.5). It is different in o-minimal structures, think of the chain $\mathbb{Q}+\lambda+\mathbb{Z}$, where λ is a singleton and $\mathbb{Q} < \lambda < \mathbb{Z}$: the closures of $De(= \mathbb{Q})$ and $Di(= \mathbb{Z})$ intersect non trivially.

4.4 Interactions between the dense and the discrete parts

In order to give some examples, we first come back to a construction presented at the end of Subsection 2.2.

Proposition 2.10 *Suppose that E is homogeneous and such that $T(E)$ has a root. Then $E \rtimes \mathbb{M}$ is C-minimal iff E and $\mathbb{M}$ are.*

Proof. The canonical tree of $\mathbb{N} := E \rtimes \mathbb{M}$ consists of $T(M)$ in which each leaf $x \in M$ is replaced by a copy T_x of $T(E)$. Let us call r_x the root of T_x. Then $T(M)$ is isomorphic to the initial subtree $\{c \in T(N); c \leq r_x \text{ for some } x \in M\}$ of $T(N)$, and the canonical projection $\pi : N \to M$

induces a one-to-one map (preserving inclusion and disjointness) between cones of N with basis in in $T(M) \setminus \{r_x; x \in M\}$ and cones of M, and between thick cones of N with basis $T(M)$ and thick cones of M.
1. Assume $E \rtimes \mathbb{M}$ is C-minimal. Then E is C-minimal too, as it is isomorphic to a thick cone of N. If $D \subseteq M$ is a definable subset, then $\pi^{-1}(D)$ is N-definable, hence a finite (disjoint) union of Swiss cheeses k of N. Now D is the union of the πk and each πk is a Swiss cheese of M, which proves C-minimality of $\mathbb{M}$.
2. We prove now the converse. Given any formula ψ of the language $\{C\}$, consider the formula ψ_* of the language $\{C, \sim\}$ such that, for any elements $x_1, \ldots, x_n \in N$, $\mathbb{N} \models \psi_*(x_1, \ldots, x_n)$ iff $[\bigwedge_{1<k\leq s}(x_{j_1} \sim x_{j_k})$ and $\psi(x_1, \ldots, x_n)$ is true in the class of the x_i]. Now, since $\mathbb{M}$ is interpretable in $\mathbb{N}$, for any formula φ of $\mathscr{L}$ there is a formula φ^* of $\mathscr{L} \cup \{\sim\}$ such that, for any $x_1, \ldots, x_n \in N$, $\mathbb{N} \models \varphi^*(x_1, \ldots, x_n)$ iff $\mathbb{M} \models \varphi(\pi(x_1), \ldots, \pi(x_n))$.

Lemma 4.10 *Any formula of $\mathscr{L} \cup \{\sim\}$ in the variables $x_1, \ldots, x_n$ is equivalent in $\mathbb{N}$ to a Boolean combination of formulas $\varphi^*(x_{i_1}, \ldots, x_{i_r})$ or $\psi_*(x_{j_1}, \ldots, x_{j_s})$.*

Proof. With the test of Shoenfield we will prove simultaneously the lemma and that $\mathbb{N}$ is axiomatized by the following sentences:
- $\sim$ is an equivalence relation, whose classes are thick cones; hence $[\neg(x \sim y \sim z) \wedge C(x; y, z) \wedge x \sim x' \wedge y \sim y' \wedge z \sim z'] \to C(x'; y', z')$;
- $[\bigwedge_1^n x_i \sim y_i \wedge R(x_1, \ldots, x_n)] \to R(y_1, \ldots, y_n)$ for any n-ary predicate symbol $R \neq C$ of $\mathscr{L}$; hence $N/\sim$ is equipped canonically with an $\mathscr{L}$-structure which we denote $\mathbb{N}/\sim$;
- $\bigwedge\{\varphi^*; \mathbb{M} \models \varphi\}$, i.e. $\mathbb{N}/\sim \equiv \mathbb{M}$;
- $\bigwedge\{\psi_*; E \models \psi\}$, i.e. each class modulo $\sim$ as a C-set with the restriction of C is elementarily equivalent to E.

Given models $\mathbb{N}$ and $\mathbb{N}'$ of these axioms, $\mathbb{N}$ countable and $\mathbb{N}'$ ω_1-saturated, substructures $B \subseteq N$ and $B' \subseteq N'$ preserving formulas φ^* and ψ_*, and f an isomorphism $B \simeq B'$ preserving these formulas too, we have to find an embedding of $\mathbb{N}$ into $\mathbb{N}'$ extending f and preserving formulas φ^* and ψ_*. Note first that, for $\varphi(x, y)$ the formula $x = y$, then $\varphi^*(x, y)$ is equivalent to $x \sim y$. Thus f preserves $\sim$ and induces $g : \pi(B) \simeq \pi(B')$, which is $\mathscr{L}$-elementary and thus extends to an elementary embedding of $\mathbb{M} := \pi(\mathbb{N})$ into $\mathbb{M}' := \pi(\mathbb{N}')$. Now, for any $m \in M$, $\pi^{-1}(m)$ is a class modulo $\sim$, *i.e.* a set of size $\leq \aleph_0$ with no other structure than that given by C (and m), hence elementary embeddable into $\pi^{-1}(g(m))$ in such a way to extend $f \restriction \pi^{-1}(m)n\ dom(f)$. So we have got an embedding of $\mathbb{N}$ into $\mathbb{N}'$, extending f and preserving formulas φ^* and ψ_*. □

If $\mathbb{M}$ is C-minimal, a formula φ^* involving a single variable defines clearly in $\mathbb{N}$ a Boolean combination of cones (see item 1. at the beginning of the proof of the proposition). Consider now a formula ψ_*. If it has parameters and is not empty, it defines a subset of a unique class modulo $\sim$, hence a Boolean combination of cones or thick cones by C-minimality of E. If it has no parameters, it is stable under $\sim$ by homogeneity of E and is therefore of the form φ^*. Thus $\mathbb{M}$ is C-minimal if E and $\mathbb{M}$ are. □

For $m \in M$, $\pi^{-1}(m)$ is a thick cone in N, and the set of bases of such cones is the maximal antichain $\{r_x; x \in M\}$ of $T(N)$. Now N is dense if E is, and this will allow us to apply Propositions 4.5 and 4.6 to functions with a dense domain, say D, and taking values in a discrete C-structure, say M, such functions will be considered as functions $D \to T(N)$ for N as above.

Combination with a connection

Let E be an homogeneous pure C-set with a root in $T(t)$ and $\mathbb{M}$ a C-structure in a relational language $\mathscr{L}$. First form the extension $\mathbb{N} := E \rtimes \mathbb{M}$ and then the connection which we denote $\mathbb{P} := \mathbb{N} \sqcup \mathbb{M}_1$ with $\mathbb{M}_1 \simeq \mathbb{M}$ in order to distinguish between the two copies of $\mathbb{M}$. The canonical projection $\pi : N \mapsto M$ is read as a partial function $\pi : P \mapsto M_1$. Consider $\mathbb{P}$ in the language consisting of C, two unary predicates for N and M_1, a symbol for π, $\mathscr{L}$ on M_1, and $\mathscr{L} \cup \{\sim\}$ on N. Structures $\langle P; C, N, M_1, \pi\rangle$ and $\langle E \rtimes M; C, \sim\rangle$ are quantifier free bi-interpretable, hence by Lemma 4.10, $\mathbb{P}$ eliminates quantifiers in the language $\{C, N, M_1, \pi\} \cup \{$formulas φ^* interpreted in $N\} \cup \{\mathscr{L}$-formulas on terms ranging in $M_1\}$. In particular, if E is homogeneous, $T(E)$ has a root, and E and $\mathbb{M}$ are C-minimal, then $\mathbb{P}$ is C-minimal too. In the structure $\mathbb{P}$, the range of π consists now of (not only imaginary but) "real" elements.

Examples. 1. Given an algebraically closed non trivially valued field (K, v), first consider it as a two-sorted structure with sorts for K and $\underline{vK} := vK \cup \{\infty\}$, and then the connection $\mathbb{K} := K \sqcup \underline{vK}$. Now $De\mathbb{K} = K$, $Di\mathbb{K} = vK$ and the valuation is a definable function, locally constant and onto, from $De\mathbb{K} \setminus \{0\}$ to $Di\mathbb{K}$.
2. Let $\mathbb{M}$ be a dense C-minimal structure, $\mathbb{N}$ its extension by some C-structure trivial and different from a singleton, $\pi : N \to M$ the canonical projection and $\mathbb{P} := \mathbb{M} \sqcup \mathbb{N}$ enriched with π as above.In this way $\mathbb{P}$ is a

C-minimal structure with $DiP = N$ and $DeP = M$, and π is a definable function from DiP onto DeP.
3. Consider an homogeneous pure C-set E such that $T(E)$ has two branches at each node and branches isomorphic to $\mathbb{Q}$ plus endpoints. A routine back-and-forth in the language $\{C, \perp\}$, where $x \perp y$ iff $x \wedge y$ is the root of $T(E)$, shows that E is $\aleph_0$-categorical and eliminates quantifiers in the language $\{C, \perp\}$. In particular E is C-minimal and homogeneous. And it is dense, as $E \rtimes M$ is for any C-structure $\mathbb{M}$.

Proposition 4.11 *In a C-minimal structure, any definable partial function from the dense part to the discrete one is locally constant except in finitely many points.*

Proof. We will use the local description of definable functions in dense C-minimal structures by Haskell and Macpherson. We could reread their proofs line by line and make sure that density is used in fact only on the domain of the function; we can also use the previous construction as we explain now. Let $\mathbb{M}$ be a C-minimal structure and $f : G \mapsto H$ a definable partial function from the dense part to the discrete one. Assume for a contradiction that there are infinitely many points where f is not locally constant. Then there is a cone (in G) of such points and this cone must contain a cone of M. Wlog we may assume that G is this last cone and $H = f(G)$ is infinite (if $f(G)$ is finite, some of the $f^{-1}(h)$ for $h \in H$ is infinite, thus contains a cone, contradiction).
Consider E as in Example 3 above, $N := E \rtimes M$ and $\mathbb{P} = \langle G \sqcup N, C, G, N, f \rangle$; $\mathbb{P}$ is dense and C-minimal, f may be seen as a definable function $f' : G \mapsto T(N)$, and f is continuous iff f' is. Proposition 4.6 applies to f'. Now, a continuous map to a discrete space is necessarily locally constant, contradiction. □

We will prove in the next section that, in a C-minimal structure *with exchange*, any definable function from the dense part to the discrete one has a finite range.

5 Geometric dense C-minimal structures and the question of the trichotomy

Strongly minimal structures and o-minimal structures have a good dimension theory, originating in the exchange property; a property not necessarily shared by dense C-minimal structures. The assumption of this property allows us to consider the pregeometry associated with the algebraic closure operator and to investigate which information it carries about the whole structure, as it is classical to do for strongly minimal structures. We end with some other consequences of the exchange property.

5.1 Pregeometries and Zilber conjecture

We recall that the algebraic closure $aclA$ of some set of parameters A in a structure $\mathbb{M}$, is the union of all finite A-definable subsets of M (see [C] 2.13). For general properties of algebraic closure, and everything we will mention about pregeometries or strongly minimal structures, we refer to [Mar].

Definition 5.1 The algebraic closure has the exchange property in a structure $\mathbb{M}$, or $\mathbb{M}$ has the exchange property if, for any $A \subseteq M$, b and c in M, if $b \in acl(A \cup \{c\} \setminus aclA)$, then $c \in acl(A \cup \{b\}$.
In this situation, the pregeometry associated to $\mathbb{M}$ consists of M and the operator which maps $A \subseteq M$ to $aclA$; A is said to be independent if, for any $a \in A$, $a \notin acl(A \setminus \{a\})$; $B \subseteq A$ is a basis of A iff B is independent and generates A in the sense that $A \subseteq aclB$.

As for linear independence in linear algebra or for algebraic independence in field theory, we have:

Fact 5.2 *A is independent iff it admits an enumeration $A = \{x_\alpha; \alpha < \alpha_0\}$ such that, for any $\alpha < \alpha_0$, x_α is not algebraic over $\{x_\alpha; \beta < \alpha\}$. Any two bases of A have the same cardinality.* □

Definition 5.3 We call the cardinality of any basis of A the dimension $dimA$ of A.
For $C \subseteq M$, the structure $\mathbb{M}_C$ consisting of $\mathbb{M}$ enriched by constant symbols for the elements from C, has still the exchange property. It is called the localisation of $\mathbb{M}$ at C. Its algebraic closure acl_C is given by $acl_C A = acl(A \cup C)$. Its dimension satisfies $\dim_C A + \dim C = \dim A$.

Examples.
1. Strongly minimal and o-minimal structures have the exchange property, see [Mar] and [PiSte1]. In particular, algebraically closed fields do, and the dimension coincides in this case with the transcendence degree.
2. Dense C-minimal structures need not have the exchange property, see [MacSte] example 3.2. But:

3. Algebraically closed valued fields have the exchange property, the corresponding dimension being the transcendence degree. Indeed, the algebraic closure of a set of parameters A, in the model theoretic sense, for both the field and the valued field structure, is the same thing as the algebraic closure, in the sense of field theory, of the subfield generated by A.
4. Dense C-minimal pure C-structures have the exchange property, see [D].
5. Haskell and Macpherson have given a criterion for the exchange property in dense C-minimal structures ([HMac1] Proposition 6.1 and its proof): A dense C-minimal structure $\mathbb{M}$ has the exchange property iff, for any integer n, any locally constant definable partial function from M to $\binom{M}{n}$, the set of finite subsets of M of size n, has a finite range, iff there is no definable C-isomorphism from some cone of M to $T(M) \setminus F(M)$.

Note: in the above definition of dimension, A is a set of parameters, living in some model, and staying the same in any bigger model. At the opposite, in the definition below, D is a definable set, which means that, except if it is finite, its size depends on the model in which it is considered and it is legitimate to consider it in any model containing the parameters used in its definition. A and D are just not objects of the same nature! This remark resolves the apparent contradiction between Definition 5.3 and Definition 5.4 below.

Definition 5.4 Let $D \subseteq M^n$ be definable with parameters a.
The dimension of D is $\dim D := \max\{\dim(x; a); x \in D\}$.
Let $A \subseteq M$, $A \ni a$; a point $m \in D$ is said to be generic in D over A if $\dim(m; A) = \dim D$; "m generic in D" means "m generic in D over a".

In dense C-minimal structures with exchange, this dimension coincides with the topological dimension defined in section 4:

Proposition 5.5 *Let $\mathbb{M}$ be a dense C-minimal structure with exchange, κ^+-saturated if κ is the cardinality of the language, and $D \subseteq M^n$ a definable subset. Then*

$$\operatorname{topdim} D = \max\{\dim\{x_1, \ldots, x_n\}; (x_1, \ldots, x_n) \in D\}.$$

Proof. [HMac1], Proposition 6.3. □

This may be rephrased as follows: $m \in D$ is generic in D iff it is contained in some open subset of D.

Definition 5.6 A pregeometry (M, acl) is said to be trivial if for any $A \subseteq M$, $aclA = \bigcup\{acl(\{a\}); a \in A\}$.
It is modular if, for any $A, B \subseteq M$, A and B finite dimensional, $\dim(A \cup B) = \dim A + \dim B - \dim(acl(A \cap B))$.
It is locally modular if its localisation at some element from M is modular.

Examples. 1. A strongly minimal group is not trivial: consider two independent generics x and y; then $x.y \in acl(x, y) \setminus acl(x) \setminus acl(y)$.
2. A strongly minimal field (in fact it must be algebraically closed) is not locally modular: given any a, consider three independent generics over a, say x, y and z, i.e. $tr(a, x, y, z) = 4$; then $acl(a, x, y) \cap acl(a, z, xy + z) = acl(a)$, thus $dim_a(acl(a, x, y) \cap acl(a, z, xy + z)) = 0$, but $\dim_a(x, y) = \dim_a(z, xy + z) = 2$ and $\dim_a(x, y, z) = 3$.

Boris Zilber hoped that these examples had a general character, precisely that a non trivial strongly minimal structure "contains" in some sense an infinite group, and a non locally modular structure a field. His intuition was correct concerning locally modular structures, as Hrushovski has shown: Let $\mathbb{M}$ be a strongly minimal structure, κ^+-saturated if κ is the cardinality of the language; suppose its pregeometry is locally modular and non trivial; then there is an infinite group interpretable in $\mathbb{M}$ ([Hr]). Disappointingly, a non locally modular strongly minimal structure, even sufficiently saturated, need not interpret an infinite group; it becomes true under additional conditions, called Zariski geometries, elaborated by Hrushovski and Zilber: a non locally modular Zariski geometry interprets indeed a field ([HrZ]).

Astonishingly, $\aleph_1$-saturated non locally modular dense o-minimal structures do interpret, not exactly a field but, let us say, a bounded part of a field. More precisely, Peterzil and Starchenko have proven what we called here the *trichotomy*, namely, first: either the structure is trivial, or it defines on some convex subset an ordered group; secondly either it is locally modular or it defines on some convex subset a real closed field ([PeSta]). Crucial facts in their proof is that o-minimal dense structures are *geometric* (from [HrPi1]), which means that algebraic closure has exchange and that they are *algebraically bounded* (given a formula $\varphi(x, y)$ there is an integer n such that, for any a, $\varphi(x, a)$ has either less than n solutions or infinitely many). And furthermore the dimension associated with algebraic closure coincides with the topological dimension (for the topology given by the order presently). Now dense

C-minimal structures with exchange share these properties ([HMac1] Lemma 2.4). Do they satisfy the trichotomy? Mixing techniques used in strongly minimal and o-minimal cases as well as features distinguishing C-relations, Fares Maalouf has proven the two following results.

Proposition 5.7 ([Maa1] and [Maa2]) *A locally modular non trivial dense C-minimal structure with exchange defines an infinite group.*

Proposition 5.8 ([Maa4]) *Let $\mathbb{F} = (F, +, \cdot, v)$ be an algebraically closed valued field of characteristic 0, and $\mathbb{B}$ the reduct of $\mathbb{F}$ to a language containing C and the language of vector spaces over F. Then if $\mathbb{B}$ is not locally modular, the multiplication of F is definable in $\mathbb{B}$ on an open neighbourhood of 0.*

5.2 More about the dense/discrete partition

Let us come back to the example of algebraically closed non trivially valued fields. Such a structure has the exchange property when considered as a one-sorted structure, for elements from the field ("$\mathscr{L}_{\text{div}}$" or "language 2" in Example 4 from [C], 1.5), but not as a two-sorted structure, for elements from the field and the valuation group. Indeed if $x \in K$ is generic in K then $v(x)$ is definable over x but x is not algebraic over $v(x)$. Now, in this 2-sorted structure, the valuation is a locally constant definable map which is onto vK. Is there a connection between exchange and the existence of such functions? We will indeed show that in a C-minimal structure with exchange, any definable application from the dense part to the discrete one has a finite range.

Proposition 5.9 *In a C-minimal structure with exchange, any definable application from the dense part to the discrete one has a finite range.*

Lemma 5.10 *Let $\mathbb{M}$ be a C-minimal structure with exchange, M_1, $M_2 \subseteq M$, definable, M_1 dense and $f : M_1 \to M_2$ partial, definable and locally constant. Then the range of f is finite.*

Proof. If the range is infinite, it contains a generic element y, *i.e.* not algebraic over all parameters occurring; $f^{-1}(y)$ contains a cone, which contains an x generic over y, but y is definable over x, contradiction. □

Proof of Proposition 5.9. The result follows from proposition 4.11 and the previous lemma. □

Remark. By what we have just seen, if a function $f : De \to Di$ is A-definable then each point in its range is A-algebraic; it may not be A-definable, as shows the following example. Let $\mathbb{M}$ be a C-minimal dense structure, $\mathbb{M}_1$ and $\mathbb{M}_2$ two copies of $\mathbb{M}$, $\mathbb{N}_0 := \mathbb{M}_1 \sqcup \mathbb{M}_2 \sqcup \{a_1, a_2\}$, f the map $M_1 \cup M_2 \to \{a_1, a_2\}$ sending M_i to a_i for $i = 1, 2$, and $\mathbb{N} := \langle N_0, C, f\rangle$. So $\mathbb{N}$ is C-minimal, $DeN = M_1 \cup M_2$, $DiN = \{a_1, a_2\}$, f is $\emptyset$-definable and there are automorphisms of $\mathbb{N}$ mapping a_1 to a_2.

Example of a definable function from the discrete part to the dense one with an infinite range (with $Di \subsetneq aclDe$ and $acl\emptyset = \emptyset$). Let us go back to the second example before Proposition 4.11, $\mathbb{P} = \langle E \rtimes \mathbb{M} \sqcup \mathbb{M}, \pi\rangle$, with M dense. Assume furthermore E finite, M with exchange and such that $acl_{\mathbb{M}}\emptyset = \emptyset$. For tuples $x \in M$ and $y \in N$, we have $acl_{\mathbb{P}}(x, y) = acl_{\mathbb{M}}(\pi(y), x) \cup \pi^{-1}(acl_{\mathbb{M}}(\pi(y), x))$, which implies that $\mathbb{P}$ has exchange and $acl_{\mathbb{P}}\emptyset = \emptyset$. The function π is onto from $N = Di(P)$ to $M = De(P)$.

The next proposition shows that this example is "locally canonical".

Proposition 5.11 *Let $\mathbb{M}$ be a C-minimal structure with exchange, $H \subseteq DiM$, $D \subseteq DeM$, and $f : H \twoheadrightarrow D$, H, D and f definable. Then there exist $y_1, \ldots, y_s \in D$ and finite homogeneous[14] C-sets $E_1, \ldots, E_r$ such that, for any $y \in D$, $y \neq y_1, \ldots, y_s$, and any $x \in f^{-1}(y)$, there exist a cone Γ (in H) containing x, a cone Δ (in D) containing y, and $j \in \{1, \ldots, r\}$, such that Γ is C-isomorphic to $E_j \rtimes \Delta$ with $f \restriction \Gamma$ sent to the canonical projection.*

Proof. By exchange, only finitely many $f^{-1}(y)$, $y \in D$, are infinite. Remove the corresponding y. The other $f^{-1}(y)$ have a bounded size. We cut what remains from D into finitely many pieces such that, on each of them, all $f^{-1}(y)$ are isomorphic. In this way wlog $f : H \to D$ is n-to-1, and has therefore an inverse $g : D \to \binom{H}{n}$, $g(y) := f^{-1}(y)$. By Proposition 4.6 there is a partition of D into a finite subset F and finitely many definable open subsets such that on each of them g is reducible to a finite family of functions g_i, which must be local multi-isomorphisms since, by definition of g, $g(y)$ and $g(y')$ are disjoint if $y \neq y'$. Any $y \in H$ is in the image of one of the g_i, say g_{i_0}. Now $g(y) = \bigcup_i g_i(y)$, and the form of the statement ("$\forall x \in f^{-1}(y)$ etc.") allows us to replace g by g_{i_0}, *i.e.* to assume that g itself is a local multi-isomophism. By definition

[14] "homogeneous" is *a priori* in the sense of C only (cf. footnote 8) but, by the present proposition, the E_j appear finally as embedded in $\mathbb{M}$ and they can be taken homogeneous in the sense of $\mathbb{M}$.

there is a local strong multi-isomophism $\hat{g} : D \to \binom{T(H)}{m}$ such that $g(y) = \{z \in D; z \geq t$ for some $t \in \hat{g}(y)\}$. Fix some $y_0 \in D \setminus F$, and $x_0 \in \hat{g}(y_0)$. There are cones Δ of D and Γ_0 of $im\, \hat{g}$ such that $y_0 \in \Delta$, $x_0 \in \Gamma_0$, and for any $y \in \Delta$, $\hat{g}(y) \cap \Gamma_0$ is a singleton, and the function induced from Δ to Γ_0 is a C-isomorphism. Now, $g(y)$ is a finite tree, say E_y, constant up to isomorphism. The only possibility is that $\hat{g}(y)$ is the root of E_y and E_y the set $\{x \in H; x \geq \hat{g}(y)\}$. It remains to prove that E_y is homogeneous. The E_y are uniformly definable in $T(H)$. If their branches did not all have same type, there would exist some C-formula φ satisfied by some $x \in H$ but not by all of them. Then φ as well as its negation would non trivially intersect each E_y, which contradicts Proposition 4.7. □

References

[AN] Samson A. Adeleke and Peter M. Neumann, Relations Related to Betweenness: Their Structure and Automorphisms, Mem. AMS 623, 1998.

[Adl] Hans Adler, Theories controlled by formulas of Vapnik-Chervonenkis codimension 1, manuscript, 2008.

[C] Zoé Chatzidakis, Introductory notes on the model theory of valued fields, this volume.

[ClLoe1] Raf Cluckers and François Loeser, b-minimality, JML 7 (2007), 195–227.

[ClLoe2] Raf Cluckers and François Loeser, Integration through cell decomposition and b-minimality, this volume.

[D] Françoise Delon, Geometries of pure C-minimal structures, submitted, 2009.

[DMo] Françoise Delon and Marie-Hélène Mourgues, Classification of $\aleph_0$-categorical pure C-minimal structures, manuscript.

[DSi] Françoise Delon and Patrick Simonetta, Classification of C-minimal valued Abelian groups, manuscript.

[HHrMac] Deirdre Haskell, Ehud Hrushovski and Dugald Macpherson,Stable Domination and Independence in Algebraically Closed Valued Fields, LNL 30, ASL, Cambridge University Press, Cambridge, 2007.

[HMac1] Deirdre Haskell and Dugald Macpherson, Cell decompositions of C-minimal structures, APAL 66 (1994), 113–162.

[HMac2] Deirdre Haskell and Dugald Macpherson, A version of o-minimality for the p-adics, JSL 62 (1997), 1075–1092.

[Ho] Jan Élise Holly, Canonical forms for definable subsets of algebraically closed and real closed valued fields. JSL 60 (1995), 843–860.

[Hr] Ehud Hrushovski, Locally modular regular types, in Classification theory (Ed J.T.Baldwin), 132-164, LNM 1292, Springer, Berlin, 1987.

[HrK] Ehud Hrushovski and David Kazhdan, Integration in valued fields, in Algebraic geometry and number theory, 261–405, Progr. Math. 253, Birkhäuser Boston, Boston (2006).

[HrPi1] Ehud Hrushovski and Anand Pillay, Groups definable in local fields and pseudo-finite fields, Israel J. of M. 85 (1994), 203–262.

[HrPi2] Ehud Hrushovski and Anand Pillay, On NIP and invariant measures, ArXiv.

[HrZ] Ehud Hrushovski and Boris Zilber, Zariski geometries, Journal of the AMS 9 (1996), 1–56.

[KiPi] Byunghan Kim and Anand Pillay, From stability to simplicity, BSL 4 (1998), 17–36.

[LR] Leonard Lipshitz and Zacharie Robinson, One dimensional fibers of rigid subanalytic sets, JSL 63 (1998), 83–88.

[Maa1] Fares Maalouf, Construction d'un groupe dans les structures C-minimales, JSL 73 (2008), 957–968.

[Maa2] Fares Maalouf, Type-definable groups in C-minimal structures, CRAS, 1348 (2010), 709–712.

[Maa3] Fares Maalouf, Espaces vectoriels C-minimaux, JSL, 75 (2010), 741–758.

[Maa4] Fares Maalouf, Additive reducts of algebraically closed valued fields, preprint.

[Mac] Dugald Macpherson, Notes on o-minimality and variations, in Model Theory, Algebra, and Geometry, Haskell, Pillay Steinhorn (editors), MSRI Publications 39, Cambridge University Press (2000), 97–130.

[MacSte] Dugald Macpherson and Charlie Steinhorn, On variants of o-minimality, APAL 79 (1996), 165–209.

[Mar] David Marker, Model Theory: An Introduction, Graduate Texts in Mathematics, Springer (2002).

[PeSta] Ya'acov Peterzil and Sergei Starchenko, Geometry, Calculus and Zil'ber Conjecture, BSL 2 (1996), 77–83.

[Pi] Anand Pillay, An introduction to Stability Theory, Oxford University Press, Oxford (1983).

[PiSte1] Anand Pillay and Charlie Steinhorn, Discrete o-minimal structures, APAL 34 (1987), 275–290.

[PiSte2] Anand Pillay and Charlie Steinhorn, Definable sets in ordered structures 3, TAMS 309 (1988), 469–476.

[S] Saharon Shelah, Classification Theory and the number of non-isomorphic models, North-Holland, Amsterdam (1978).

[Si1] Patrick Simonetta, An example of a C-minimal group which is not abelian-by-finite, Proceedings of the AMS 131 (2003), 3913–3917.

[Si2] Patrick Simonetta, On non-abelian C-minimal groups, APAL 122 (2003), 263–287.

11

Trees of definable sets in $\mathbb{Z}_p$

Immanuel Halupczok

The author was supported by the Fondation Sciences mathématiques de Paris.

Abstract

Understanding definable sets in the ring of p-adic integers amounts to understanding certain trees associated to the definable sets. A conjectural description of these trees is given in the article *Trees of definable sets over the p-adics* by the author. I will explain this conjecture and show how it is related to the rationality of some Poincaré series.

1 Introduction

The metric on the p-adic integers $\mathbb{Z}_p$ induces a metric on subsets of $\mathbb{Z}_p^n$. Suppose we have two such subsets X and X' and we would like to know whether there exists an isometry between them. This would be easiest if we had some invariants such that an isometry exists if and only the invariants are equal. A first step in this direction is the following: to each set $X \subset \mathbb{Z}_p^n$ we will associate a (rooted) tree $\mathrm{T}(X)$ such that if $X, X' \subset \mathbb{Z}_p^n$ are closed in the p-adic topology, then isometries $X \to X'$ correspond exactly to isomorphisms of trees $\mathrm{T}(X) \xrightarrow{\sim} \mathrm{T}(X')$.

Checking whether two arbitrary trees are isomorphic is not much easier than checking whether two sets are in isometry, so now it would be

Motivic Integration and its Interactions with Model Theory and Non-Archimedean Geometry (Volume II), ed. Raf Cluckers, Johannes Nicaise, and Julien Sebag. Published by Cambridge University Press.

helpful if we had an explicit description of the class of trees which we can obtain from subsets of $\mathbb{Z}_p^n$. For arbitrary subsets $X \subset \mathbb{Z}_p^n$, the trees may be almost arbitrary, too, but now let us restrict ourselves to sets X which are algebraic. In that case, it turns out that the possible trees are very particular ones. The goal of these notes is to present the main conjecture from [3], which gives a precise description of the class of these trees. This class will be sufficiently small and explicit so that checking whether two of its trees are isomorphic should not be a difficult problem anymore.

These notes are organized as follows. First we will introduce the trees (Section 1.1) and present the conjecture and some variants (Section 2), but without actually defining the class of trees which conjecturally can appear (the "trees of level *d*"). Then we give another motivation to consider the trees, namely a connection to Poincaré series (Section 3).

The longest part of the notes will then be the definition of the trees of level d. The complete definition is rather technical, so we will start by considering only the "easier half" of it (Section 4) and verify it on some examples (Section 5); in particular, we will prove the conjecture for algebraic sets which are smooth in a weak sense. Finally, we will describe the second half of the definition (Section 6) and see how this helps understanding the Poincaré series (Section 7).

1.1 Associating trees to subsets of $\mathbb{Z}_p^n$

Let us first fix some notation.

Notation 1.1 • Fix once and for all a prime p.

- $\mathbb{Q}_p$ is the field of p-adic numbers and $\mathbb{Z}_p$ are the p-adic integers. We write $v\colon \mathbb{Q}_p \to \mathbb{Z} \cup \{\infty\}$ for the valuation. It will be useful to define the valuation of a tuple as the minimum of the valuations of the coordinates: $v(\underline{a}) := \min_i v(a_i)$ for $\underline{a} = (a_1, \dots, a_n) \in \mathbb{Q}_p^n$.
- For sets $X, X' \subset \mathbb{Q}_p^n$, we say that $f\colon X \to X'$ is an isometry if for all $\underline{x}_1, \underline{x}_2 \in X$, we have $v(f(\underline{x}_1) - f(\underline{x}_2)) = v(\underline{x}_1 - \underline{x}_2)$. By our definition of the valuation of a tuple, this means that we are considering the maximum norm on $\mathbb{Q}_p^n$. This is indeed the most natural norm to use on $\mathbb{Q}_p^n$.
- The set of all balls in $\mathbb{Q}_p^n$ will play a crucial role. If $\underline{a} \in \mathbb{Q}_p^n$ and $\lambda \in \mathbb{Z}$, we write $B(\underline{a}, \lambda) = \underline{a} + p^\lambda \mathbb{Z}_p^n = \{\underline{x} \in \mathbb{Q}_p^n \mid v(\underline{x} - \underline{a}) \geq \lambda\}$ for the ball around $\underline{a}$ of "valuative radius" λ. (By radius, we will always mean valuative radius.)

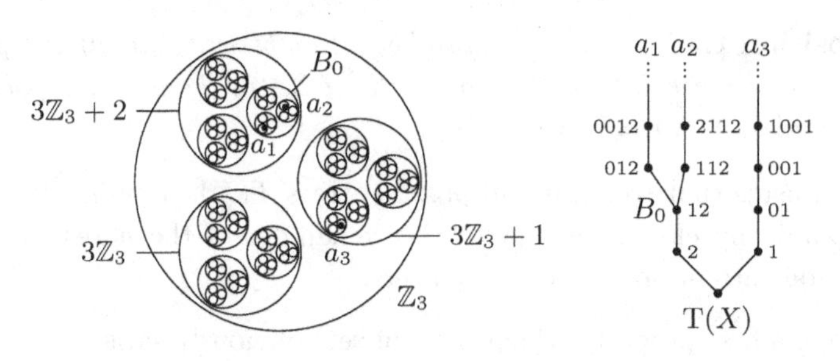

Figure 11.1 A three-point set $X = \{a_1, a_2, a_3\}$ in $\mathbb{Z}_3$ and its tree. Each node in the tree of X corresponds to fixing some 3-adic digits of the numbers a_i.

The set of balls contained in $\mathbb{Z}_p^n$ forms a tree under inclusion: each ball of radius λ consists of exactly p^n balls of radius $\lambda + 1$; the root of this tree is the ball $\mathbb{Z}_p^n$ itself. Another way to describe this tree is the following: a ball of radius λ contains all elements of $\mathbb{Z}_p^n$ where the first λ p-adic digits of each coordinate have been fixed; going up one step in the tree means fixing one more digit; the infinite paths[1] in the tree correspond exactly to the elements of $\mathbb{Z}_p^n$.

We now define the tree associated to a subset of $\mathbb{Z}_p^n$:

Definition 1.2 Suppose $X \subset \mathbb{Z}_p^n$. Then the *tree of* X consists of all those balls in $\mathbb{Z}_p^n$ that intersect X:

$$\mathrm{T}(X) := \{B = B(\underline{a}, \lambda) \mid B \cap X \neq \emptyset\} = \{B(\underline{a}, \lambda) \mid \underline{a} \in X, \lambda \geq 0\}.$$

The tree structure is given by inclusion of balls.

Example 1.3
- As we already noted, the full tree $\mathrm{T}(\mathbb{Z}_p^n)$ is the one where each node has exactly p^n children.
- If $B = B(\underline{a}, \lambda) \subset \mathbb{Z}_p^n$ is a ball, then the tree $\mathrm{T}(B)$ consists of a path of length λ (from the root to the ball B itself), with a tree isomorphic to $\mathrm{T}(\mathbb{Z}_p^n)$ attached to its end. (By "a tree $\mathscr{T}$ is attached to a node v of another tree", we mean that the root of $\mathscr{T}$ is identified with v.)
- If $X = \{\underline{a}_1, \ldots, \underline{a}_l\}$ consists of l points, then $\mathrm{T}(X)$ consists of l infinite paths, each one corresponding to one of the points of X. The paths can have a common segment at the beginning; the paths corresponding to $\underline{a}_i$ and $\underline{a}_j$ separate at height $v(\underline{a}_i - \underline{a}_j)$.

More generally, one easily verifies:

[1] By an *infinite path*, we mean a totally ordered subset P of (the nodes of) the tree consisting of one node at each height.

Proposition 1.4 *There is a natural bijection between the infinite paths of* $\mathrm{T}(X)$ *and the points of the closure* $\bar{X}$ *of* X *(in the p-adic topology). In particular,* $\mathrm{T}(X) = \mathrm{T}(\bar{X})$.

This means that to obtain all possible trees $\mathrm{T}(X)$, it suffices to consider p-adically closed sets X. In the remainder of the notes, we will restrict our attention to such closed sets.

Here is a first property which trees of sets obviously satisfy:

Proposition 1.5 *For any* $X \subset \mathbb{Z}_p^n$, *the tree* $\mathrm{T}(X)$ *has no leaves, i.e., no nodes without children.*

Let us now verify that isometries of sets correspond to isomorphisms of trees.

Proposition 1.6 *For any two sets* $X \subset \mathbb{Z}_p^n$ *and* $X' \subset \mathbb{Z}_p^{n'}$ *which are closed in p-adic topology, we have a natural bijection*

$$\{\textit{isometries } X \longrightarrow X'\} \overset{1:1}{\longleftrightarrow} \{\textit{isomorphisms } \mathrm{T}(X) \overset{\sim}{\longrightarrow} \mathrm{T}(X')\}.$$

Sketch of proof Suppose $f\colon X \to X'$ is an isometry. Then for $\underline{x} \in X$ define $f_{\text{tree}}\colon B(\underline{x}, \lambda) \mapsto B(f(\underline{x}), \lambda)$. The isometry condition implies that this yields a well-defined map $\mathrm{T}(X) \to \mathrm{T}(X')$. Use f^{-1} to find an inverse of f_{tree}.

On the other hand, a map $\mathrm{T}(X) \to \mathrm{T}(X')$ defines a map between the sets of infinite paths of the trees, and these correspond to the points of X and X', respectively. □

2 The goal

Our goal is to understand what the tree of an algebraic set $X \subset \mathbb{Z}_p^n$ (a zero-set of polynomials) can look like. More precisely:

Question 2.1 For which abstract trees $\mathscr{T}$ does there exist an algebraic set $X \subset \mathbb{Z}_p^n$ (for some suitable n) such that $\mathscr{T} \cong \mathrm{T}(X)$?

Just to see that the answer is not trivial (i.e., that there are indeed trees which do not come from algebraic sets), consider the tree of Figure 11.2. The corresponding set would have infinitely many isolated points, which is impossible.

Instead of considering only algebraic subsets of $\mathbb{Z}_p^n$, we might generalize the question to some other sets. Let me present two such generalizations.

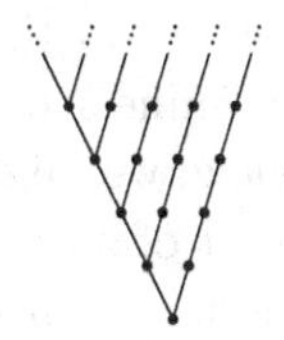

Figure 11.2 Not the tree of any algebraic set.

2.1 Algebraic generalization

In the language of algebraic geometry, an algebraic subset of $\mathbb{Z}_p^n$ is a set of the form $V(\mathbb{Z}_p)$, where $V \subset \mathbb{A}^n_{\mathbb{Z}_p}$ is an affine embedded variety (not necessarily irreducible). Using this notation, we can define $\mathrm{T}(X)$ in a more algebraic way. For $\lambda \in \mathbb{N}$, consider the following canonical maps:

$$\begin{array}{ccc} V(\mathbb{Z}_p) & \xrightarrow{\pi_{\lambda+1}} & V(\mathbb{Z}_p/p^{\lambda+1}\mathbb{Z}_p) \\ & \searrow^{\pi_\lambda} & \downarrow \sigma_\lambda \\ & & V(\mathbb{Z}_p/p^{\lambda}\mathbb{Z}_p) \end{array}.$$

Assume that V is the set of zeros of some polynomials $f_i \in \mathbb{Z}_p[x_1, \ldots, x_n]$. Since balls of radius λ are just cosets of the ball $(p^\lambda\mathbb{Z}_p)^n$, we can identify $(\mathbb{Z}_p/p^\lambda\mathbb{Z}_p)^n$ with the set of balls of radius λ in $\mathbb{Z}_p^n$ and get

$$V(\mathbb{Z}_p/p^\lambda\mathbb{Z}_p) = \{B(\underline{x}, \lambda) \mid \underline{x} \in \mathbb{Z}_p^n, \forall i\, f_i(\underline{x}) \equiv 0 \mod p^\lambda\}.$$

The image of the map π_λ inside this set consists of those balls which do contain a point $\underline{x}$ satisfying $f_i(\underline{x}) = 0$, i.e., it is exactly the set of nodes of $\mathrm{T}(V(\mathbb{Z}_p))$ at height λ. Thus an alternative definition of the tree $\mathrm{T}(V(\mathbb{Z}_p))$ is

$$\mathrm{T}(V(\mathbb{Z}_p)) = \dot{\bigcup_{\lambda=0}^{\infty}} \operatorname{im} \pi_\lambda.$$

The tree structure on this is given by the maps σ_λ: they map each small ball into the bigger ball which contains the small one.

This new definition of $\mathrm{T}(V(\mathbb{Z}_p))$ has the advantage that it does not depend on the embedding of V into $\mathbb{A}^n_{\mathbb{Z}_p}$ and that it works for non-affine V. So here is an algebraic generalization of the question:

Question 2.2 For which abstract trees $\mathscr{T}$ does there exist a variety V over $\mathbb{Z}_p$ such that $\mathscr{T} \cong \mathrm{T}(V(\mathbb{Z}_p))$?

(In fact, "variety" can even be replaced by "scheme of finite type".)

It turns out that essentially no new trees appear in this generalization. This is due to the fact that whether a tree comes from an algebraic set is essentially a local question. Hence in the remainder of the notes we will restrict ourselves to trees of affine sets.

From the algebraic point of view, we might also consider another tree associated to V: the one whose set of nodes of height λ is the whole set $V(\mathbb{Z}_p/p^\lambda\mathbb{Z}_p)$, instead of only the image of the map π_λ. These trees are quite similar in nature to the other ones; for these notes let us stick to the first ones.

2.2 Model-theoretic generalization

From a model theoretic point of view, it is also natural to replace algebraic subsets of $\mathbb{Z}_p^n$ by definable ones (say, using the two-sorted language consisting of the field $\mathbb{Q}_p$ with the ring language, the value group $\mathbb{Z}$ with the ordered group language, and the valuation map v). These definable sets are well understood; in particular, we have the following quantifier elimination result, which may also serve as definition of definable sets for those readers who are not familiar with model theory.

Proposition 2.3 (see [2], [4]) *The definable sets are exactly the* semi-algebraic *ones, i.e., boolean combinations of sets of the following types:*[2]

- *algebraic sets (zero-sets of polynomials);*
- *for each polynomial $f \in \mathbb{Q}_p[x_1, \dots, x_n]$ and each $r \in \mathbb{N}$, the inverse image under f of the set of r-th powers: $f^{-1}(\{x^r \mid x \in \mathbb{Q}_p\})$.*

So here is a model-theoretic version of our question:

Question 2.4 For which abstract trees $\mathscr{T}$ does there exist a definable set $X \subset \mathbb{Z}_p^n$ (for some suitable n) such that $\mathscr{T} \cong \mathrm{T}(X)$?

2.3 The conjecture

To state the main conjecture, we need a notion of dimension for the subsets of $\mathbb{Z}_p^n$ we are considering. For algebraic subsets, such a notion of course exists, but we have to be a little bit careful: an affine variety $V \subset \mathbb{A}^n_{\mathbb{Z}_p}$ might have components which have no points over $\mathbb{Z}_p$ (e.g.

[2] Over $\mathbb{R}$, the "classical" definition is that a semi-algebraic set is a boolean combination of polynomial inequalities. However, since in $\mathbb{R}$, the set of r-th powers is either everything (if r is odd) or $\mathbb{R}_{\geq 0}$ (if r is even), over $\mathbb{R}$ our definition is equivalent to the classical one.
The reason to generalize it to $\mathbb{Q}_p$ in the above way is that this yields precisely the smallest class of sets which contains the algebraic sets and which is closed under boolean combinations and projections.

when V is given by $X^2 = a$ and a has no square root in $\mathbb{Q}_p$). However, we want our dimension to depend only on the set of $\mathbb{Z}_p$-valued points of V. The following definition yields such a notion of dimension; moreover, it works even for definable sets.

Definition 2.5 For $X \subset \mathbb{Z}_p^n$ definable, $\dim X$ is defined as the dimension of the Zariski closure of X in $\tilde{\mathbb{Q}}_p^n$ (which is an algebraic set), where $\tilde{\mathbb{Q}}_p$ is the algebraic closure of $\mathbb{Q}_p$.

An equivalent definition would be: $\dim X$ is the maximal integer d such that there exists a coordinate projection π onto d coordinates such that the image $\pi(X)$ contains an open ball. (For definable sets, this has been proven in [5]).

Most of the work of these notes will be to define a class of (abstract) trees called *trees of level d*. The main conjecture is then:

Conjecture 2.6 *A tree* $\mathscr{T}$ *is of level d if and only if there exists a definable set* $X \subset \mathbb{Z}_p^n$ *(for some suitable n) of dimension d such that* $\mathrm{T}(X) \cong \mathscr{T}$.

The class of level d trees will be surprisingly small. For this reason, the difficult direction of the conjecture is "$\Leftarrow$". Indeed, "$\Rightarrow$" can be proven by explicitly constructing, for any given tree $\mathscr{T}$ of level d, a corresponding definable set X.

The conjecture does not say when we can choose X to be algebraic. However, we will see on an example that trees of algebraic sets are not essentially simpler than trees of definable sets.

Concerning "$\Leftarrow$", we will see a proof when X is smooth (in a weak sense) in Section 5.2. Other cases in which I can prove the conjecture (see [3]) are the following: when X is one-dimensional (in that case, the theorem of Puiseux gives a sufficiently explicit description of X), and when X is an arbitrary definable subset of $\mathbb{Z}_p^2$ (use cell decomposition and apply the theorem of Puiseux to the cell centers).

3 Trees and Poincaré series

Before coming to the definition of level d trees, let me mention one possible application of the conjecture, which was also one of the motivations to study the trees.

The *Poincaré series* of a set $X \subset \mathbb{Z}_p^n$ is defined as follows:

$$P_X(Z) := \sum_{\lambda=0}^{\infty} N_\lambda Z^\lambda \in \mathbb{Z}[[Z]],$$

where N_λ is the number of balls of radius λ in $\mathbb{Z}_p^n$ intersecting X; in terms of trees, N_λ is the number of nodes of $\mathrm{T}(X)$ at height λ.

If X is a definable set, then this series is known to be a rational function in Z, i.e., $P_X(Z) \in \mathbb{Q}(Z)$ (see [1]). This is of course a strong condition on the coefficients of $P_X(Z)$ and hence on $\mathrm{T}(X)$, which somehow should be reflected in the structure of $\mathrm{T}(X)$. Indeed, the definition of level d trees will be restrictive enough so that rationality of the Poincaré series is easily implied.

And even better: a level d tree can be described by a finite amount of data, and from this data one can directly compute the Poincaré series as a rational function; we will see this on an example at the end of the notes. Hence one can hope that our conjecture will help getting a better understanding of Poincaré series.

4 Definition of trees of level d: part I

The definition of trees of level d is inductive; we start by defining trees of level $d = 0$.

4.1 Trees of level 0

Definition 4.1 $\mathscr{T}$ is of level 0 iff it has no leaves and only finitely many bifurcations (= nodes with more than one child); in other words, it is the union of finitely many infinite paths, possibly having some segments in common at the beginning.

The 0-dimensional definable sets are exactly the finite ones. As we already saw, finite sets have trees satisfying this definition. Conversely, for any tree $\mathscr{T}$ of level 0 it is easy to construct a finite set whose points have distances given by the bifurcation heights of $\mathscr{T}$.

4.2 Trees of level d: preliminaries

For $d > 0$, instead of defining trees of level exactly d, it will be easier to define the class of trees of level $\leq d$. Then of course we can define a tree to be of level exactly d if it is of level $\leq d$ but not of level $\leq d - 1$.

The following notation will be needed in the definition:

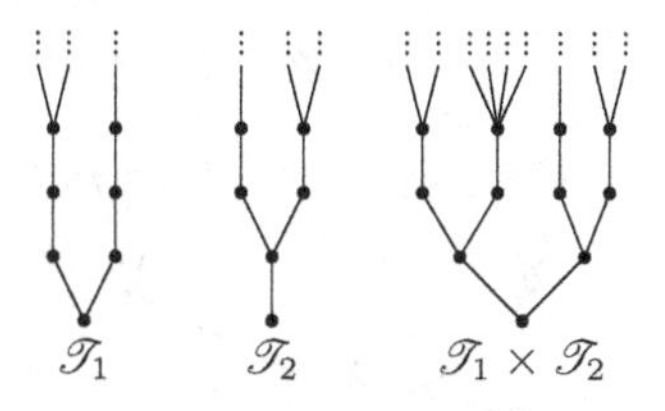

Figure 11.3 Two trees and their product.

Notation 4.2 If $\mathscr{T}_1$ and $\mathscr{T}_2$ are two trees, we write $\mathscr{T}_1 \times \mathscr{T}_2$ for the tree whose set of nodes at height λ is the product of the corresponding sets of nodes of $\mathscr{T}_1$ and $\mathscr{T}_2$, and where a node (v_1', v_2') is a child of (v_1, v_2) if v_i' is a child of v_i in $\mathscr{T}_i$ for $i = 1, 2$ (see Figure 11.3).

One easily checks: $\mathrm{T}(X_1 \times X_2) \cong \mathrm{T}(X_1) \times \mathrm{T}(X_2)$. We will mainly need this notation in the case $\mathscr{T} \times \mathrm{T}(\mathbb{Z}_p)$.

4.3 Trees of level $\leq d$: the smoothness condition

Intuitively, the definition of trees of level $\leq d$ consists of two parts: a "smoothness condition" (S) and a "uniformity condition when approaching singularities" (U). A complete and precise statement of (U) automatically also includes (S); however, this is long and technical, and (S) by itself already gives a good idea on how the trees will look like. For this reason, we postpone the details of (U) to Section 6.

Definition 4.3 A tree $\mathscr{T}$ is of level $\leq d$ if there exists a finite set P_0 of infinite paths (the "singular paths") in $\mathscr{T}$ such that the following two conditions holds:

(S) For $\lambda \in \mathbb{N}$, consider the tree $\tilde{\mathscr{T}}_\lambda$ obtained from $\mathscr{T}$ by removing some full subtrees[3] as follows: for each path $\mathscr{P} \in P_0$, remove the full subtree whose root is the node at height λ on $\mathscr{P}$ (see Figure 11.4). We require $\tilde{\mathscr{T}}_\lambda$ to satisfy the following condition (for every λ):

 (∇) $\tilde{\mathscr{T}}_\lambda$ consists of a finite tree $\mathscr{F}$ with trees $\mathscr{T}_i' \times \mathrm{T}(\mathbb{Z}_p)$ attached to its leaves, where each $\mathscr{T}_i'$ is a tree of level $\leq d - 1$.

(U) For each path $\mathscr{P} \in P_0$, a uniformity condition on the side branches of $\mathscr{T}$ on $\mathscr{P}$ (see Section 6).

[3] By *subtree*, we mean a subset which is connected when viewed as a graph; a *full subtree* is a subtree consisting of one node and everything above that node.

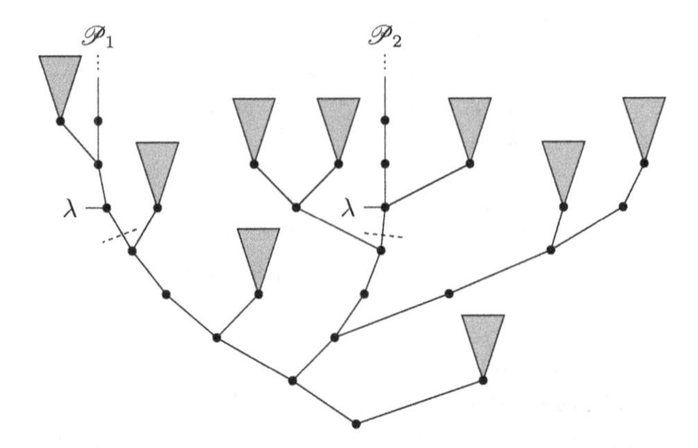

Figure 11.4 Condition (S) on trees of level $\leq d$: P_0 consists of the two paths $\mathscr{P}_1, \mathscr{P}_2$. Each triangle stands for a tree of the form $\mathscr{T}' \times \mathrm{T}(\mathbb{Z}_p)$, where $\mathscr{T}'$ is of level $\leq d-1$. After cutting the paths $\mathscr{P}_i$ at any height (say, at the dotted lines), the remaining tree $\tilde{\mathscr{T}}_\lambda$ satisfies (∇).

Condition (S) may sound complicated, but it has an easy geometric interpretation. Suppose that $\mathscr{T} = \mathrm{T}(X)$ is the tree of a set X which is closed in p-adic topology. I claim that (S) is equivalent to the following condition on X:

(S') There exists a finite set $S_0 \subset X$ such that the following holds: around each $x \in X \setminus S_0$, there exists a neighbourhood $B = B(x, \mu)$ such that $X \cap B$ is isometric to a set of the form $X' \times p^\mu \mathbb{Z}_p$, where X' has a tree of level $\leq d-1$.

Proof of the claim Let S_0 be the set of points of X corresponding to the paths P_0.

"(S) $\Rightarrow$ (S')": Let $x \in X \setminus S_0$ be given. Choose $\lambda \in \mathbb{N}$ such that $\lambda > v(x-s)$ for all $s \in S_0$ and consider the corresponding tree $\tilde{\mathscr{T}}_\lambda$ defined in (S). This tree $\tilde{\mathscr{T}}_\lambda$ is just the tree of the set $\tilde{X}_\lambda$ obtained from X by removing a ball of radius λ around each point of S_0:

$$\tilde{X}_\lambda = X \setminus \bigcup_{\underline{s} \in S_0} B(\underline{s}, \lambda).$$

Condition (∇) translates as follows. There are finitely many disjoint balls $B_i = B(\underline{x}_i, \mu_i)$ (corresponding to the leaves of $\mathscr{F}$) such that $\tilde{X}_\lambda$ is contained in the union of those balls, and each intersection $X_i := \tilde{X}_\lambda \cap B_i$ has a tree starting with a path of length μ_i, with $\mathscr{T}_i' \times \mathrm{T}(\mathbb{Z}_p)$ attached to its end. In other words, X_i is isometric to a set of the form $X_i' \times p^{\mu_i}\mathbb{Z}_p$, where X_i' has a tree of level $\leq d-1$. (Here, we use that adding or removing a path at the root of a tree does not change whether that tree is of level $d-1$.) By our choice of λ, x is contained in $\tilde{X}_\lambda$, so it is contained in one of the balls B_i, and (S') follows.

"(S) $\Leftarrow$ (S')": Suppose first that the sets P_0 and S_0 are empty. Then the balls from (S') cover the whole set X. As X is a closed subset of $\mathbb{Z}_p^n$, it is compact, so finitely many balls suffice to cover X. Moreover, any two balls are either disjoint or contained in one another, so by keeping only the largest balls of our cover, we may suppose that they are all disjoint. Now we obtain that $\mathrm{T}(X)$ satisfies ($\triangledown$) by letting $\mathscr{F}$ be the finite tree whose leaves are exactly those balls.

If we permit a non-empty singular set S_0 in condition (S'), then we do not get a similar finite cover of $X \setminus S_0$, since it is not compact. However, for any radius λ, the set $\tilde{X}_\lambda := X \setminus \bigcup_{\underline{s} \in S_0} B(\underline{s}, \lambda)$ is compact, so we may apply the above arguments to $\tilde{X}_\lambda$, which implies ($\triangledown$) for the trees $\mathrm{T}(\tilde{X}_\lambda) = \tilde{\mathscr{T}}_\lambda$. □

5 Examples

Before getting to the missing part of the definition of level $\leq d$ trees, let us consider a few examples. In particular, we will prove the conjecture for smooth algebraic sets; in that case, the set P_0 can be chosen to be empty and condition (U) becomes trivial.

5.1 The full set $\mathbb{Z}_p^n$

The tree of $\mathbb{Z}_p$ can be written as $\mathrm{T}(\{0\}) \times \mathrm{T}(\mathbb{Z}_p)$. As $\mathrm{T}(\{0\})$ is of level 0, $\mathrm{T}(\mathbb{Z}_p)$ is of level ≤ 1. Inductively, one gets that $\mathrm{T}(\mathbb{Z}_p^n)$ is of level $\leq n$.

To see that $\mathrm{T}(\mathbb{Z}_p^n)$ is exactly of level n, we use induction. It is clear that for $n > 0$, $\mathrm{T}(\mathbb{Z}_p^n)$ is not of level 0. Now suppose that $\mathrm{T}(\mathbb{Z}_p^n)$ is of level $\leq m$ with $n > m \geq 1$ and consider condition (S). As there are only finitely many singular paths, it implies that somewhere in the tree $\mathrm{T}(\mathbb{Z}_p^n)$ there is a full subtree of the form $\mathscr{T}' \times \mathrm{T}(\mathbb{Z}_p)$, where $\mathscr{T}'$ is of level $\leq m-1$. Now every full subtree of $\mathrm{T}(\mathbb{Z}_p^n)$ is isomorphic to $\mathrm{T}(\mathbb{Z}_p^n)$ itself, so we get $\mathscr{T}' \times \mathrm{T}(\mathbb{Z}_p) \cong \mathrm{T}(\mathbb{Z}_p^n)$, which implies $\mathscr{T}' \cong \mathrm{T}(\mathbb{Z}_p^{n-1})$; however, $\mathrm{T}(\mathbb{Z}_p^{n-1})$ is not of level $\leq m-1$ by induction.

With a bit of additional work, this can be turned into a general argument showing that $\mathrm{T}(X)$ can not be a tree of level d if $d < \dim X$.

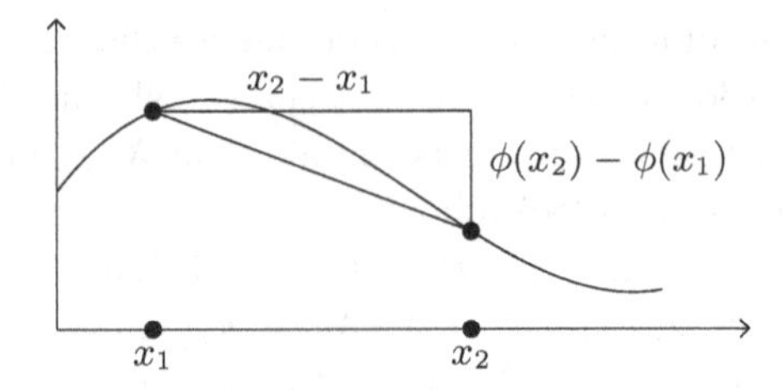

Figure 11.5 The graph of a 1-Lipschitz map: the vertical distance is dominated by the horizontal one, so p-adically the diagonal distance is equal to the horizontal one; hence the graph is isometric to the horizontal line.

5.2 Smooth algebraic sets

Now suppose that $X \subset \mathbb{Z}_p^2$ and that X is a smooth curve. Here, we use "smooth curve" in the following naive sense: X is the set of zeros of a polynomial $f \in \mathbb{Z}_p[x, y]$ such that $\nabla f = (\partial f/\partial x, \partial f/\partial y)$ vanishes nowhere on X. (Note that this notion of smoothness is weaker than the usual scheme theoretic notion of smoothness: if we write $X = V(\mathbb{Z}_p)$ with $V \subset \mathbb{A}^2_{\mathbb{Z}_p}$, then our condition is equivalent to: for any $\mathbb{Z}_p$-valued point x: **Spec** $\mathbb{Z}_p \to V$, V is smooth at $x(\eta)$, where η is the generic point of **Spec** $\mathbb{Z}_p$.)

We would like to verify that X satisfies the geometric condition (S'), i.e., that X is piecewise isometric to straight lines almost everywhere. It may sound surprising that this is possible: over the reals, a curved line is never isometric to a straight one. In the p-adics however, this is perfectly possible; the reason is that we are using the metric induced by the maximum norm. Suppose first that $X = \{(x, \phi(x)) \mid x \in \mathbb{Z}_p\}$ is the graph of a function ϕ which is 1-Lipschitz with respect to the p-adic metric, i.e., ϕ satisfies

$$v(\phi(x_1) - \phi(x_2)) \geq v(x_1 - x_2)$$

for any $x_1, x_2 \in \mathbb{Z}_p$. Then one easily verifies that the map

$$\mathbb{Z}_p \to X, x \mapsto (x, \phi(x))$$

is an isometry (see Figure 11.5). By Proposition 1.6, we get $\mathrm{T}(X) \cong \mathrm{T}(\mathbb{Z}_p)$, so in particular it is of level 1.

Now if X is an arbitrary smooth curve, then by using a p-adic version of the implicit function theorem and choosing coordinates adequately, we can write it piecewise as graphs of 1-Lipschitz functions, thus its tree $\mathrm{T}(X)$ is of level 1. The same idea also works for higher-dimensional smooth algebraic sets.

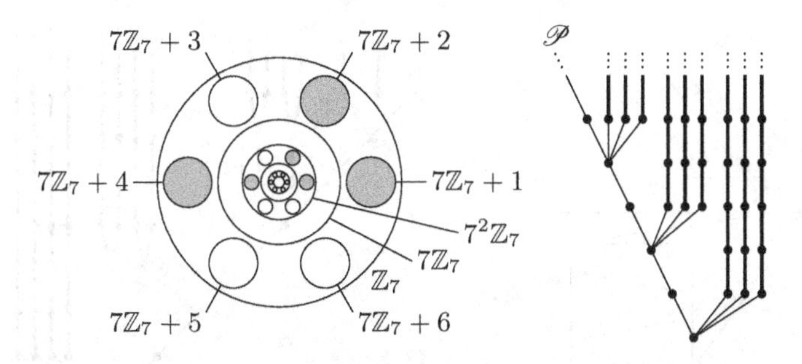

Figure 11.6 The set of squares in $\mathbb{Z}_7$ (grey area) and its tree. To simplify the picture of the tree, the subtrees isomorphic to $\mathrm{T}(\mathbb{Z}_7)$ have simply been drawn as thick paths. In other words, a thick subtree in the picture stands for $\mathrm{T}(\mathbb{Z}_p)$ times that thick subtree.

If X is an algebraic set which has only isolated singularities, then we let S_0 be the set of these singularities; in this way, we obtain that X satisfies Condition (S'). However, the real difficulty in proving that X has a tree of level $\dim X$ lies in verifying the uniformity condition (U) at each singularity.

5.3 The set of squares

Now let us consider a concrete example of a tree with a singular path. Let $X = \{x^2 \mid x \in \mathbb{Z}_p\}$ be the set of squares. (Recall that this is semi-algebraic.) For simplicity, suppose $p \neq 2$. An element of $\mathbb{Q}_p$ is a square if and only if it has even valuation and its angular component is a square in the residue field $\mathbb{F}_p$. (If x has odd valuation, then a root of x would have non-integer valuation. If x has valuation 2μ, then write it as $x = (p^\mu)^2 x_0$, where $v(x_0) = 0$. By Hensel's Lemma, x_0 is a square if and only if its residue is a square.)

In other words, X is a disjoint union of balls $B(p^{2\mu}a, 2\mu + 1)$, where μ runs through $\mathbb{N}$ and a runs through a set A of representatives of the non-zero squares in the residue field. As the multiplicative group of $\mathbb{F}_p$ is cyclic, exactly half of its elements are squares, so $|A| = \frac{p-1}{2}$. Thus the tree of X is obtained in the following way (see Figure 11.6):

- Start with an infinite path $\mathscr{P}$ (corresponding to $0 \in X$).
- At each node of $\mathscr{P}$ of even height λ, add $\frac{p-1}{2}$ additional children (the balls $B(p^\lambda a, \lambda + 1)$ for $a \in A$).
- Attach a copy of $\mathrm{T}(\mathbb{Z}_p)$ to each of these additional children.

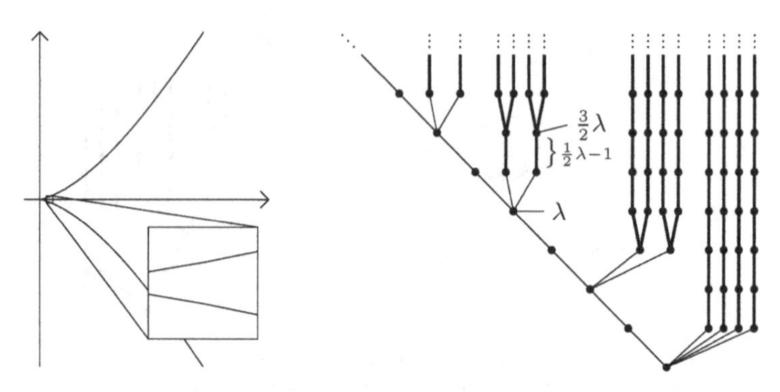

Figure 11.7 The cusp curve over $\mathbb{R}$, a suggestive closeup, and its tree in $\mathbb{Z}_5$. Again, a thick subtree in the picture stands for $\mathrm{T}(\mathbb{Z}_p)$ times that tree.

This tree does satisfy Condition (S) if we set $P_0 := \{\mathscr{P}\}$. Moreover, we get a first idea of how the uniformity condition (U) should look like. In this example, after fixing the height modulo 2, all side branches attached to $\mathscr{P}$ are the same. In the next example, we will see that the general situation is a bit more complicated. Moreover, we will see that complicated trees can already arise from algebraic sets, and not only from semi-algebraic ones.

5.4 The cusp curve

Consider the cusp curve $X = \{(x, y) \in \mathbb{Z}_p^2 \mid x^3 = y^2\}$; again suppose $p \neq 2$. The condition $x^3 = y^2$ implies that x has a square root, so the projection of X to the first coordinate is just our set of squares from the previous example. However, for each square x, there are two corresponding points $(x, \pm x\sqrt{x})$ in X. The picture of the cusp curve over $\mathbb{R}$ suggests that close to 0, these two points lie on two almost straight, almost parallel curves which are close together (see Figure 11.7). This intuition turns out to be correct; more precisely, for each ball $B_0 = B(p^{2\mu}a, 2\mu+1)$ of our set of squares (where a is a square with $v(a) = 0$), the preimage $X \cap (B_0 \times \mathbb{Z}_p)$ of that ball is isometric to two horizontal lines at (valuative) distance 3μ. (The distance 3μ comes from $v(x) = 2\mu$ for $x \in B_0$, which implies $v(y) = \frac{1}{2}v(y^2) = \frac{3}{2}v(x) = 3\mu$.) In this way, one finally obtains that $\mathrm{T}(X)$ can be constructed as follows:

- Start with an infinite path $\mathscr{P}$ (corresponding to $0 \in X$). For each even λ, attach the following tree ("side branch") $\mathscr{B}_\lambda$ to the node at height λ:

- If $\lambda = 0$, then $\mathscr{B}_\lambda$ consists of a root with $p-1$ children, above each of which the tree is isomorphic to $\mathrm{T}(\mathbb{Z}_p)$.
- If $\lambda > 0$, then $\mathscr{B}_\lambda$ consists of a root with $\frac{p-1}{2}$ children, above each of which the tree is isomorphic to $\mathscr{T}_\lambda \times \mathrm{T}(\mathbb{Z}_p)$, where $\mathscr{T}_\lambda$ is the tree consisting of two infinite paths separating at height $\frac{\lambda}{2} - 1$. (Counted from the root of the whole tree, this separation is at height $\frac{3}{2}\lambda$; i.e., the distance between the two almost parallel curves.)

6 Definition of trees of level d: part II

I will now state the uniformity condition (U) which was still missing in the definition of trees of level $\leq d$. To get an idea of what it should be, let us look again at the cusp example: for $\lambda \geq 2$ and even, each side branch $\mathscr{B}_\lambda$ consists of the same finite tree, with copies of $\mathscr{T}_\lambda \times \mathrm{T}(\mathbb{Z}_p)$ attached to its leaves, where $\mathscr{T}_\lambda$ are trees of level 0 which are, in a certain sense, uniform in λ. This is roughly what will also happen in general. In particular, to state condition (U) for level $\leq d$, we will need a notion of *uniform families* of trees of level $\leq d-1$; this means that we have to completely rewrite the definition to include such uniform families.

The uniform families of trees of level $\leq d-1$ which we need are parametrized by subsets of $\mathbb{N}$. To define those families, we will then need uniform families of trees of level $\leq d-2$ parametrized by subsets of $\mathbb{N}^2$, etc. Therefore, we will consider, right from the beginning, families parametrized by subsets $M \subset \mathbb{N}^m$ for any $m \in \mathbb{N}$. (For model theorists: M is definable in the value group.)

6.1 Uniform families of side branches

Before I give the definition of uniform families of trees of level $\leq d$, let me define how this then yields "uniform families of side branches"; these side branches will then be attached to the singular paths (see Figure 11.8).

Definition 6.1 Suppose $M \subset \mathbb{N}^m$. A family of trees $(\mathscr{B}_{\underline{\kappa}})_{\underline{\kappa} \in M}$ is a *uniform family of side branches* (of level $\leq d$) if the following holds:

- Each tree $\mathscr{B}_{\underline{\kappa}}$ satisfies (∇), i.e., it consists of a finite tree $\mathscr{F}_{\underline{\kappa}}$ with trees $\mathscr{T}'_{\underline{\kappa},1} \times \mathrm{T}(\mathbb{Z}_p), \ldots, \mathscr{T}'_{\underline{\kappa},\ell} \times \mathrm{T}(\mathbb{Z}_p)$ attached to its leaves (where ℓ is the number of leaves of $\mathscr{F}_{\underline{\kappa}}$).
- The finite trees $\mathscr{F}_{\underline{\kappa}}$ are all equal.
- For each $i \in \{1, \ldots, \ell\}$, the trees $(\mathscr{T}'_{\underline{\kappa},i})_{\underline{\kappa}}$ form a uniform family of trees of level $\leq d-1$.

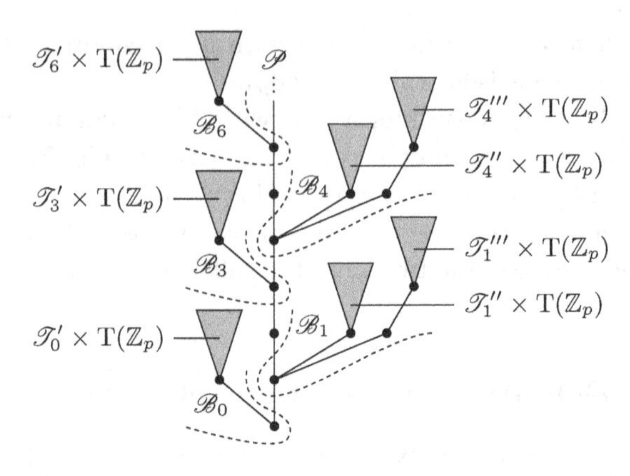

Figure 11.8 A tree which will satisfy condition (U): to the (single) singular path $\mathscr{P}$, two uniform families of side branches (see Definition 6.1) have been attached: $(\mathscr{B}_{3\mu})_{\mu\in\mathbb{N}}$ and $(\mathscr{B}_{3\mu+1})_{\mu\in\mathbb{N}}$. For the families of side branches to be uniform, each of the three families $(\mathscr{T}'_{3\mu})_\mu$, $(\mathscr{T}''_{3\mu+1})_\mu$ and $(\mathscr{T}'''_{3\mu+1})_\mu$ has to be a uniform family of trees of level $\leq d-1$.

Note that the side branches of the cusp example do satisfy this definition (for $\lambda \geq 2$ and even), assuming that the trees $\mathscr{T}_\lambda$ appearing there form a uniform family of level 0 trees (which we did not yet define).

6.2 Uniform families of trees of level 0

Now we define uniform families of trees of level $d = 0$. Recall that a tree is of level 0 if it has finitely many bifurcations and no leaves. In the uniform version, we require that the different trees of the family differ only in the lengths of the segments between the bifurcations, and moreover that these lengths are linear in the parameter $\underline{\kappa}$. In other words:

Definition 6.2 Suppose $M \subset \mathbb{N}^m$. A family of trees $(\mathscr{T}_{\underline{\kappa}})_{\underline{\kappa}\in M}$ is *uniformly of level* 0 if in each tree $\mathscr{T}_{\underline{\kappa}}$ we can choose finitely many nodes $v_{\underline{\kappa},0,\dots,}v_{\underline{\kappa},\nu}$ (called *joints*) such that:

- $v_{\underline{\kappa},0}$ is the root of $\mathscr{T}_{\underline{\kappa}}$.
- The number ν of joints does not depend on $\underline{\kappa}$.
- The tree $\mathscr{T}_{\underline{\kappa}}$ is obtained from the set of joints by adding some finite paths between pairs of joints and attaching some infinite paths to some of them. (All these paths will be called *bones*.)

- For each $j, j' \leq \nu$, whether or not $v_{\underline{\kappa},j}$ and $v_{\underline{\kappa},j'}$ are connected by a bone does not depend on $\underline{\kappa}$, and if a bone exists, then its length is a linear function in $\underline{\kappa}$.
- For each $j \leq \nu$, the number of infinite bones attached to $v_{\underline{\kappa},j}$ does not depend on $\underline{\kappa}$.

See Figure 11.9 for an example (for the moment, ignore the small indices 0 and 1 in the figure). In this example, joint c is useless: we could as well have attached an infinite bone directly to joint a. However, such useless joints will become useful later.

The family of trees $\mathscr{T}_\lambda$ ($\lambda \in M := \{4, 6, 8, \dots\}$) from the cusp example obviousely fits into Definition 6.2, with one joint for the root and one joint for the bifurcation (for $\lambda = 2$, the two joints fall together). Note that the linear function yielding the length of a bone (in this case the bone from the root to the bifurcation) can have rational coefficients, as long as the value is an integer for any parameter in M.

6.3 Uniform families of level d

Suppose now that $d \geq 1$ and recall the set P_0 of singular paths from Definition 4.3. The union of those paths forms a tree of level 0; let us call this union the *skeleton* $\mathscr{S}$ of $\mathscr{T}$. We can consider $\mathscr{T}$ as being obtained by starting with $\mathscr{S}$ and attaching side branches to it. We will adopt this point of view in the definition of uniform families of trees. For a family of trees $\mathscr{T}_{\underline{\kappa}}$ to be uniformly of level $\leq d$, we will then require that the skeletons $\mathscr{S}_{\underline{\kappa}}$ form a uniform family of trees of level 0, and that the attached side branches also come in certain uniform families (in the sense of Definition 6.1).

Note that we now want to impose *two* kinds of uniformity on the side branches: uniformity for different side branches in the same tree (after all, this was the initial reason to introduce uniform families), and uniformity in $\underline{\kappa}$.

To define uniformity inside a single tree, we assume that in the skeletons $\mathscr{S}_{\underline{\kappa}}$, we have choosen families of joints as in Definition 6.2. The uniformity of the side branches inside $\mathscr{T}_{\underline{\kappa}}$ will be required separately for each bone of $\mathscr{S}_{\underline{\kappa}}$. Moreover, recall that in the cusp example, we only had uniformity separately for side branches at even heights and for side branches at odd heights. In general, we will require uniformity separately depending on the height of the side branch modulo some integer ρ.

For the uniformity in $\underline{\kappa}$, recall that the joints and bones of a family of trees $\mathscr{S}_{\underline{\kappa}}$ also come in families (consisting of one joint resp. bone in each of the trees $\mathscr{S}_{\underline{\kappa}}$). We will require the side branches to be uniform inside each of these families.

We now present a complete definition of uniform families of trees of level $\leq d$. Note that we do not need to require the smoothness condition (S) anymore: using that each side branch satisfies (∇) (by Definition 6.1), we also get (∇) for the trees $\tilde{\mathscr{T}}_\lambda$ appearing in (S).

Definition 6.3 A family $(\mathscr{T}_{\underline{\kappa}})_{\underline{\kappa}\in M}$ is a *uniform family of trees of level* $\leq d$ if there exists

- a subtree $\mathscr{S}_{\underline{\kappa}}$ of $\mathscr{T}_{\underline{\kappa}}$ (the *skeleton*)
- a natural number ρ

such that the skeletons $(\mathscr{S}_{\underline{\kappa}})_{\underline{\kappa}\in M}$ form a uniform family of level 0 trees and $\mathscr{T}_{\underline{\kappa}}$ is obtained from $\mathscr{S}_{\underline{\kappa}}$ by attaching one side branch to each node of $\mathscr{S}_{\underline{\kappa}}$. We require that there is a choice of joints for $(\mathscr{S}_{\underline{\kappa}})_{\underline{\kappa}\in M}$ as in Definition 6.2 such that the side branches satisfy the following:

- For each family of joints of the trees $\mathscr{S}_{\underline{\kappa}}$: denote by $\mathscr{B}_{\underline{\kappa}}$ the side branch attached to the joint of $\mathscr{S}_{\underline{\kappa}}$ in that family; the family $(\mathscr{B}_{\underline{\kappa}})_{\underline{\kappa}\in M}$ is required to be a uniform family of side branches of level $\leq d-1$.
- For each family of bones of the trees $\mathscr{S}_{\underline{\kappa}}$ and for each congruence class $C = c + \rho\mathbb{Z}$: let N be the set of those pairs $(\underline{\kappa}, \lambda) \in M \times C$ for which the bone of $\mathscr{S}_{\underline{\kappa}}$ in our family has a node at height λ (not counting the joint(s) at the end(s) of the bone). For $(\underline{\kappa}, \lambda) \in N$, denote by $\mathscr{B}_{\underline{\kappa},\lambda}$ the side branch attached to the corresponding node of $\mathscr{S}_{\underline{\kappa}}$. We require $(\mathscr{B}_{\underline{\kappa},\lambda})_{(\underline{\kappa},\lambda)\in N}$ to be a uniform family of side branches of level $\leq d-1$.

See Figure 11.9 for an example of how the side branches are grouped into families.

This was a complete definition of uniform families of trees of level $\leq d$; to get the definition of a single tree of level $\leq d$, simply choose M to be a one-element set.

To formally fit the tree of the cusp curve into this definition, note the following:

- The skeleton is of course the infinite path $\mathscr{P}$ corresponding to the point 0, and of course we choose $\rho = 2$. However, as the family $\mathscr{T}_\lambda$ is uniform only for $\lambda > 2$, we have to insert a joint at height 2.
- Not attaching anything to the nodes of $\mathscr{P}$ at odd height is possible by attaching a side branch consisting only of a root.

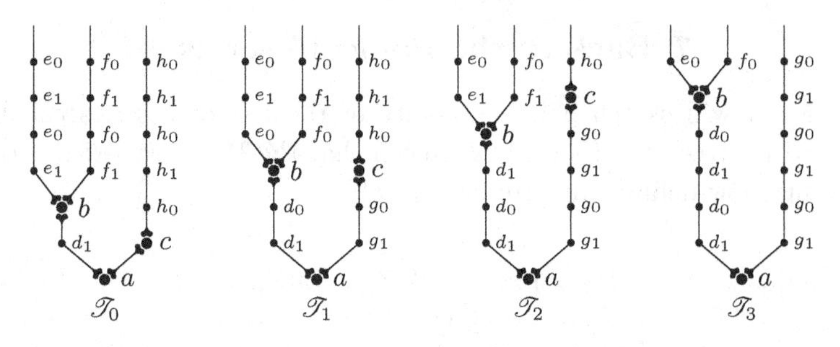

Figure 11.9 A uniform family of level 0 trees $\mathscr{T}_\lambda$ consisting of three joints a, b, c, a bone d of length $\lambda + 2$, a bone g of length $2\lambda + 1$, and three infinite bones e, f, h. If we choose $\rho = 2$ in Definition 6.3, then each label (including the indices) corresponds to one uniform family of side branches.

The congruence condition in the above definition might seem somewhat unnatural: should we really consider the absolute height of the side branch, or would it make more sense to consider the height relative to the lower end of the bone? (In the example of Figure 11.9, bone e starts with a side branch of type e_1 for even λ and with a side branch of type e_0 for odd λ.) Or what about the upper end of the bone, for those bones which have finite length? The answer is: it doesn't matter. More precisely, it *does* change the notion of uniform families, but it *does not* change the notion of (single) level $\leq d$ trees. We may even additionally require that in a family, the length of each bone is constant modulo ρ (for all $\kappa \in M$). In that case, all the above variants become equivalent.

Indeed: Suppose for example that $\mathscr{T}$ is a tree using the family of trees from Figure 11.9 in some side branches: $\mathscr{T}_\lambda \times \mathrm{T}(\mathbb{Z}_p)$ appears in a side branch at height λ of $\mathscr{T}$. Then set $\rho = 2$ for $\mathscr{T}$ (don't confuse this ρ with the one from the family $\mathscr{T}_\lambda$). This means that the trees $\mathscr{T}_\lambda$ for even and for odd λ are now two independent uniform families; in each of these families, the bone lengths are indeed constant modulo 2.

From a model theoretic point of view, one might also wonder whether Definition 6.3 is too restrictive: in the skeletons, what about allowing the bone lengths to be given by definable functions (in the value group) instead of linear ones? And concerning the uniform families of side branches on bones: what about replacing the congruence condition on λ in Definition 6.3 by an arbitrary definable condition on $\underline{\kappa}$ and λ? Again, the answer is: it does not matter for the notion of single (unparametrized) level $\leq d$ trees. (This can be proven using a cell decomposition theorem in the value group.)

7 Back to the Poincaré series

To finish, I will sketch how rationality of the Poincaré series can be obtained for trees $\mathscr{T}$ of level $\leq d$. Recall that the Poincaré series is the following (now defined directly for a tree):

$$P_{\mathscr{T}}(Z) := \sum_{\lambda=0}^{\infty} N_\lambda Z^\lambda \in \mathbb{Z}[[Z]]$$

where N_λ is the number of nodes of $\mathscr{T}$ at height λ.

For any subset A of the nodes of $\mathscr{T}$, we can define a similar series $P_A(Z)$: use the same definition, but let N_λ count only nodes inside A. For any finite partition $(A_i)_i$ of the nodes of $\mathscr{T}$, we then get the total series $P_{\mathscr{T}}(Z)$ as sum of the series $P_{A_i}(Z)$, so to get rationality of $P_{\mathscr{T}}(Z)$, it suffices to prove that all $P_{A_i}(Z)$ are rational.

The idea is now that our definition of level d trees naturally yields a finite partition of $\mathscr{T}$ into sets each of which is easy to describe. The rough idea is the following: each node of $\mathscr{T}$ lies in one of the side branches attached to the skeleton (nodes of the skeleton itself are roots of side branches). Start by partitioning the nodes of $\mathscr{T}$ such that each part contains all nodes of all side branches of one family; note that there are only finitely many such families. Now consider one such family $\mathscr{B}_\lambda$ and suppose that $\mathscr{T}'_\lambda$ is a family of trees of level $\leq d-1$ appearing in the construction of $\mathscr{B}_\lambda$. All side branches of all $\mathscr{T}'_\lambda$ again come in finitely many families; use this to further partition the nodes of the trees $\mathscr{T}'_\lambda \times \mathrm{T}(\mathbb{Z}_p)$ appearing in $\mathscr{B}_\lambda$. Do this recursively until we arrive at trees of level 0, which are partitioned according to their joints and bones.

Each of the resulting sets will have a series which can be written as nested geometric series, and such nested series are rational. Let me show this on an example. Figure 11.10 shows (a simplified version of) the partition one obtains for the cusp curve. The series of the sets A and B are just geometric series: $P_A(Z) = \sum_{\lambda=0}^{\infty} Z^\lambda = \frac{1}{1-Z}$ and $P_B(Z) = \sum_{\lambda=0}^{\infty}(p-1)p^\lambda Z^{\lambda+1} = \frac{(p-1)Z}{1-pZ}$. For the sets C and D, we can write the series using an outer sum which runs over the different side branches (below, summand μ corresponds to the side branch starting at height 2μ) and an inner sum which runs over the different heights inside the side branch. The series obtained in this way are

$$P_C(Z) = \frac{p-1}{2} \sum_{\mu=1}^{\infty} \sum_{\nu=0}^{\mu-1} p^\nu Z^{2\mu+1+\nu} = \frac{(p-1)Z^3}{2(1-Z^2)(1-pZ^3)}$$

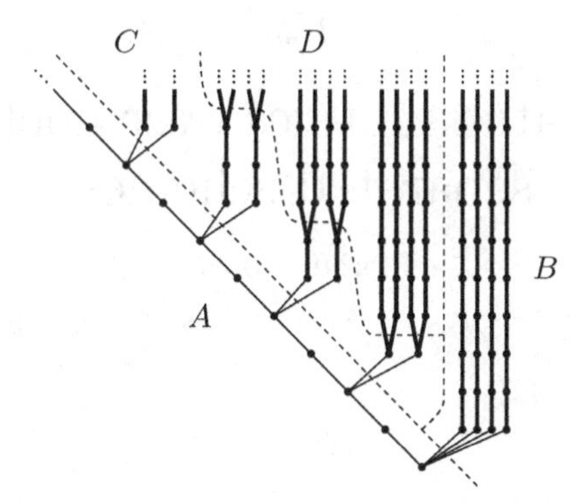

Figure 11.10 Partitioning the nodes of the tree of the cusp curve to compute its Poincaré series (don't forget that the thick lines stand for: take the product of that tree with $\mathrm{T}(\mathbb{Z}_p)$).

and

$$P_D(Z) = (p-1)\sum_{\mu=1}^{\infty}\sum_{\nu=\mu}^{\infty} p^{\nu} Z^{2\mu+1+\nu} = \frac{(p-1)pZ^4}{(1-pZ)(1-pZ^3)}.$$

(Doing these computations is left to the reader.) This yields the following total series for the cusp curve:

$$P_{\mathscr{T}}(Z) = \frac{2p^2Z^5 + (p^2-p)Z^4 + (1-3p)Z^3 - 2pZ^2 + 2}{2(1-pZ)(1-Z^2)(1-pZ^3)}.$$

References

[1] J. Denef, *The rationality of the Poincaré series associated to the p-adic points on a variety*, Invent. Math., 77 (1984), pp. 1–23.

[2] ———, *p-adic semi-algebraic sets and cell decomposition*, J. Reine Angew. Math., 369 (1986), pp. 154–166.

[3] I. Halupczok, *Trees of definable sets over the p-adics.* To appear in J. Reine Angew. Math.

[4] A. Macintyre, *On definable subsets of p-adic fields*, J. Symbolic Logic, 41 (1976), pp. 605–610.

[5] P. Scowcroft and L. van den Dries, *On the structure of semialgebraic sets over p-adic fields*, J. Symbolic Logic, 53 (1988), pp. 1138–1164.

12

Triangulated motives over noetherian separated schemes

Florian Ivorra

Abstract The main purpose of this paper is to provide a detailed exposition of Voevodsky's construction of triangulated motives over a general basis. We mainly give a thorough review of the theory of relative cycles due to A. Suslin and V. Voevodsky [16] and then provide the definition and basic properties of finite correspondences over a noetherian separated scheme as well as the definition of the associated triangulated categories of (effective) motives.

Introduction

In [19], V. Voevodsky has given the definition of the triangulated category $DM^{\mathrm{eff}}_{gm}(k)$ of geometrical (effective) motives over a perfect field k. Moreover using sheaves with transfers for the Nisnevich topology, he has constructed a larger category $DM^{\mathrm{eff}}_{-}(k)$ and shown that $DM^{\mathrm{eff}}_{gm}(k)$ is a strictly full triangulated subcategory of $DM^{\mathrm{eff}}_{-}(k)$. This result known as the embedding theorem, embeds a category with a rather nice and compact definition into a category more suited to computation (*e.g.* see [22]). The construction of both $DM^{\mathrm{eff}}_{gm}(k)$ and $DM^{\mathrm{eff}}_{-}(k)$ relies on the notion of finite correspondences. Over a field the definition of a finite correspondence, between two smooth k-schemes of finite type X and Y, is simple: it is a linear combination of closed integral subschemes of $X \times_k Y$ which are finite and surjective over a connected component of X. Moreover the composition of finite correspondences is provided by t he classical intersection theory.

Motivic Integration and its Interactions with Model Theory and Non-Archimedean Geometry (Volume II), ed. Raf Cluckers, Johannes Nicaise, and Julien Sebag.
Published by Cambridge University Press.

Over a noetherian base scheme the notion of finite correspondences still makes sense even for non smooth schemes, though the definition and the composition formula are more involved and require the full theory of relative cycles developed by A. Suslin and V. Voevodsky [16]. This generality was not available in [19] although the composition formula do appear briefly in the litterature [21, §2]. The purpose of this paper is to provide a survey of the theory of relative cycles and then detail the properties of finite correspondences over a noetherian scheme. These properties are the cornerstone of the various constructions of motivic triangulated categories[1]. The theory of relative algebraic cycles plays also a central role in [9] and a multiplicity oriented treatment of the theory can be found in the appendices[2] of this book.

Let us give a brief overview of the content of this paper. The first section is devoted to the main operation on relative cycles: the base change morphism which bears an important intersection theoretic meaning. Given a relative cycle α on X over S and a morphism $\theta : T \to S$, the idea is to define a relative cycle $\theta^{\circledast}\alpha$ on X_T over T in such a way that it coincides with the flat base changes whenever they are defined: for flat morphisms and flat cycles (see §1.4.2). It happens that there exists a unique way to encompass the two flat base changes into one theory, however two subtleties appear: the base change is not necessarily defined for all relative cycles, a condition has to be taken into account which can be phrased using either fat points or abstract blow ups, and even when defined some denominators may appear.

The second section deals with the explicit computation of the multiplicities that appear by base change. It allows to show in particular that the general base change is also compatible over regular schemes with the operation defined using the Tor-formula and thus generalizes the classical intersection theory obtained via Serre's multiplicities. In the third section we recall the basic operations on relative cycles and their compatibilities with base change. The fourth section is devoted to the finite correspondences and the definition of geometrical motives while the last one is devoted to the embedding theorem.

Conventions All schemes are assumed to be noetherian and separated. A morphism of schemes $\theta : T \to S$ is said to be maximal if any

[1] The category of geometrical motives over a noetherian scheme forms the backbone of [7] where an ℓ-adic realization functor is provided for those categories and a fine study of finite correspondences from the local point of view is given.

[2] However the theory without explicit multiplicities apply more generally: there is no need to assume the base scheme to be geometrically unibranch.

irreductible component of T dominates an irreductible component of S. If X is a S-scheme it will be convenient to denote by X_T the scheme $T \times_S X$ and by θ_X the morphism $X_T \to X$ obtained by base change.

In this paper a relative cycle on X over S is any linear combination of closed subschemes of X that are of finite type over S and dominate an irreducible component of S. In [16] this term is reserved for the cycles that admit well-defined base changes, those cycles are called universally rational instead. The group of relative/universally rational cycles is denoted by $Cycl(X/S, n)$ in [16] and by $\mathscr{Z}(X/S, n)$ here. The base change morphism is denoted by $cycl(\theta)$ in [16] while we adopt the more compact notation $\theta^{\circledast}$ here.

1 Relative cycles

1.1 Let X be a scheme. An algebraic cycle on X is a linear combination of closed integral subschemes of X, in other words an algebraic cycle is an element of the free abelian group $Z(X)$ generated by the closed integral subschemes of X. Given such a subscheme Z of X, we denote by $[Z]$ its image in $Z(X)$. More generally any closed subscheme Z of X defines an algebraic cycle using the length function:

$$[Z] := \sum_{z} \lg(\mathscr{O}_{Z,z}) \left(\left[\overline{\{z\}}\right]\right)$$

where the sum is taken over the generic points z of the irreductible components of Z. Given a cycle α on X and a closed integral subscheme Z of X we denote by $m(Z, \alpha)$ the multiplicity of Z in α, that is to say the coefficient of $[Z]$ in α. Algebraic cycles are contravariant with respect to flat morphisms and covariant with respect to proper morphisms. In the sequel the following remark is used endlessly:

Remark 1.2 *The pullback map along a flat surjective morphism is a monomorphism.*

To develop a well behaved theory of base change for cycles, it is crucial to look at a special class of cycles.

Definition 1.3 Let X be a S-scheme. The group $Z(X/S)$ of relative algebraic cycles is the free abelian group generated by the closed integral subschemes Z of X which are of *finite type* over S and *dominates an irreductible component* of S.

The relative dimension considered here is defined as in [3, chapter 20]. Namely if Z is an integral S-scheme of finite type, its dimension over S is the nonnegative integer

$$\dim_S(Z) = \operatorname{degtr}_{\kappa(s)} \kappa(z) - \dim(\mathscr{O}_{S,s})$$

where z is the generic point of Z and s its image in S. When Z dominates an irreducible component of S as in definition 1.3, the ring $\mathscr{O}_{S,s}$ is artinian hence zero dimensional and $\dim_S(Z)$ is simply the dimension of the generic fiber of X. More generally the dimension of a scheme of finite type over S is the supremum of the dimensions of its irreductible components endowed with their reduced structure. A relative cycle $\alpha \in Z(X/S)$ is said to be of dimension n if all the closed integral subschemes Z such that $m(Z, \alpha)$ is non zero are of relative dimension n over S. This provides a graduation

$$Z(X/S) = \bigoplus_{n \in \mathbb{N}} Z(X/S, n)$$

where $Z(X/S, n)$ denotes the subgroup of relative cycles of dimension n over S.

1.4 The base change of a relative cycle α in $Z(X/S, n)$ along a flat morphism $\theta : T \to S$ is defined by

$$\theta^{\circledast}_{\text{flat}} \alpha = \theta^*_X \alpha = \sum_Z m(Z, \alpha) \big[Z_T\big].$$

the sum being taken over the closed integral subschemes of X of dimension n over S. It is simply the flat pullback of α along the flat morphism θ_X. Since a flat morphism is maximal, lemmas 5.10 and 5.11 assure that we get a relative cycle $\theta^{\circledast}_{\text{flat}} \alpha \in Z(X_T/T, n)$.

There is another flat base change defined for an arbitrary morphism but on a special class of cycles which are assumed to be flat in some sense over S. Namely let us denote by $\mathrm{Hilb}(X/S, n)$ the set of closed subschemes of X which are flat and equidimensional of dimension n over S. Taking the associated algebraic cycle, we get a morphism

$$[-] : \mathbb{Z}\mathrm{Hilb}(X/S, n) \to Z_{\text{equi}}(X/S, n)$$

and we let $Z_{\text{Hilb}}(X/S, n)$ be its image. Since flat and equidimensional morphisms of relative dimension n are stable by base change, we have a map

$$T \times_S - : \mathbb{Z}\mathrm{Hilb}(X/S, n) \to \mathbb{Z}\mathrm{Hilb}(X_T/T, n).$$

To see that this map is compatible with the associated cycle, the next proposition is crucial.

Proposition 1.5 *Let X be a S-scheme and $\theta : T \to S$ a morphism of schemes. Given α and β in $\mathbb{Z}\mathrm{Hilb}(X/S, n)$ such that $[\alpha] = [\beta]$, one has also $[T \times_S \alpha] = [T \times_S \beta]$.*

It shows that we get a well defined morphism

$$\theta^{\circledast}_{\mathrm{Hilb}} : Z_{\mathrm{Hilb}}(X/S, n) \to Z_{\mathrm{Hilb}}(X_T/T, n)$$

by letting $\theta^{\circledast}_{\mathrm{Hilb}}(\alpha) = \sum_{\mathscr{Z}} \alpha_{\mathscr{Z}}[T \times_S \mathscr{Z}]$ for an element

$$\alpha = \sum_{\mathscr{Z} \in \mathrm{Hilb}(X/S,n)} \alpha_{\mathscr{Z}}[\mathscr{Z}]$$

in $Z_{\mathrm{Hilb}}(X/S, n)$. In particular we have a commutative square

$$\begin{array}{ccc} \mathbb{Z}\mathrm{Hilb}(X/S,n) & \xrightarrow{[-]} & Z_{\mathrm{Hilb}}(X/S,n) \\ \downarrow{\scriptstyle T\times_S -} & & \vdots{\scriptstyle \theta^{\circledast}_{\mathrm{Hilb}}} \\ \mathbb{Z}\mathrm{Hilb}(X_T/T,n) & \xrightarrow{[-]} & Z_{\mathrm{Hilb}}(X_T/T,n). \end{array}$$

The rest of the paragraph is devoted to the proof of proposition 1.5. For a finite and flat S-scheme X, the $\mathscr{O}_S$-module $\pi_{X/S*}\mathscr{O}_X$ is locally free of finite rank, we denote by $\deg_s(X/S)$ its rank on a neighborhood of a maximal point s.

Remark 1.6 Let X be an S-scheme finite and flat over S:

1. given any maximal point s in S, one has
$$\deg_s(X/S)\,\mathrm{lg}_{\mathscr{O}_{S,s}}(\mathscr{O}_{S,s}) = \sum_{x \in X_s} \mathrm{lg}_{\mathscr{O}_{X,x}}(\mathscr{O}_{X,x})[\kappa(x) : \kappa(s)];$$
2. given a morphism of schemes $\theta : T \to S$, a maximal point t in T and a maximal generization s of the image of t in S, one has
$$\deg_s(X/S) = \deg_t(X_T/T).$$

Proof of proposition 1.5 The result holds when θ is flat since in that case $[T \times_S \alpha] = \theta^*_X\alpha$, where θ^*_X is the morphism provided by the functoriality of cycles with respect to flat maps. Consider now the general case. By linearity it enough to check that for any α such that $[\alpha] = 0$ one has also $[T \times_S \alpha] = 0$. The cycle α may be written

$$\alpha = \sum_{i=1}^{r} \alpha_i \mathscr{Z}_i$$

where the $\mathscr{Z}_i$'s are closed subschemes of X which are flat and equidimensional of dimension n over S. One may replace X by the union of the $\mathscr{Z}_i$'s, and assume without restriction that X is equidimensional of dimension n over S. Let W be a closed integral subscheme of X_T. We have to show that

$$m(W, [T \times_S \alpha]) = 0.$$

For this we may assume that W is of dimension n over T and dominates an irreductible component of T. Let w be the generic point of W and z its image in X. The equality we have to proves is local for the Zariski topology around z, so we may replace X by any convenient open neighborhood of z. Since the point w is maximal in its fibre, z is also maximal in its fibre, and by lemma 5.13, we may assume that the projection of X over S is a composition $X \xrightarrow{q} \mathbb{A}^n_S \to S$ where q is a flat and quasi-finite morphism. Since $T \times_S \alpha = \mathbb{A}^n_T \times_{\mathbb{A}^n_S} \alpha$, we may assume $n = 0$.

Let t be the generic point of the irreductible component of T dominated by W. The local ring $\mathscr{O}_{T,t}$ is then artinian and

$$m(W, [T \times_S \alpha]) = m(\overline{W}, [\mathrm{Spec}(\mathscr{O}_{T,t}) \times_S \alpha])$$

where $\overline{W}$ denotes the closure of w in $\mathrm{Spec}(\mathscr{O}_{T,t}) \times_S X$. One may therefore assume that T is an artinian local scheme with closed point t. Consider now a strict henselization $u : T' \to T$ of T at its closed point. The scheme T' is also artinian[3], and the morphism u being faithfully flat, we get

$$[T' \times_S \alpha] = [T' \times_T (T \times_S \alpha)] = u_X^*[T \times_S \alpha]$$

and so by remark 1.2, we may assume that T is a strictly henselian local artinian scheme. Let now $\bar{s}$ be the geometric point of S defined by θ. There exists a commutative triangle

$$\begin{array}{ccc} & & \mathrm{Spec}(\mathscr{O}^{sh}_{S,\bar{s}}) \\ & \overset{\vartheta}{\nearrow} & \downarrow {\scriptstyle \mathfrak{i}^{sh}_{S,\bar{s}}} \\ T & \xrightarrow{\theta} & S, \end{array}$$

where $\mathrm{Spec}(\mathscr{O}^{sh}_{S,\bar{s}})$ denotes the spectrum of the strict henselization of S at the geometric point $\bar{s}$. The canonical morphism $\mathfrak{i}^{sh}_{S,\bar{s}}$ being flat, the

[3] A strict henselization of a local artinian ring is also a local artinian ring. If follows from [5, **IV**,18.6.9.1] and [5, **IV**,18.8.8] since any essentially étale $\mathscr{O}$-algebra is artinian.

equality

$$[T \times_S \alpha] = \left[T \times_{\mathrm{Spec}(\mathscr{O}^{\mathrm{sh}}_{S,\bar{s}})} (\mathrm{Spec}(\mathscr{O}^{\mathrm{sh}}_{S,\bar{s}}) \times_S \alpha)\right]$$

allows to assume that S is a strictly henselian local scheme. In that case, by [5, **IV**, 18.5.11], we may assume that X is a local S-scheme flat and finite over S. Moreover we may assume (if necessary by taking an open neighborhood of z) that z is the unique point of X_s. This assumption assures that $(X_s)_{\mathrm{red}} = \mathrm{Spec}(\kappa(z))$ which implies that

$$\begin{aligned}(X_T)_{\mathrm{red}} = ((X_T)_t)_{\mathrm{red}} &= \left(\mathrm{Spec}(\kappa(t)) \times_{\mathrm{Spec}(\kappa(s))} X_s\right)_{\mathrm{red}} \\ &= \left(\mathrm{Spec}(\kappa(t)) \times_{\mathrm{Spec}(\kappa(s))} (X_s)_{\mathrm{red}}\right)_{\mathrm{red}} \\ &= \left(\mathrm{Spec}(\kappa(t) \otimes_{\kappa(s)} \kappa(z))\right)_{\mathrm{red}}\end{aligned}$$

Since S is strictly henselian, $\kappa(z)/\kappa(s)$ is a purely inseparable extension, which implies that the artinian ring $\kappa(t) \otimes_{\kappa(s)} \kappa(z)$ is local and therefore that X_T is a local artinian scheme reduced to the sole point w. Fix a maximal generization η of s. Using remark 1.6, one gets

$$\begin{aligned}m(W, [T \times_S \alpha]) &= \sum_{i=1}^{r} \alpha_i \lg(\mathscr{O}_{T \times_S \mathscr{Z}_i, w}) \\ &= \frac{\lg(\mathscr{O}_{T,t})}{[\kappa(w) : \kappa(t)]} \left[\sum_{i=1}^{r} \alpha_i \deg_t((\mathscr{Z}_i)_T/T)\right] \\ &= \frac{\lg(\mathscr{O}_{T,t})}{[\kappa(w) : \kappa(t)]} \left[\sum_{i=1}^{r} \alpha_i \deg_\eta(\mathscr{Z}_i/S)\right] \\ &= \frac{\lg(\mathscr{O}_{T,t})}{[\kappa(w) : \kappa(t)]} \left[\sum_{i=1}^{r} \alpha_i \left(\sum_{x \in X_\eta} \lg(\mathscr{O}_{\mathscr{Z}_i,x}) \frac{[\kappa(x) : \kappa(\eta)]}{\lg(\mathscr{O}_{S,\eta})}\right)\right].\end{aligned}$$

This implies that $m(W, [T \times_S \alpha])$ is equal to

$$\frac{\lg(\mathscr{O}_{T,t})}{[\kappa(w) : \kappa(t)] \lg(\mathscr{O}_{S,\eta})} \left[\sum_{x \in X_\eta} [\kappa(x) : \kappa(\eta)] \left(\sum_{i=1}^{r} \alpha_i \lg(\mathscr{O}_{\mathscr{Z}_i,x})\right)\right]$$

and since for any point x above η, we know by hypothesis that

$$m(\overline{\{x\}}, [\alpha]) = \sum_{i=1}^{r} \alpha_i \lg(\mathscr{O}_{\mathscr{Z}_i,x}) = 0,$$

the proposition is proved. □

1.7 Relative algebraic cycles have the advantage over arbitrary cycles to be completely determined by what happens over the generic points of the irreductible components of S. This property is essential and the next remark provides a precise formulation.

Remark 1.8 Let $s_1, \dots, s_r$ be the generic points of the irreductible component of S. There is an isomorphism $Z(X/S, n) = Z(S_{\mathrm{red}} \times_S X/S_{\mathrm{red}}, n)$ and the morphism

$$Z(S_{\mathrm{red}} \times_S X/S_{\mathrm{red}}, n) \to \oplus_{i=1}^{r} Z(X_{s_i}/\kappa(s_i), n)$$

given by the morphisms $(s_i)^{\circledast}_{\mathrm{flat}}$ is injective.

Definition 1.9 A presheaf of relative cycles over X with coefficient in a commutative ring A is the data for any S-scheme T of a A-submodule $\mathfrak{Z}(X_T/T, n)$ of $Z(X_T/T, n)_A$ and for any morphism of S-schemes $T' \to T$ of a morphism of A-modules

$$\theta^{\circledast}_{\mathfrak{Z}} : \mathfrak{Z}(X_T/T, n) \to \mathfrak{Z}(X_{T'}/T', n)$$

such that the following conditions are satisfied:

1. if θ is a flat morphism then $\theta^{\circledast}_{\mathfrak{Z}}$ coincides with the flat base change $\theta^{\circledast}_{\mathrm{flat}}$;
2. if α belongs to $Z_{\mathrm{Hilb}}(X/S, n)$ then $\theta^{\circledast}_{\mathfrak{Z}}$ is given by
$$\theta^{\circledast}_{\mathfrak{Z}}\alpha = \theta^{\circledast}_{\mathrm{Hilb}}\alpha;$$
3. if $\theta : T_{\mathrm{red}} \to T$ is the natural morphism, then $\theta^{\circledast}_{\mathfrak{Z}}\alpha = \alpha$ for any α in $\mathfrak{Z}(X_T/T, n)$;
4. for any morphisms of S-schemes $\vartheta : T'' \to T', \theta : T' \to T$ and any element α in $\mathfrak{Z}(X_T/T, n)$, one has
$$(\theta \circ \vartheta)^{\circledast}_{\mathfrak{Z}}\alpha = \vartheta^{\circledast}_{\mathfrak{Z}}(\theta^{\circledast}_{\mathfrak{Z}}\alpha).$$

Thus a presheaf of relative cycles is a way to define, for a specific class of cycles, a functorial theory of base change which coincides with the two possible flat pullbacks. In particular for a flat morphism $\theta : T' \to T$ and an element $\alpha \in \mathfrak{Z}(X_T/T, n)$, one has the equality

$$\theta^{\circledast}_{\mathfrak{Z}}\alpha = \sum_Z \alpha_Z \left(\sum_W m(W, Z_{T'})[W] \right)$$

where the second sum is taken over the irreductible components W of $Z_{T'}$. Since we work with relative cycles, this formula holds for the base change by any maximal morphism:

Lemma 1.10 *Let $\mathfrak{Z}$ be a presheaf of relative cycles over X with coefficient in a commutative ring A , $\theta : T' \to T$ a maximal morphism of S-schemes and α an element in $\mathfrak{Z}(X_T/T, n)$. One has the equality*

$$\theta_{\mathfrak{Z}}^{\circledast}\alpha = \sum_Z \alpha_Z \left(\sum_W m(W, Z_{T'})[W] \right)$$

where the second sum is taken over the irreductible components W of $Z_{T'}$ that dominate an irreductible component of T'.

Proof Let β be the relative cycle in $Z(X_{T'}/T', n)$

$$\beta = \sum_Z \alpha_Z \left(\sum_W m(W, Z_{T'})[W] \right).$$

We may assume T and T' integral. By remark 1.8, it is enough to show that for any maximal point t' in T' the cycles $\theta_{\mathfrak{Z}}^{\circledast}\alpha$ and β coincide after application of the flat base change $(t')^{\circledast}_{\text{flat}}$. Our assumption assures that the image t of t' is a maximal point in T, it is then enough to remark by functoriality that the first condition of definition 1.9 ensures that

$$(t')^{\circledast}_{\text{flat}}\theta_{\mathfrak{Z}}^{\circledast}\alpha = \sum_Z \alpha_Z[(Z_t)_{\kappa(t')}] = \sum_Z \alpha_Z[(Z_{T'})_{t'}] = (t')^{\circledast}_{\text{flat}}\beta.$$ □

The reason for considering only relative cycles appears more clearly over an integral regular S of dimension one. Indeed under this assumption, we have the equality

$$Z(X/S, n) = Z_{\text{Hilb}}(X/S, n) \tag{12.1}$$

as a result of [5, **IV**, 2.8] and [5, **IV**, 14.2.2]. The main idea to obtain a reasonable base change for relative cycles is to reduce the definition to this case: this is where the notion of fat points appears. Let K be a field and $\underline{s}$ a K-point of S. A thickening of $\underline{s}$ is the data of a spectrum $\mathscr{O}$ of a DVR and a factorization

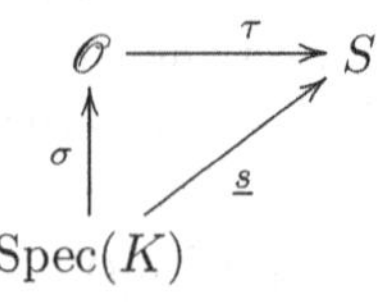

where the image of σ is the closed point of $\mathscr{O}$ and τ is a birational morphism over an irreductible component of S containing the image of $\underline{s}$. The point $\underline{s}$ is said to be fat if it possesses a thickening. We do not impose that K is the residue field of $\mathscr{O}$, *a priori* it is simply an extension of it. In other words σ is simply a K-point of $\mathcal{O}$ localized at the closed point.

Remark 1.11 The locus s of a fat K-point $\underline{s}$ cannot be a maximal point of S. An arbitrary K-point $\underline{s}$ of S is not necessarily flat. However if its locus is not a maximal point of S, there exists always an extension E of K such that the E-point $\underline{s}_E$ obtained by extension to E is fat. Indeed [5, **II**, 7.1.9] implies that for any maximal generization η of the locus s of $\underline{s}$, there exists a spectrum $\mathcal{O}$ of a DVR and a morphism

$$\tau : \mathcal{O} \to S$$

sending the closed point of $\mathcal{O}$ to s and inducing a birational morphism between $\mathcal{O}$ and the irreductible component $\overline{\{\eta\}}$. The residue field k of $\mathcal{O}$ is an extension of $\kappa(s)$, and choosing a composite extension E of K and k over $\kappa(s)$, we get a thickening

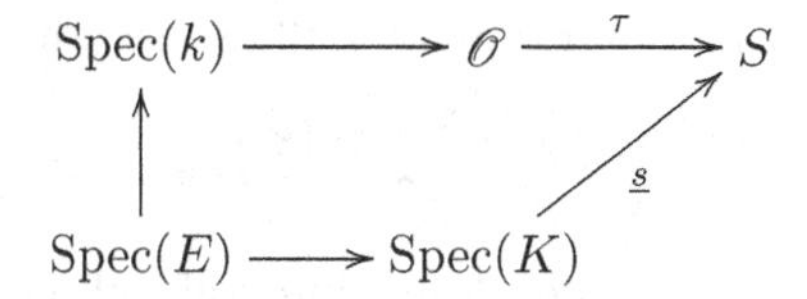

of the E-point $\underline{s}_E$.

In the sequel, given a thickening $(\mathcal{O}, \tau, \sigma)$ of a fat K-point of S, we denote by ρ the generic point of the irreductible component of S dominated by $\mathcal{O}$.

Remark 1.12 Let $\underline{s}$ be a fat K-point and $(\mathcal{O}, \tau, \sigma)$ a thickening of this fat point. Then for any integral S-scheme of finite type Z, that is of dimension n over S and that dominates $\overline{\{\rho\}}$, the scheme $Z_{\mathcal{O}}$ has a unique irreducible component $\mathcal{Z}$ that dominates $\mathcal{O}$. It enjoys the following properties:

1. $\mathcal{Z}$ is the unique closed subscheme of $Z_{\mathcal{O}}$ flat over $\mathcal{O}$ and generically isomorphic to $Z_{\mathcal{O}}$;
2. $\mathcal{Z}$ is a $\mathcal{O}$-scheme flat and equidimensional of dimension n;
3. $\mathcal{Z}$ has multiplicity one in $Z_{\mathcal{O}}$.

Let $\underline{s}$ be a fat K-point of S and fix a thickening $(\mathcal{O}, \tau, \sigma)$ of $\underline{s}$. Let X be a S-scheme and $\alpha \in Z(X/S, n)$ be a relative cycle. By remark 1.12, we may consider the cycle

$$\sum_Z \alpha_Z [\mathcal{Z}] \tag{12.2}$$

the sum being taken over the closed integral subschemes Z of X, of dimension n over S and that dominate $\overline{\{\rho\}}$. Remark 1.12 implies moreover that (12.2) belongs to $Z_{\mathrm{Hilb}}(X_{\mathscr{O}}/\mathscr{O}, n)$, and so we may consider the cycle

$$(\mathscr{O}, \tau, \sigma)^{\circledast}\alpha := \sigma_{\mathrm{Hilb}}^{\circledast}\left(\sum_Z \alpha_Z[\mathscr{Z}]\right) = \sum_Z \alpha_Z[\mathrm{Spec}(K) \times_{\mathscr{O}} \mathscr{Z}]. \quad (12.3)$$

However for a fat K-point $\underline{s}$ of S, the choice of the thickening is not unique and so the cycles (12.3) may differ for different thickenings. Nevertheless we have the following lemma:

Lemma 1.13 *Let X be a S-scheme and $\mathfrak{Z}$ a presheaf of relative cycles with rational coefficients over X. Then for any fat K-point $\underline{s}$ of S and any thickening $(\mathscr{O}, \tau, \sigma)$ one has*

$$\underline{s}_{\mathfrak{Z}}^{\circledast}(\alpha) = (\mathscr{O}, \tau, \sigma)^{\circledast}\alpha.$$

Proof Using lemma 1.10 and remark 1.12, we see that

$$\tau_{\mathfrak{Z}}^{\circledast}\alpha = \sum_Z \alpha_Z[\mathscr{Z}]$$

the sum being taken over the closed integral subschemes Z of X of dimension n over S and that dominate $\overline{\{\rho\}}$. It is then enough to use functoriality and the second condition of definition 1.9 to get

$$\underline{s}_{\mathfrak{Z}}^{\circledast}\alpha = \sigma_{\mathrm{Hilb}}^{\circledast}\left(\sum_Z \alpha_Z[\mathscr{Z}]\right) = (\mathscr{O}, \tau, \sigma)^{\circledast}\alpha.$$

□

Lemma 1.13 implies in particular that if we can apply any kind of reasonable base change to a relative cycle $\alpha \in Z(X/S, n)$ then the cycles (12.3) have to be independant of the chosen thickenings.

Definition 1.14 A relative cycle $\alpha \in Z(X/S, n)$ is said to be universally rational if for any fat K-point $\underline{s}$ of S the cycles

$$(\mathscr{O}, \tau, \sigma)^{\circledast}\alpha$$

do not depend of the chosen thickening $(\mathscr{O}, \tau, \sigma)$. The universally rational cycles form a subgroup $\mathscr{Z}(X/S, n)$ of $Z(X/S, n)$.

In the sequel if α is a universally rational cycle, the common value of the cycles (12.3) is denoted by $\underline{s}^{\circledast}\alpha$.

1.15 Fat points are used to produce flat cycles out of a given relative cycle. This can be also achieved using blow-ups by [13] and leads to another description of universally rational algebraic cycles in terms of blow-ups.

Definition 1.16 An abstract blow-up $\theta : T \to S$ is a morphism of schemes such that

1. θ is proper and maximal ;
2. there exists an open subscheme U in S such that the morphism $(U_T)_{\text{red}} \to U_{\text{red}}$ is an isomorphism.

An abstract blow-up is said to be generic if there exists some dense open subscheme satisfying the last condition.

Let $\theta : T \to S$ be an abstract blow-up, U be the biggest open subset satisfying the last condition of definition 1.16 and V be the open subscheme $V = U \times_S X$. The strict transform Z_T^{st} of a closed subscheme $Z \subset X$ is the closed subscheme of Z_T obtained by taking the schematic closure of the subscheme $Z_T \cap V_T$ in X_T. Given a relative cycle $\alpha = \sum_{i=1}^r \alpha_i [Z_i]$ one may then define its strict transform as

$$\theta_{\text{st}}^{\circledast} \alpha = \sum_{i=1}^{r} \alpha_i \left[(Z_i)_T^{\text{st}} \right].$$

A abstract blow-up is said to flatten a cycle when the strict transforms of the Z_i's are flat over T. In that case we know by [5, **IV**, 14.2.2] that those strict transforms are equidimensional of dimension n over T and thus that $\theta_{\text{st}}^{\circledast} \alpha$ belongs to $Z_{\text{Hilb}}(X_T/T, n)$.

Proposition 1.17 *Let X be a S-scheme and $\alpha \in Z(X/S, n)$ a relative cycle. There exists a generic abstract blow-up $\theta : T \to S$ that flattens the cycle α.*

Proof Let $\vartheta : S_{\text{red}} \to S$ be the canonical morphism. This morphism is a generic abstract blow-up and the canonical isomorphism

$$Z(S_{\text{red}} \times_S X / S_{\text{red}}, n) = Z(X/S, n)$$

of remark 1.8 coincides with $\vartheta_{\text{st}}^{\circledast}$. In particular we may assume that S is reduced. Write α as

$$\alpha = \sum_{i=1}^{r} \alpha_i [Z_i]$$

where the Z_i's are closed integral subschemes of X of dimension n over S that dominate an irreductible component of S. Since S is reduced, [5, **II**, 6.9.1] implies by induction that there exists a dense open subset U in S such that the schemes $U \times_S Z_i$ are flat over U. The proposition then follows from[13, 5.2.2]. □

One of the key observations needed to define the base change is the following remark:

Remark 1.18 Assume that $\theta : T \to S$ is a proper birational morphism. Consider the diagram

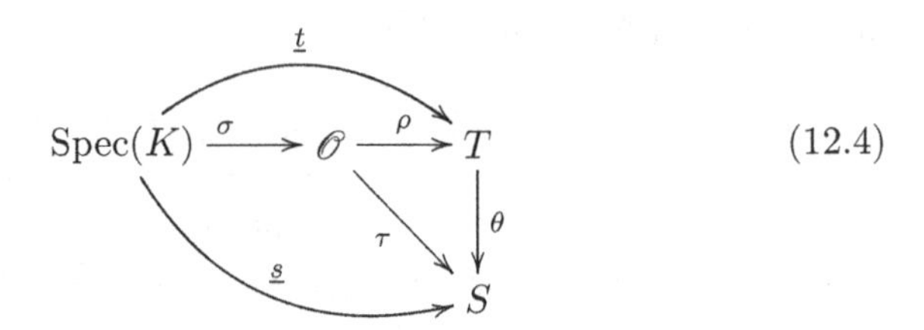

(12.4)

where $\mathscr{O}$ is the spectrum of a DVR. Then τ is a birational morphism over some irreductible component of S containing the image of $\underline{s}$ if and only if ρ is also a birational morphism over an irreductible component of T containing the image of $\underline{t}$. The valuative criterion of properness [5, **II**, 7.3.8] implies then that, given any such τ, there always exists a unique ρ fitting into this commutative diagram. Therefore $\underline{t}$ is fat if and only if its image $\underline{s}$ is fat and

$$(\mathscr{O}, \rho, \sigma) \mapsto (\mathscr{O}, \tau, \sigma)$$

defines a 1:1 correspondences between their thickenings. Assuming moreover that θ is a generic abstract blow-up that flattens α, we see that

$$(\mathscr{O}, \tau, \sigma)^{\circledast}\alpha = (\mathscr{O}, \rho, \sigma)^{\circledast}(\theta^{\circledast}_{\mathrm{st}}\alpha) = \underline{t}^{\circledast}_{\mathrm{Hilb}}(\theta^{\circledast}_{\mathrm{st}}\alpha).$$

Indeed write α as

$$\alpha = \sum_{i=1}^{r} \alpha_i [Z_i]$$

where the Z_i's are closed integral subschemes of X of dimension n over S that dominate an irreductible component of S. The strict transform $(Z_i)^{\mathrm{st}}_T$ is a closed subscheme of X_T which is flat over T and generically isomorphic to $(Z_i)_T$. This implies that $\mathscr{O} \times_T (Z_i)^{\mathrm{st}}_T$ is a closed subscheme of $\mathscr{O} \times_S X$ flat over $\mathscr{O}$ and generically isomorphic to $\mathscr{O} \times_S Z_i$. Therefore

$$(\mathscr{O}, \tau, \sigma)^{\circledast}\alpha = \sum_{i=1}^{r} \alpha_i \Big[\operatorname{Spec}(K) \times_{\mathscr{O}} \big(\mathscr{O} \times_T (Z_i)_T^{\mathrm{st}}\big) \Big] = (\mathscr{O}, \rho, \sigma)^{\circledast}(\theta_{\mathrm{st}}^{\circledast}\alpha)$$
$$= \sum_{i=1}^{r} \alpha_i \Big[\operatorname{Spec}(K) \times_T (Z_i)_T^{\mathrm{st}} \Big] = \underline{t}_{\mathrm{Hilb}}^{\circledast}(\theta_{\mathrm{st}}^{\circledast}\alpha).$$

From this the following proposition which provide a characterization of universally rational cycles in terms of blow-ups may be easily deduced:

Proposition 1.19 *Let X be an S-scheme and α a cycle in $Z(X/S, n)$. Then the following conditions are equivalent:*

1. *given a fat K-point $\underline{s}$, the cycles $(\mathscr{O}, \tau, \sigma)^{\circledast}\alpha$ do not depend on the thickening $(\mathscr{O}, \tau, \sigma)$;*
2. *given a generic abstract blow-up $\theta : T \to S$ that flattens α and two K-points $\underline{t}_1, \underline{t}_2$ such that $\theta \circ \underline{t}_1 = \theta \circ \underline{t}_2$, one has*

$$\underline{t}_{1,\mathrm{Hilb}}^{\circledast}\theta_{\mathrm{st}}^{\circledast}\alpha = \underline{t}_{2,\mathrm{Hilb}}^{\circledast}\theta_{\mathrm{st}}^{\circledast}\alpha.$$

Proof Assume the first condition and let $\theta : T \to S$ be a generic abstract blow-up that flattens α. Let t_1, t_2 be the loci of the K-points $\underline{t}_1$ and $\underline{t}_2$ and s the locus of the K-point $\underline{s}$ image of $\underline{t}_1, \underline{t}_2$ by θ. The abstract blow-up θ being a birational morphism, we may assume that s is not maximal. In that case using remark 1.11, one can choose an extension E/K such that the E-points $(\underline{t}_1)_E$, $(\underline{t}_2)_E$ of T are fat. Fix for each of them a thickening $(\mathscr{O}_1, \tau_1, \sigma_1)$ and $(\mathscr{O}_2, \tau_2, \sigma_2)$ such that we have a commutative diagram

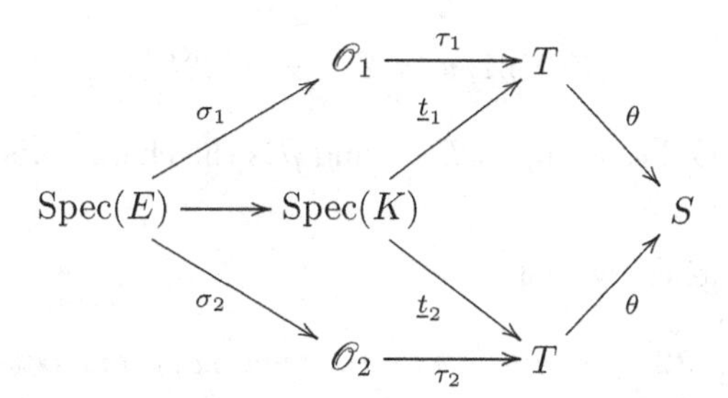

since θ is birational, the triplets $(\mathscr{O}_1, \theta \circ \tau_1, \sigma_1)$ and $(\mathscr{O}_2, \theta \circ \tau_2, \sigma_2)$ are two thickening of the same E-point and our hypothesis implies that

$$(\mathscr{O}_1, \theta \circ \tau_1, \sigma_1)^{\circledast}\alpha = (\mathscr{O}_2, \theta \circ \tau_2, \sigma_2)^{\circledast}\alpha.$$

Applying remark 1.18, we get

$$(\mathscr{O}_1, \theta\circ\tau_1, \sigma_1)^{\circledast}\alpha = \left(\underline{t}_{1,\mathrm{Hilb}}^{\circledast}\left[\theta_{\mathrm{st}}^{\circledast}\alpha\right]\right)_E, (\mathscr{O}_2, \theta\circ\tau_2, \sigma_2)^{\circledast}\alpha = \left(\underline{t}_{2,\mathrm{Hilb}}^{\circledast}\left[\theta_{\mathrm{st}}^{\circledast}\alpha\right]\right)_E$$

and the second assertion is finally a consequence of remark 1.2.

Assume now the second assertion to be satisfied et take two thickenings $(\mathscr{O}_1, \tau_1, \sigma_1)$ and $(\mathscr{O}_2, \tau_2, \sigma_2)$ of a same fat K-point $\underline{s}$. By proposition 1.17 there exists a generic abstract blow-up $\theta : T \to S$ that flattens the cycle α. The image of the generic point of $\mathscr{O}_1$ being a maximal point and since θ is birational, the valuative criterion of properness [5, **II**, 7.3.8] applied to the proper morphism θ provides a canonical lifting

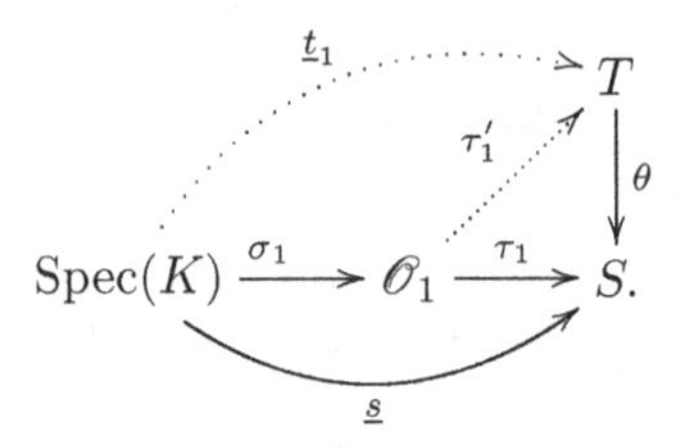

Applying remark 1.18, we get $(\mathscr{O}_1, \tau_1, \sigma_1)^{\circledast}\alpha = \underline{t}_1^{\circledast}\left[\theta_{\mathrm{st}}^{\circledast}\alpha\right]$. Similarly one gets $(\mathscr{O}_2, \tau_2, \sigma_2)^{\circledast}\alpha = \underline{t}_2^{\circledast}\left[\theta_{\mathrm{st}}^{\circledast}\alpha\right]$ and the proposition follows. □

1.20 In this section we define (see theorem 1.23) the base change of a cycle $\alpha \in \mathscr{Z}(X/S, n)_{\mathbb{Q}}$ by a morphism $\theta : T \to S$. The fact that denominators may appear when one performs the base change is explained by the remark below:

Remark 1.21 Let E/K be a finite normal extension and X a K-scheme of finite type, then the extension of scalars induces an isomorphism

$$Z(X)[1/p] = Z(X_E)[1/p]^G$$

where G is the Galois group of E/K and p is the characteristic exponent of K.

As a consequence, we get:

Proposition 1.22 *Let s be a non maximal point and α be an element in $\mathscr{Z}(X/S, n)$. Then there exists a unique cycle*

$$\beta \in \mathscr{Z}(X_s/\kappa(s), n)[1/p]$$

satisfying for all fat K-point $\underline{s}$ with image s the relation $\underline{s}^{\circledast}\alpha = \beta_K$.

The cycle β obtained in the proposition above will be denoted by $s^{\circledast}\alpha$ in the sequel. It follows from the definition that, given a generic abstract blow-up $\theta : T \to S$ that flattens α and a K-point $\underline{t}$ in T above s, we have

$$\underline{t}^{\circledast}_{\mathrm{Hilb}}\left(\theta^{\circledast}_{\mathrm{st}}\right) = (s^{\circledast}\alpha)_K. \tag{12.5}$$

Proof Let us fix a generic abstract blow-up $\theta : T \to S$ that flattens α. Since θ is surjective of finite type we may find a finite normal extension $E/\kappa(s)$ and a E-point $\underline{u}$ such that such that $\underline{u} = s_E$. Each element g in the Galois group of E/K provides two E-points $\underline{u}$ and $\underline{u} \circ g$ which coincide after composition with θ. Proposition 1.19 implies then that $\underline{u}^{\circledast}_{\mathrm{Hilb}}\theta^{\circledast}_{\mathrm{st}}\alpha$ is invariant under Galois and thus, by remark 1.21, we know that it is the scalar extension of a cycle

$$\beta \in \mathscr{Z}(X_s/\kappa(s), n)[1/p].$$

It is enough to check that this cycle has the required property and for this we may assume that K is an extension of E. Let $\underline{s}$ be our fat K-point and $(\mathscr{O}, \tau, \sigma)$ a thickening. According to remark 1.18, we have the diagram (12.4) and the relation

$$(\mathscr{O}, \tau, \sigma)^{\circledast}\alpha = \underline{t}^{\circledast}_{\mathrm{Hilb}}\theta^{\circledast}_{\mathrm{st}}\alpha.$$

The desired relation then follows from proposition 1.19 since $\underline{u}_K$ and $\underline{t}$ coincide after composition with θ. □

Theorem 1.23 *Let X be a S-scheme, $\theta : T \to S$ be a morphisms of schemes and α an element in $\mathscr{Z}(X/S, n)_{\mathbb{Q}}$. There exists a unique cycle*

$$\beta \in \mathscr{Z}(X \times_S T/T, n)_{\mathbb{Q}}$$

such that for any maximal or fat K-point $\underline{t}$ of T whose image $\underline{s}$ by θ is a maximal or fat K-point of S, we have

$$\underline{t}^{\circledast}\beta = \underline{s}^{\circledast}\alpha. \tag{12.6}$$

The cycle β obtained in the theorem is called the base change of α along θ and denoted by $\theta^{\circledast}\alpha$. If α is in $Z_{\mathrm{Hilb}}(X/S, n)$ we have $\theta^{\circledast}\alpha = \theta^{\circledast}_{\mathrm{Hilb}}\alpha$, similarly if θ is flat $\theta^{\circledast}\alpha = \theta^{\circledast}_{\mathrm{flat}}\alpha$. In the sequel there will be no need to keep the indices indicating the kind of flat base change performed.

Proof Denote by $T_1, \ldots, T_r$ the irreducible components of T, by t_i the generic point of T_i and by s_i its image in S. Denote also by K_i the residue field of T at the point t_i. Assume that such a cycle β exists.

The condition imposed prescribes the value of β at the maximal points $t_1, \ldots, t_r$ of T and therefore the uniqueness of β is a consequence of remark 1.8. Moreover one knows also, writing β as

$$\beta = \sum_Z \beta_Z[Z]$$

where the sum is taken over the closed integral subschemes Z of X_T of dimension n over T that dominate an irreductible component of T, that the multiplicity β_Z of Z in β is given by

$$\beta_Z = m\big(Z_{t_i}, (s_i^{\circledast}\alpha)_{K_i}\big)$$

where T_i is the irreductible component of T dominated by Z. So it is enough to check that this cycle β satisfies the prescribed conditions. Since it has been defined to satisfy (12.6) when $\underline{t}$ is maximal, one may assume that $\underline{t}$ is a fat K-point. Let us fix a thickening $(\mathscr{O}', \tau', \sigma')$ of this fat point as well as a generic abstract blow-up $\vartheta : S' \to S$ that flattens α.

The image of the generic point of $\mathscr{O}'$ being a maximal point of T, one may assume it is the generic point of T_1. The function field of $\mathscr{O}'$ is then the extension of finite type K_1 of $\kappa(s)$. The morphism ϑ being surjective of finite type, there exists an extension of finite type E of $\kappa(s)$ such that the E-point $(s_1)_E$ lifts to S', if necessary one can replace E by a composite extension of K_1 and E over $\kappa(s)$ oand assume that E is an extension of finite type of K_1. According to [5, **II**, 7.1.7], there exists a spectrum $\mathscr{O}''$ of a DVR with function field E and a dominant morphism

$$\varrho : \mathscr{O}'' \to \mathscr{O}'$$

sending the closed point to the closed point and inducing at the level of the function fields the extension E/K_1. By the valuative criterion of properness [5, **II**, 7.3.8] one gets then a commutative diagram

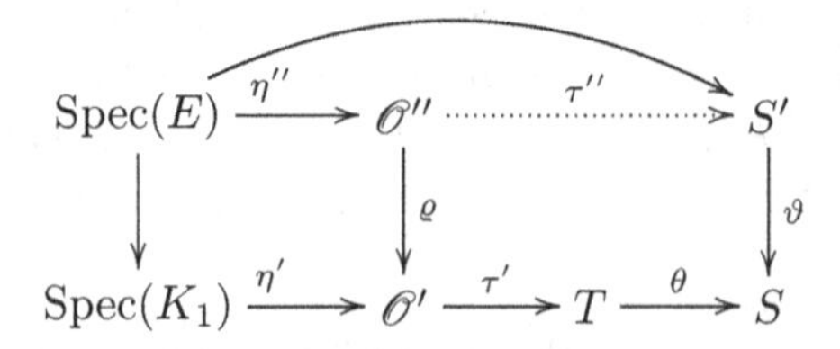

$$\begin{array}{ccccccc} \mathrm{Spec}(E) & \xrightarrow{\eta''} & \mathscr{O}'' & \overset{\tau''}{\dashrightarrow} & & & S' \\ \downarrow & & \downarrow \varrho & & & & \downarrow \vartheta \\ \mathrm{Spec}(K_1) & \xrightarrow{\eta'} & \mathscr{O}' & \xrightarrow{\tau'} & T & \xrightarrow{\theta} & S \end{array}$$

in which η' and η'' are the generic points of $\mathscr{O}'$ and $\mathscr{O}''$. The field K and the residue field $\mathscr{O}''$ at the closed point are two extensions of the residue field of $\mathscr{O}'$ at its closed point, one can fix a composite extension F

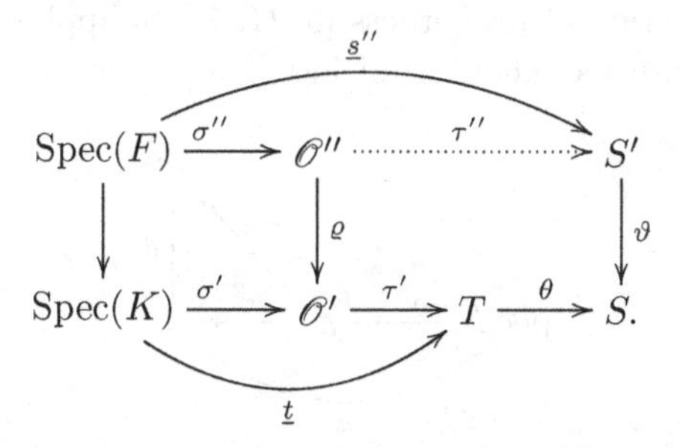

For any closed integral subscheme Z of X_T of dimension n over T and that dominates T_1, the scheme $\mathscr{O}' \times_T Z$ has according to remark 1.12 a unique irreductible component $\mathscr{Z}$ that dominates $\mathscr{O}'$. Let γ be the relative cycle

$$\gamma = \sum_Z \beta_Z[\mathscr{Z}]$$

the sum being taken over the closed integral subschemes Z of X_T of dimension n over T and that dominate T_1. The construction of β assures that $(s^{\circledast}\alpha)_{K_1} = (\eta')^{\circledast}_{\text{flat}}\gamma$, using (12.5) one therefore gets

$$\begin{aligned}(\eta'')^{\circledast}_{\text{flat}}\left[(\tau'')^{\circledast}_{\text{Hilb}}(\theta^{\circledast}_{\text{st}}\alpha)\right] &= (\tau'' \circ \eta'')^{\circledast}_{\text{Hilb}}(\vartheta^{\circledast}_{\text{st}}\alpha) = (s^{\circledast}\alpha)_E \\ &= ((\eta')^{\circledast}_{\text{flat}}\gamma)_E = (\eta'')^{\circledast}_{\text{flat}}(\varrho^{\circledast}_{\text{Hilb}}\gamma)\end{aligned}$$

which yields thanks to remark 1.8 the equality

$$(\tau'')^{\circledast}_{\text{Hilb}}(\vartheta^{\circledast}_{\text{st}}\alpha) = \varrho^{\circledast}_{\text{Hilb}}\gamma.$$

Using proposition 1.5 and the definition of the base change along a fat point, one sees that

$$\begin{aligned}(\underline{s}'')^{\circledast}_{\text{Hilb}}(\vartheta^{\circledast}_{\text{st}}\alpha) &= (\sigma'')^{\circledast}_{\text{Hilb}}(\varrho^{\circledast}_{\text{Hilb}}\gamma) = \left[(\sigma')^{\circledast}_{\text{Hilb}}\gamma\right]_F \\ &= (\mathscr{O}', \tau', \sigma')^{\circledast}\beta = (\underline{t}^{\circledast}\beta)_F.\end{aligned}$$

Now assume that $\underline{s}$ is maximal. The morphism ϑ being birational, one can see $\underline{s}$ as a maximal K-point $\underline{s}'$ of S'. The F-points $\underline{s}''$ and $\underline{s}'_F$ induce then the same F-point of S and proposition 1.19 assures that

$$(\underline{s}^{\circledast}\alpha)_F = (\underline{s}'_F)^{\circledast}_{\text{Hilb}}(\vartheta^{\circledast}_{\text{st}}\alpha) = (\underline{s}'')^{\circledast}_{\text{Hilb}}(\vartheta^{\circledast}_{\text{st}}\alpha) = (\underline{t}^{\circledast}\beta)_F.$$

Remark 1.2 ensures finally that $\underline{t}^{\circledast}(\beta) = \underline{s}^{\circledast}\alpha$, which is the desired equality.

Assume on the contrary that $\underline{s}$ is a fat K-point and choose a thickening $(\mathscr{O}, \tau, \sigma)$ of this fat point. Since the image of the generic point of

$\mathscr{O}$ is a maximal point and since the abstract blow-up ϑ is birational, the valuative criterion of properness [5, **II**, 7.3.8] applied to the proper morphism ϑ provides a canonical lifting

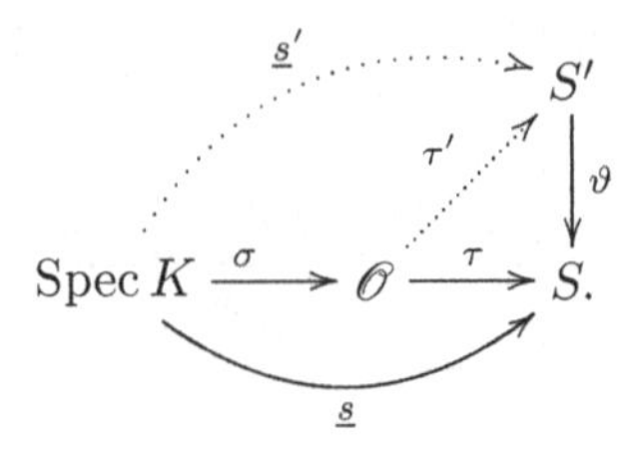

Using remark 1.18 we have

$$\underline{s}^{\circledast}\alpha = (\mathscr{O}, \tau, \sigma)^{\circledast}\alpha = (\mathscr{O}, \tau', \sigma)^{\circledast}[\vartheta^{\circledast}_{\mathrm{st}}\alpha] = (\underline{s}')^{\circledast}[\vartheta^{\circledast}_{\mathrm{st}}\alpha].$$

As before the F-points $\underline{s}''$ and $\underline{s}'_F$ induce the same F-point of S and proposition 1.19 gives the equality

$$(\underline{s}^{\circledast}\alpha)_F = (\underline{s}'_F)^{\circledast}_{\mathrm{Hilb}}(\vartheta^{\circledast}_{\mathrm{st}}\alpha) = (\underline{s}'')^{\circledast}_{\mathrm{Hilb}}(\vartheta^{\circledast}_{\mathrm{st}}\alpha) = (\underline{t}^{\circledast}\beta)_F$$

and to conclude it is enough to apply remark 1.2 . □

Remark 1.24 Given an element $\alpha \in \mathscr{Z}(X/S, n)$ and a morphism $\theta : T \to S$, the relative cycle $\theta^{\circledast}\alpha$ may not have integral coefficients, however one sees from its construction that the prime factors that may appear in the denominators are the residual characteristics of S at the images of the maximal points of T. In particular if the characteristic of S is zero, then the cycle $\theta^{\circledast}\alpha$ has integral coefficients.

Theorem 1.23 provides a presheaf or relative cycles $\underline{\mathscr{Z}}(X/S, n)_{\mathbb{Q}}$ with rational coefficients over X. It maps a S-scheme T to the $\mathbb{Q}$-vector space

$$\underline{\mathscr{Z}}(X/S, n)_{\mathbb{Q}}(T) = \mathscr{Z}(X \times_S T/T, n)_{\mathbb{Q}}$$

and a morphism $T' \to T$ of S-schemes to the $\mathbb{Q}$-linear morphism

$$\theta^{\circledast} : \mathscr{Z}(X \times_S T/T, n)_{\mathbb{Q}} \to \mathscr{Z}(X \times_S T'/T', n)_{\mathbb{Q}}.$$

Lemma 1.13 implies the following result which shows that the base change theory developed by A. Suslin and V. Voevodsky is the only possible functorial extension of the two flat base changes:

Corollary 1.25 *Let X be a S-scheme:*

1. *$\underline{\mathscr{Z}}(X/S, n)_{\mathbb{Q}}$ is a presheaf of algebraic cycles ;*
2. *any presheaf of relative algebraic cycles with rational coefficients over X is a subpresheaf of $\underline{\mathscr{Z}}(X/S, n)_{\mathbb{Q}}$.*

In the sequel we shall say that a relative cycle α is universally integral if it is universally rational and if moreover for any morphism of schemes $\theta : T \to S$ the base change $\theta^{\circledast}\alpha$ has integral coefficients. We denote by $z(X/S, n)$ the subgroup of $\mathscr{Z}(X/S, n)$ formed by universally integral relative cycles.

Proposition 1.26 *Let $\theta : T \to S$ be a morphism of schemes and $\alpha \in \mathscr{Z}(X/S, n)$ be a relative cycle. One has the inclusion*

$$\mathrm{supp}(\theta^{\circledast}\alpha) \subset \theta_X^{-1}(\mathrm{supp}(\alpha)).$$

Proof Since $\theta^{\circledast}\alpha$ is a relative cycle, one can reduce the proof to the case where θ is a morphism associated to a point s in S. Moreover the result being true if s is a maximal point, we may assume it is not the case and take, thanks to remark 1.11, an extension $K/\kappa(s)$ such that the K-point

$$\mathrm{Spec}(K) \longrightarrow \mathrm{Spec}(\kappa(s)) \xrightarrow{s} S \quad (\text{composite: } \underline{s})$$

is fat. It is then enough to check the result of the point $\underline{s}$. Fix a thickening $(\mathscr{O}, \tau, \sigma)$ and write α as

$$\alpha = \sum_{i=1}^{r} \alpha_i [Z_i]$$

where the Z_i's are closed integral subschemes of X of dimension n over S that dominate an irreductible component of S. Up to renumbering the Z_i's, we may assume that the first k are those that dominate the irreductible component of S with generic point the image by τ of the generic point of $\mathscr{O}$. Then we get

$$\underline{s}^{\circledast}\alpha = \sum_{i=1}^{k} [\mathrm{Spec}(K) \times_{\mathscr{O}} \mathscr{Z}_i]$$

where $\mathscr{Z}_i$ is the unique irreductible component of $\mathscr{O} \times_S Z_i$ that dominates $\mathscr{O}$. From this one deduces that

$$\begin{aligned} \mathrm{supp}(\underline{s}^{\circledast}\alpha) &\subset \sigma_X^{-1}\left(\bigcup_{i=1}^{k} \mathscr{Z}_i\right) \subset \sigma_X^{-1}\left(\bigcup_{i=1}^{k} (\mathscr{O} \times_S Z_i)_{\mathrm{red}}\right) \\ &\subset \theta_X^{-1}\left(\bigcup_{i=1}^{r} Z_i\right) = \theta_X^{-1}\,\mathrm{supp}(\alpha) \end{aligned}$$

and the proposition is shown. □

Denote by $\mathscr{Z}_{\mathrm{equi}}(X/S,n)$ the group of equidimensional universally rational relative cycles on X, by $z_{\mathrm{equi}}(X/S,n)$ the subgroup formed by universally integral cycles and let

$$\begin{aligned}\mathscr{C}(X/S,n) &= C(X/S,n) \cap \mathscr{Z}(X/S,n)\\ \mathscr{C}_{\mathrm{equi}}(X/S,n) &= C(X/S,n) \cap \mathscr{Z}_{\mathrm{equi}}(X/S,n)\\ c(X/S,n) &= C(X/S,n) \cap z(X/S,n)\\ c_{\mathrm{equi}}(X/S,n) &= C(X/S,n) \cap z_{\mathrm{equi}}(X/S,n).\end{aligned}$$

Proposition 1.26 provides then the following corollary:

Corollary 1.27 *The above abelian groups, as well as the abelian groups $\mathscr{Z}_{\mathrm{equi}}(X/S,n)_{\mathbb{Q}}$ and $z_{\mathrm{equi}}(X/S,n)$, are stable under base change.*

Proof Fix a morphism of schemes $\theta : T \to S$. By construction the support of $\theta^{\circledast}\alpha$ is maximal over T and of dimension n over T. Proposition 1.26 implies the inequality

$$\dim(\mathrm{supp}(\theta^{\circledast}\alpha)/T) \leqslant \dim(\mathrm{supp}(\alpha)/S) = n$$

on the other hand Chevalley's semi-continuity theorem implies the reverse inequality

$$\dim(\mathrm{supp}(\theta^{\circledast}\alpha)/T) \geqslant n.$$

This proves that the function $\dim(\mathrm{supp}(\theta^{\circledast}\alpha)/T)$ is constant and shows that $\theta^{\circledast}\alpha$ is an equidimensional cycle. The case of proper support follows also from proposition 1.26. □

Corollary 1.27, assures that we get a whole family of presheaves on the category Sch_S of S-schemes, by defining for a S-scheme T:

$$\begin{array}{ll}\underline{\mathscr{Z}}(X/S,n)_{\mathbb{Q}}(T) = \mathscr{Z}(X_T/T)_{\mathbb{Q}} & \underline{\mathscr{Z}}_{\mathrm{equi}}(X/S,n)_{\mathbb{Q}}(T) = \mathscr{Z}_{\mathrm{equi}}(X_T/T,n)_{\mathbb{Q}}\\ \underline{z}(X/S,n)(T) = z(X_T/T,n) & \underline{z}_{\mathrm{equi}}(X/S,n)(T) = z_{\mathrm{equi}}(X_T/T,n)\\ \underline{\mathscr{C}}(X/S,n)_{\mathbb{Q}}(T) = \mathscr{C}(X_T/T)_{\mathbb{Q}} & \underline{\mathscr{C}}_{\mathrm{equi}}(X/S,n)_{\mathbb{Q}}(T) = \mathscr{C}_{\mathrm{equi}}(X_T/T,n)_{\mathbb{Q}}\\ \underline{c}(X/S,n)(T) = c(X_T/T,n) & \underline{c}_{\mathrm{equi}}(X/S,n)(T) = c_{\mathrm{equi}}(X_T/T,n).\end{array}$$

Lemma 1.28 *Assume S geometrically unibranch, then for any closed suscheme Z of X which is equidimensional over S of dimension n, the relative cycle $[Z]$ associated to Z belongs to $\mathscr{Z}_{\mathrm{equi}}(X/S,n)$.*

Proof By replacing Z by one of its irreducible components, we may assume that Z is integral. According to proposition 1.17, there exists a generic abstract blow-up $\theta : T \to S$ that flattens the cycle $[Z]$ i.e. such that the strict transform Z_T^{st} is flat over T. Fix then two K-points $\underline{t}_1$ and $\underline{t}_2$ of T such that $\theta \circ \underline{t}_1 = \theta \circ \underline{t}_2$. By proposition 1.19 it is enough to check that

$$\underline{t}_{1,\mathrm{Hilb}}^{\circledast}[Z_T^{\mathrm{st}}] = \underline{t}_{2,\mathrm{Hilb}}^{\circledast}[Z_T^{\mathrm{st}}]$$

which amounts to prove that

$$[\mathrm{Spec}(K) \times_{\underline{t}_1} Z_T^{\mathrm{st}}] = [\mathrm{Spec}(K) \times_{\underline{t}_2} Z_T^{\mathrm{st}}].$$

Since S is geometrically unibranch and Z is equidimensional, the scheme $T \times_S Z$ is also equidimensional. The closed immersion of Z_T^{st} in $T \times_S Z$ being an isomorphism above the generic points of T, one sees that this immersion is defined by a nilpotent ideal. The closed subscheme $\mathrm{Spec}(K) \times_S Z_T^{\mathrm{st}}$ of

$$\mathrm{Spec}(K) \times_S Z_T = [\mathrm{Spec}(K) \times_S T] \times_{\mathrm{Spec}(K)} [\mathrm{Spec}(K) \times_S Z]$$

is thus defined by a nilpotent ideal.

Moreover the morphism θ being proper and dominant, [5, **III**, 4.3.5] and [5, **IV**, 18.8.15] imply that the scheme $\mathrm{Spec}(K) \times_S T$ is geometrically connected. Using remark 1.2, one may take a suitable extension of K and assume that K is algebraically closed. The scheme $\mathrm{Spec}(K) \times_S T$ is then connected and we may consider only the case where it is integral. Let $Z_1, \ldots, Z_r$ be the irreductible components of $\mathrm{Spec}(K) \times_S Z$ with their reduced structure. Since K is algebraically closed and $\mathrm{Spec}(K) \times_S T$ is integral, [5, **IV**, 4.5.8] implies that

$$[\mathrm{Spec}(K) \times_S T] \times_{\mathrm{Spec}(K)} Z_i \tag{12.7}$$

is irreductible and therefore integral by [5, **IV**, 4.6.5]. The irreducible components of the scheme $\mathrm{Spec}(K) \times_S Z_T$ are therefore precisely the schemes (12.7) and the cycle associated to $\mathrm{Spec}(E) \times_S Z_T^{\mathrm{st}}$ is then given by a linear combination

$$[\mathrm{Spec}(K) \times_S Z_T^{\mathrm{st}}] = \sum_{i=1}^{r} n_i \left[[\mathrm{Spec}(K) \times_S T] \times_{\mathrm{Spec}(K)} Z_i\right].$$

The result is finally a consequence of proposition 1.5. ☐

This lemma implies the following corollary:

Corollary 1.29 *Assume S geometrically unibranch, then:*

1. *the abelian group $\mathscr{Z}_{\mathrm{equi}}(X/S,n)$ is the free abelian group generated by the closed integral subschemes of X equidimensional over S of dimension n;*
2. *the group $\mathscr{C}_{\mathrm{equi}}(X/S,n)$ is the free abelian group generated by the closed integral subschemes of X proper over S and equidimensional over S of dimension n.*

Assuming moreover that S is regular, the preceding corollary may be refined:

Corollary 1.30 *Assume S regular, then:*

1. *the abelian group $z_{\mathrm{equi}}(X/S,n)$ is the free abelian group generated by the closed integral subschemes of X equidimensional over S of dimension n;*
2. *the group $c_{\mathrm{equi}}(X/S,n)$ is the free abelian group generated by the closed integral subschemes of X proper over S and equidimensional over S of dimension n.*

1.31 The groups of relative cycles behave quite nicely with respect to base change along flat and surjective morphisms.

Proposition 1.32 *Let X be a S-scheme and $\theta : T \to S$ be a flat and surjective morphism, then:*

1. *the squares*

$$\begin{array}{ccccc} z(X/S,n) & \longrightarrow & \mathscr{Z}(X/S,n) & \longrightarrow & Z(X/S,n) \\ \downarrow{\scriptstyle \theta^{\circledast}} & & \downarrow{\scriptstyle \theta^{\circledast}} & & \downarrow{\scriptstyle \theta^{\circledast}} \\ z(X\times_S T/T,n) & \to & \mathscr{Z}(X\times_S T/T,n) & \to & Z(X\times_S T/T,n) \end{array}$$

   are cartesian;
2. *similarly the squares*

$$\begin{array}{ccccc} c(X/S,n) & \longrightarrow & \mathscr{C}(X/S,n) & \longrightarrow & C(X/S,n) \\ \downarrow{\scriptstyle \theta^{\circledast}} & & \downarrow{\scriptstyle \theta^{\circledast}} & & \downarrow{\scriptstyle \theta^{\circledast}} \\ c(X\times_S T/T,n) & \longrightarrow & \mathscr{C}(X\times_S T/T,n) & \longrightarrow & C(X\times_S T/T,n) \end{array}$$

   are cartesian.

Proof The second assertion follows from the first one since the squares

$$\begin{array}{ccccc} c(X/S,n) & \longrightarrow & \mathscr{C}(X/S,n) & \longrightarrow & C(X/S,n) \\ \downarrow & & \downarrow & & \downarrow \theta^{\circledast} \\ z(X/S,n) & \longrightarrow & \mathscr{Z}(X/S,n) & \longrightarrow & Z(X/S,n) \end{array}$$

are cartesian by definition. Let α be an element in $Z(X/S,n)$ such that $\theta^{\circledast}\alpha$ belongs to $\mathscr{Z}(X/S,n)$. Fix a generic abstract blow-up $\vartheta : S' \to S$ that flattens α as well as two K-points $\underline{s}_1, \underline{s}_2$ of S' such that $\vartheta \circ \underline{s}_1 = \vartheta \circ \underline{s}_2$. By proposition 1.19, it is enough to check that $\underline{s}_1^*(\vartheta_{\mathrm{st}}^*\alpha)$ and $\underline{s}_2^*(\vartheta_{\mathrm{st}}^*\alpha)$ are equal. In the cartesian square

$$\begin{array}{ccc} T' & \xrightarrow{\vartheta'} & T \\ {\scriptstyle \text{flat, surjective}}\ \downarrow \theta' & \square & \theta \downarrow\ {\scriptstyle \text{flat, surjective}} \\ S' & \xrightarrow{\vartheta} & S \end{array} \tag{12.8}$$

the morphism ϑ' is a generic abstract blow-up that flattens $\theta^{\circledast}\alpha$. Moreover

$$(\theta')^{\circledast}(\vartheta_{\mathrm{st}}^{\circledast}\alpha) = (\vartheta')_{\mathrm{st}}^{\circledast}(\theta^{\circledast}\alpha).$$

Since θ' is surjective, there exists an extension F/K and two F-points $\underline{t}_1, \underline{t}_2$ of T' for which the squares

$$\begin{array}{ccc} \mathrm{Spec}(F) & \xrightarrow{\underline{t}_1} & T' \\ \downarrow & & \downarrow \theta' \\ \mathrm{Spec}(K) & \xrightarrow{\underline{s}_1} & S' \end{array} \qquad \begin{array}{ccc} \mathrm{Spec}(F) & \xrightarrow{\underline{t}_2} & T' \\ \downarrow & & \downarrow \theta' \\ \mathrm{Spec}(K) & \xrightarrow{\underline{s}_2} & S' \end{array}$$

commute. By assumption $\theta^{\circledast}\alpha$ is in $\mathscr{Z}(X/S,n)$, so by proposition 1.19, we have the equality

$$\underline{t}_1^{\circledast}\left((\vartheta')_{\mathrm{st}}^{\circledast}(\theta^{\circledast}\alpha)\right) = \underline{t}_2^{\circledast}\left((\vartheta')_{\mathrm{st}}^{\circledast}(\theta^{\circledast}\alpha)\right)$$

which implies that the cycles $\underline{s}_1^*(\vartheta_{\mathrm{st}}^*\alpha)$ and $\underline{s}_2^*(\vartheta_{\mathrm{st}}^*\alpha)$ are equal after extension to F. Remark 1.2 implies then the desired equality.

Assume now that $\theta^{\circledast}\alpha$ is in $z(X \times_S T/T, n)$ and consider a morphism of schemes $\vartheta : S' \to S$. We use the notation of square (12.8). The first part of the lemma implies that α is in $\mathscr{Z}(X/S,n)$, which implies by functoriality of base change that

$$(\theta')^{\circledast}\vartheta^{\circledast}\alpha = (\tau')^{\circledast}\theta^{\circledast}\alpha$$

in $\mathscr{Z}(X \times_S T'/T', n)_{\mathbb{Q}}$. Using corollary 1.25 and the fact that $\theta^{\circledast}\alpha$ is assumed to be universally integral , we see that the cycles in the previous equality have integral coefficients. The flatness of θ' implies that $\vartheta^{\circledast}\alpha$ has integral coefficients and therefore α is universally integral. □

2 Multiplicities

2.1 When the base scheme is assumed to be regular, the intersection multiplicities defined by Serre suffice to define a suitable notion of base change for relative cycles. Indeed if S is regular and Z is a scheme equidimensional of dimension n over S, then:

- Z is of finite Tor-dimension over S ;
- Z_T is equidimensional of dimension n over T.

It is thus possible to define an element $\theta^{\circledast}_{\mathrm{Tor}}[Z]$ in $Z_{\mathrm{equi}}(X_T/T, n)$ using the Tor-formula

$$\theta^{\circledast}_{\mathrm{Tor}}[Z] := \sum_W \lg_{\mathscr{O}_{Z_T,w}} (\mathsf{L}\theta^*_X(\mathscr{O}_Z)_w)\,[W]$$

the sum being taken over the irreductible component W of Z_T with generic point w. By linearity we get a morphism

$$\theta^{\circledast}_{\mathrm{Tor}} : Z_{\mathrm{equi}}(X/S, n) \to Z_{\mathrm{equi}}(X_T/T, n)$$

and the associativity of Serre's intersection multiplicities implies that $\vartheta^{\circledast}_{\mathrm{Tor}} \circ \theta^{\circledast}_{\mathrm{Tor}} = (\theta \circ \vartheta)^{\circledast}_{\mathrm{Tor}}$ for morphisms $\vartheta : U \to T$ and $\theta : T \to S$ with T and S regular. In fact as we shall see, this extension of the flat pullbacks is already encompassed in the theory elaborated by Suslin and Voevodsky. Their strategy to prove this compatibility is to compute explicitly the multiplicities obtained by base change and to compare them to the multiplicities defines by Serre.

2.2 We refer to [10] for the definition and the properties of the multiplicity $e_I(\mathscr{O})$ of a $\mathfrak{m}$-primary ideal I in a local noetherian ring $\mathscr{O}$. We assume S to be geometrically unibranch and integral. We fix in this section a point s in S and we choose a $\mathfrak{m}_{S,s}$-primary ideal I in $\mathscr{O}_{S,s}$.

Lemma 2.3 *Let X be an equidimensional S-scheme and s be a point in S. Then for any generic point w of an irreductible component of the fibre X_s, the ideal $\mathfrak{m}_{S,s}\mathscr{O}_{X,w}$ is a $\mathfrak{m}_{X,w}$-primary ideal of $\mathscr{O}_{X,w}$.*

Proof One may assume that S is affine and the spectrum of an integral ring A and denote by $\mathfrak{p}$ the prime ideal of A corresponding to s. By proposition 5.12, we may assume, after taking a small enough neighborhood of w, that the structural morphism from X to S is given by a composition $X \xrightarrow{q} \mathbb{A}^n_S \to S$ where n is a nonnegative integer and q is a maximal quasi-finite morphism. Since q is quasi-finite, one knows from [5, **0**, 7.4.4] that $\mathfrak{m}_{\mathbb{A}^n_S,q(w)}\mathcal{O}_{X,w}$ is a $\mathfrak{m}_{X,w}$-primary ideal of $\mathcal{O}_{X,w}$. It is then enough to check that

$$\mathfrak{m}_{S,s}\mathcal{O}_{\mathbb{A}^n_S,q(w)} = \mathfrak{m}_{\mathbb{A}^n_S,q(w)}. \tag{12.1}$$

Since q is equidimensional of dimension 0 and $\mathbb{A}^n_S$ is geometrically unibranch, the morphism p in the cartesian squares

$$\begin{array}{ccccc} X_s & \xrightarrow{p} & \mathbb{A}^n_{\kappa(s)} & \longrightarrow & \mathrm{Spec}(\kappa(s)) \\ \downarrow & \square & \downarrow & \square & \downarrow \\ X & \xrightarrow{q} & \mathbb{A}^n_S & \longrightarrow & S \end{array}$$

is also equidimensional of dimension 0. In particular $p(w)$ is the generic point of $\mathbb{A}^n_{\kappa(s)}$. This implies that the prime ideal $\mathfrak{q}$ of $A[x_1, \dots, x_n]$ corresponding to $q(w)$ is minimal among the prime ideals $\mathfrak{r}$ of $A[x_1, \dots, x_n]$ such that $\mathfrak{r} \cap A = \mathfrak{p}$. One thus gets $\mathfrak{q} = \mathfrak{p}A[x_1, \dots, x_n]$ and the equality (12.1) follows. □

Let s be a point in S and I a $\mathfrak{m}_{S,s}$-primary ideal in $\mathcal{O}_{S,s}$. Given an integral scheme X equidimensional over S and an irreductible component W of the fiber X_s with generic point w. Lemma 2.3 implies in particular that for any $\mathfrak{m}_{S,s}$-primary ideal I of $\mathcal{O}_{S,s}$ the ideal $I\mathcal{O}_{X,w}$ is a $\mathfrak{m}_{X,w}$-primary ideal of $\mathcal{O}_{X,w}$, therefore, we have a well-defined rational number:

$$\frac{e_{I\mathcal{O}_{X,w}}(\mathcal{O}_{X,w})}{e_I(\mathcal{O}_{S,s})}.$$

Definition 2.4 Let Z be an integral closed subscheme of X that is equidimensional of dimension n over S. One sets

$$s^{\circledast}_{I,\mathrm{mult}}[Z] = \sum_W \frac{e_{I\mathcal{O}_{Z,w}}(\mathcal{O}_{Z,w})}{e_I(\mathcal{O}_{S,s})}[W]$$

the sum being taken over the irreductible component W of Z_s with generic point w.

Remark 2.5 When $\mathscr{O}_{Z,w}$ is an $\mathscr{O}_{S,s}$-module of finite type, one has

$$m(W, s^{\circledast}_{I,\mathrm{mult}}[Z]) = \frac{e_{I\mathscr{O}_{Z,w}}(\mathscr{O}_{Z,w})}{e_I(\mathscr{O}_{S,s})} = \frac{\dim_{F(S)}(\mathscr{O}_{Z,w} \otimes_{\mathscr{O}_{S,s}} F(S))}{[\kappa(w):\kappa(s)]}$$

where $F(S)$ is the function field of S.

Since S is geometrically unibranch, we know by corollary 1.29 that $\mathscr{Z}_{\mathrm{equi}}(X/S,n)_{\mathbb{Q}} = Z_{\mathrm{equi}}(X/S,n)_{\mathbb{Q}}$ is the $\mathbb{Q}$-vector space freely generated by the closed integral subschemes of X which are equidimensional over S of dimension n. By linearity we get a morphism

$$s^{\circledast}_{I,\mathrm{mult}} : \mathscr{Z}_{\mathrm{equi}}(X/S,n)_{\mathbb{Q}} \to \mathscr{Z}_{\mathrm{equi}}(X_s/\kappa(s),n)_{\mathbb{Q}}.$$

These rational numbers depend *a priori* upon the choice of the $\mathfrak{m}_{S,s}$-primary ideal I in $\mathscr{O}_{S,s}$, we will see that this is not the case but this fact do require a proof and is not obvious. The multiplicities introduced above are additive in the usual sense:

Lemma 2.6 *Let X be an equidimensional S-scheme and s a point in S. For any irreductible component W of X_s with generic point w*

$$m(W, s^{\circledast}_{I,\mathrm{mult}}[X]) = \frac{e_{I\mathscr{O}_{X,w}}(\mathscr{O}_{X,w})}{e_I(\mathscr{O}_{S,s})}.$$

Proof Let $X_1, \ldots, X_r$ the irreducible component of X that contain w and x_i be the generic point of X_i. By definition

$$m(W, s^{\circledast}_{I,\mathrm{mult}}[X]) = \sum_{i=1}^{r} \lg_{\mathscr{O}_{X,x_i}}(\mathscr{O}_{X,x_i}) \left[\frac{e_{I\mathscr{O}_{X_i,w}}(\mathscr{O}_{X_i,w})}{e_I(\mathscr{O}_{S,s})}\right].$$

Denote by $\mathfrak{p}_i$ the minimal prime ideal of $\mathscr{O}_{X,w}$ corresponding to X_i, we have

$$\mathscr{O}_{X_i,w} = \mathscr{O}_{X,w}/\mathfrak{p}_i \qquad \mathscr{O}_{X,x_i} = (\mathscr{O}_{X,w})_{\mathfrak{p}_i}$$

and therefore

$$\begin{aligned} m(W, s^{\circledast}_{I,\mathrm{mult}}[X]) &= \sum_{i=1}^{r} \lg_{(\mathscr{O}_{X,w})_{\mathfrak{p}_i}}((\mathscr{O}_{X,x_i})_{\mathfrak{p}_i}) \left[\frac{e_{I(\mathscr{O}_{X,w}/\mathfrak{p}_i)}(\mathscr{O}_{X,w}/\mathfrak{p}_i)}{e_I(\mathscr{O}_{S,s})}\right] \\ &= \sum_{i=1}^{r} \lg_{(\mathscr{O}_{X,w})_{\mathfrak{p}_i}}((\mathscr{O}_{X,x_i})_{\mathfrak{p}_i}) \left[\frac{e_{I\mathscr{O}_{X,w}}(\mathscr{O}_{X,w}/\mathfrak{p}_i)}{e_I(\mathscr{O}_{S,s})}\right] \\ &= \frac{e_{I\mathscr{O}_{X,w}}(\mathscr{O}_{X,w})}{e_I(\mathscr{O}_{S,s})} \end{aligned}$$

which is the desired result. □

Now we may verify the independence with respect to the $\mathfrak{m}_{S,s}$-primary ideal I chosen in $\mathscr{O}_{S,s}$.

Proposition 2.7 *Let X be a S-scheme and s a point in S. For any α in $\mathscr{Z}_{\text{equi}}(X/S,n)_{\mathbb{Q}}$, the cycles $s^{\circledast}_{I,\text{mult}}\alpha$ do not depend upon the chosen $\mathfrak{m}_{S,s}$-primary I in $\mathscr{O}_{S,s}$.*

Proof It is enough to check that for any closed integral subscheme Z of X equidimensional of dimension n over S and any irreducible component W of Z_s, the rational number $m(W, s^{\circledast}_{I,\text{mult}}[Z])$ do not depend on I. Let w be the generic point of W. Since

$$m\left(W, s^{\circledast}_{I,\text{mult}}[Z]\right) = m\left(W, s^{\circledast}_{I\mathscr{O}^{\text{h}}_{S,s},\text{mult}}[\text{Spec}(\mathscr{O}^{\text{h}}_{S,s}) \times_S Z]\right),$$

we may assume that S is a henselian local scheme with closed point s. The statement we have to check is local for the Zariski topology around w, so we may assume that the structural morphism of Z over S has a factorization $Z \xrightarrow{q} \mathbb{A}^n_S \to S$ where $T = \mathbb{A}^n_S$ and q is a quasi-finite and maximal morphism. Since

$$m(W, s^{\circledast}_{I,\text{mult}}[X]) = m(V, t^{\circledast}_{I\mathscr{O}_{T,t},\text{mult}}[X])$$

where $t = q(w)$ and V is the irreductible component of X_t having w as generic point, we may also assume that $n = 0$. The scheme Z is then quasi-finite over S and, via [5, **IV**, 18.5.11], we know that $\mathscr{O}_{X,w}$ is a $\mathscr{O}_{S,s}$-module of type fini. The proposition is then a direct consequence of remark 2.5. □

There is now no reason to precise the chosen $\mathfrak{m}_{S,s}$-primary ideal I. In the sequel we denote by $s^{\circledast}_{\text{mult}}\alpha$ the common value of the cycle $s^{\circledast}_{I,\text{mult}}\alpha$.

2.8 The main result of this section, corollary 2.13, is the comparison over a regular base scheme S between the base change defined by Suslin and Voevodsky in a more general setting and the base change obtained by the Tor- formula.

Lemma 2.9 *Let $\alpha \in \mathscr{Z}_{\text{equi}}(X/S,n)$ and s be a point in S:*

1. for $\alpha \in Z_{\text{Hilb}}(X/S,n)$, one has

$$s^{\circledast}_{\text{mult}}\alpha = s^{\circledast}_{\text{Hilb}}\alpha;$$

2. for S regular, one has

$$s^{\circledast}_{\text{mult}}\alpha = s^{\circledast}_{\text{Tor}}\alpha.$$

Proof (1). By linearity we may assume that $\alpha = [Z]$ where Z is a closed integral subscheme of X flat and equidimensional over S of dimension n. Let W be an irreductible component of Z_s and w be its generic point. Since $\mathscr{O}_{Z,w}$ is a flat $\mathscr{O}_{S,s}$-module and since

$$\mathscr{O}_{Z_s,w} = \mathscr{O}_{Z,w}/\mathfrak{m}_{S,s}\mathscr{O}_{Z,w},$$

we have the equality

$$m(W, s^{\circledast}_{\mathrm{mult}}[\alpha]) = \frac{e_{I\mathscr{O}_{Z,w}}(\mathscr{O}_{Z,w})}{e_I(\mathscr{O}_{S,s})} = \lg_{\mathscr{O}_{Z_s,w}}(\mathscr{O}_{Z_s,w}) = m(W, s^{\circledast}_{\mathrm{Hilb}}\alpha).$$

(2). Let d be the dimension of the local ring $\mathscr{O}_{S,s}$. Since it is a regular local ring, its maximal idea $\mathfrak{m}_{S,s}$ is generated by a system of parameters $s_1, \ldots, s_d$. By linearity we may assume that $\alpha = [Z]$ where Z is a closed integral subscheme of X equidimensional over S of dimension n. Let W be an irreductible component of Z_s and w be its generic point. One knows by theorem 14.10 of [10] that the multiplicity of $\mathfrak{m}_{S,s}$ in $\mathscr{O}_{S,s}$ is 1. Using theorem IV.1 of [14] on sees that

$$m(W, s^{\circledast}_{\mathrm{mult}}\alpha) = e_{\mathfrak{m}_{S,s}\mathscr{O}_{Z,w}}(\mathscr{O}_{Z,w}) = \lg_{\mathscr{O}_{Z,w}}(\mathrm{K}(\underline{s}; \mathscr{O}_{Z,w}))$$

where $\underline{s}$ denotes the sequence $(s_1, \ldots, s_d)$. Since $\mathrm{K}(\underline{s}; \mathscr{O}_{S,s})$ is a complex of free $\mathscr{O}_{S,s}$-modules, we may write

$$\mathrm{K}(\underline{s}; \mathscr{O}_{Z,w}) = \mathrm{K}(\underline{s}; \mathscr{O}_{S,s}) \otimes_{\mathscr{O}_{S,s}} \mathscr{O}_{Z,w} = \mathrm{K}(\underline{s}; \mathscr{O}_{S,s}) \otimes^{\mathsf{L}}_{\mathscr{O}_{S,s}} \mathscr{O}_{Z,w}.$$

The sequence $\underline{s}$ being regular, the Koszul complex $\mathrm{K}(\underline{s}; \mathscr{O}_{S,s})$ is quasi-isomorphic to $\kappa(s)$, this implies that

$$\begin{aligned} m(W, s^{\circledast}_{\mathrm{mult}}\alpha) &= \lg_{\mathscr{O}_{Z,w}}\left[\kappa(s) \otimes^{\mathsf{L}}_{\mathscr{O}_{S,s}} \mathscr{O}_{Z,w}\right] \\ &= \lg_{\mathscr{O}_{Z_s,w}}\left[\kappa(s) \otimes^{\mathsf{L}}_{\mathscr{O}_{S,s}} \mathscr{O}_{Z,w}\right] = m(W, s^{\circledast}_{\mathrm{Tor}}\alpha). \end{aligned}$$

and proves the lemma. □

Lemma 2.10 *Let Z be equidimensional of dimension n over S. Then any closed subscheme Z' generically isomorphic to Z is equidimensional of dimension n over S and*

$$s^{\circledast}_{\mathrm{mult}}[Z] = s^{\circledast}_{\mathrm{mult}}[Z'].$$

Proof Remark first that the closed subscheme Z' of Z is defined by a nilpotent ideal in $\mathscr{O}_Z$ in other words Z and Z' have the same underlying closed subset. Indeed since Z' is generically isomorphic to Z, there exists by [5, **IV**, 8.10.5] a neighborhood U of the generic point η of S such that the immersion of Z'_U in Z_U is an isomorphism. Since Z is equidimensional over S, the maximal points $z_1, \ldots, z_k$ of Z have the same image η. Thus they all lie in Z_U and therefore in Z'. This implies that Z and Z' have the same underlying closed subset. Hence Z' is equidimensional over S of dimension n and the fibres Z_s and Z'_s have the same irreducible components. Fix such an irreducible component W and denote by w its generic point. We have to show that

$$m(W, s^{\circledast}_{\text{mult}}[Z]) = m(W, s^{\circledast}_{\text{mult}}[Z']).$$

For this choose a $\mathfrak{m}_{S,s}$-primary ideal I in $\mathscr{O}_{S,s}$. We have seen that

$$\mathscr{O}_{Z',w} = \mathscr{O}_{Z,w}/J$$

where J is a nilpotent ideal, thus contained in the intersection of the minimal prime ideals $\mathfrak{p}_1, \ldots, \mathfrak{p}_r$ of $\mathscr{O}_{Z,w}$. The minimal prime ideals in $\mathscr{O}_{Z',w}$ are therefore $\mathfrak{q}_1 = \mathfrak{p}_1/J, \ldots, \mathfrak{q}_r = \mathfrak{p}_r/J$ and one has

$$\mathscr{O}_{Z,w}/\mathfrak{p}_i = \mathscr{O}_{Z',w}/\mathfrak{q}_i.$$

We may assume that $z_1, \ldots, z_r$ are the generic points of the irreducible components of Z containing w corresponding to $\mathfrak{p}_1, \ldots, \mathfrak{p}_r$. Since Z' and Z are generically isomorphic, the rings

$$\mathscr{O}_{Z,z_i} = (\mathscr{O}_{Z,w})_{\mathfrak{p}_i} \qquad \text{and} \qquad \mathscr{O}_{Z',z_i} = (\mathscr{O}_{Z',w})_{\mathfrak{q}_i}$$

are isomorphic, and therefore

$$\begin{aligned} m(W, s^{\circledast}_{\text{mult}}[Z]) &= \sum_{i=1}^{r} e_{I(\mathscr{O}_{Z,w}/\mathfrak{p}_i)}(\mathscr{O}_{Z,w}/\mathfrak{p}_i) \lg_{\mathscr{O}_{Z,z_i}}(\mathscr{O}_{Z,z_i}) \\ &= \sum_{i=1}^{r} e_{I(\mathscr{O}_{Z',w}/\mathfrak{q}_i)}(\mathscr{O}_{Z',w}/\mathfrak{q}_i) \lg_{\mathscr{O}_{Z',z_i}}(\mathscr{O}_{Z',z_i}) \\ &= m(W, s^{\circledast}_{\text{mult}}[Z']). \end{aligned}$$

This shows the lemma. □

Lemma 2.11 *Let $T \to S$ be a maximal morphism where T is geometrically unibranch and integral. Then for all $\alpha \in \mathscr{Z}_{\text{equi}}(X/S, n)$ and all preimage $t \in T$ one has*

$$t^{\circledast}_{\text{mult}}(\theta^{\circledast}\alpha) = (s^{\circledast}_{\text{mult}}\alpha)_{\kappa(t)}$$

Proof In order to simplify the notation let k be the residue field of S at s and K be the residue field of T at t. By definition, we may assume without loss of generality that X is an integral scheme equidimensional over S of dimension n and that $\alpha = [X]$. Similarly we may assume that the schemes T and S are local schemes with closed points t and s. Fix an irreducible component W of the fiber $(X_T)_t$ and denote by w its generic point. Let Z be the irreducible component of X_s dominated by W and let z be its generic point. We have to show that

$$m(W, t^{\circledast}_{\text{mult}}[X_T]) = m(Z, s^{\circledast}_{\text{mult}}[X])m(W, [Z_K]). \tag{12.2}$$

According to [5, **IV**, 4.2.4], the local ring $\mathscr{O}_{Z_K,w}$ is given by

$$\mathscr{O}_{Z_K,w} = \Big(K \otimes_k \kappa(z)\Big)_{\mathfrak{p}_w}$$

where the minimal prime ideal $\mathfrak{p}_w$ is the kernel of the morphism from $K \otimes_k \kappa(z)$ to $\kappa(w)$. In particular

$$m(W, [Z_K]) = \lg \Big(K \otimes_k \kappa(z)\Big)_{\mathfrak{p}_w}.$$

Assume that $T = \operatorname{Spec}(\mathscr{O}^{\text{sh}}_{S,\overline{s}})$ is the strict henselization of S given by a geometric point $\overline{s}$. In that case K/k is a separable extension and the scheme Z_K is reduced by [5, **IV**, 4.6.1]. In particular $m(W, [Z_K]) = 1$ and the relation (12.2) holds since we know that

$$m(W, \overline{s}^{\circledast}_{\text{mult}}[\operatorname{Spec}(\mathscr{O}^{\text{sh}}_{S,\overline{s}}) \times_S X]) = m(Z, s^{\circledast}_{\text{mult}}[X]).$$

Now let $\overline{t}$ be geometric point with locus t and $\overline{s}$ be the geometric point of S defined by θ, using the commutative square

$$\begin{array}{ccc} \operatorname{Spec}(\mathscr{O}^{\text{sh}}_{T,\overline{t}}) & \overset{\vartheta}{\dashrightarrow} & \operatorname{Spec}(\mathscr{O}^{\text{sh}}_{S,\overline{s}}) \\ \downarrow{\scriptstyle \mathfrak{l}^{\text{sh}}_{T,\overline{t}}} & & \downarrow{\scriptstyle \mathfrak{l}^{\text{sh}}_{S,\overline{s}}} \\ T & \xrightarrow{\theta} & S, \end{array}$$

we see that we may assume S and T to be strictly henselian. If we take a small enough open neighborhood of z we may assume that the structural morphism of X is a composition $X \xrightarrow{q} \mathbb{A}^n_S \to S$ where q is an equidimensional morphism of dimension 0. Let S' be the scheme $\mathbb{A}^n_S$ and consider the commutative diagram

$$\begin{array}{ccccc} X_T & \xrightarrow{q_T} & T' & \longrightarrow & T \\ \downarrow{\scriptstyle \theta_X} & \square & \downarrow{\scriptstyle \theta'} & \square & \downarrow{\scriptstyle \theta} \\ X & \xrightarrow{q} & S' & \longrightarrow & S \end{array}$$

Denote by s', t' the images of z, w by q and q_T and denote by W' the irreducible component of $X_{t'}$ that dominates W and by Z' the irreducible component of $Z_{s'}$ that dominates Z. If z' is the generic point of Z' and w' the generic point of W', we see that

$$\left(K \otimes_k \kappa(z)\right)_{\mathfrak{p}_w} = \left(K' \otimes_{k'} \kappa(z')\right)_{\mathfrak{p}'_w}$$

where K' is the residue field at t' and k' the residue field at s'. Since

$$m(W, t^{\circledast}_{\text{mult}}[X_T]) = m(W', t'^{\circledast}_{\text{mult}}[X_{T'}]), m(Z, s^{\circledast}_{\text{mult}}[X]) = m(Z', s'^{\circledast}_{\text{mult}}[X]).$$

we may assume $n = 0$.

In that case X is quasi-finite over S and since S is strictly henselian, $\mathcal{O}_{X,z}$ is a $\mathcal{O}_{S,s}$-module of finite type. For the same reason the $\mathcal{O}_{T,t}$-module $\mathcal{O}_{X_T,w}$ is of finite type. Moreover $\kappa(z)/\kappa(s)$ is a purely inseparable extension, and [5, **IV**, 4.3.2] implies that $K \otimes_k \kappa(z)$ is a local ring with residue field $\kappa(w)$. This yields

$$m(W, [Z_K]) = \lg \left(K \otimes_k \kappa(z)\right)_{\mathfrak{p}_w} = \frac{[\kappa(z) : \kappa(s)]}{[\kappa(w) : \kappa(t)]}.$$

Since

$$\begin{aligned}\mathcal{O}_{X_T,w} \otimes_{\mathcal{O}_{T,t}} F(T) &= (\mathcal{O}_{X,z} \otimes_{\mathcal{O}_{S,s}} \mathcal{O}_{T,t}) \otimes_{\mathcal{O}_{T,t}} F(T) \\ &= (\mathcal{O}_{X,z} \otimes_{\mathcal{O}_{S,s}} F(S)) \otimes_{F(S)} F(T)\end{aligned}$$

remark 2.5 implies finally

$$\begin{aligned} m(Z, s^{\circledast}_{\text{mult}}[X]) m(W, [Z_K]) &= \frac{\dim_F(\mathcal{O}_{X,z} \otimes_{\mathcal{O}_{S,s}} F(S))}{[\kappa(z) : \kappa(s)]} \frac{[\kappa(z) : \kappa(s)]}{[\kappa(w) : \kappa(t)]} \\ &= \frac{\dim_{F(S)}(\mathcal{O}_{X,z} \otimes_{\mathcal{O}_{S,s}} F(S))}{[\kappa(w) : \kappa(t)]} \\ &= \frac{\dim_{F(T)}(\mathcal{O}_{X,w} \otimes_{\mathcal{O}_{T,t}} F(T))}{[\kappa(w) : \kappa(t)]} \\ &= m(W, t^{\circledast}_{\text{mult}}[X_T]) \end{aligned}$$

and the lemma is shown. □

Proposition 2.12 *Let X be a S-scheme and s be a point in S. Then for any cycle α in $\mathscr{Z}_{\text{equi}}(X/S, n)$, one has*

$$s^{\circledast}\alpha = s^{\circledast}_{\text{mult}}\alpha.$$

Proof First assume that s is maximal. Since S is reduced, the local artinian ring $\mathcal{O}_{S,s}$ is a field. By linearity one may assume that $\alpha = [Z]$ where Z is a closed integral subscheme of X equidimensional over S of dimension n. Let z be its generic point. We may assume that s is the image of z since otherwise $s^{\circledast}_{\mathrm{mult}}\alpha = s^{\circledast}\alpha = 0$. In that case Z_s is integral with generic point z and of dimension n. Since $\mathcal{O}_{Z,z}$ is a flat $\mathcal{O}_{S,s}$-module and that

$$\mathcal{O}_{Z_s,z} = \mathcal{O}_{Z,z}/\mathfrak{m}_{S,s}\mathcal{O}_{Z,z} = \mathcal{O}_{Z,z},$$

we have

$$m(Z_s, s^{\circledast}_{\mathrm{mult}}\alpha) = \frac{e_{I\mathcal{O}_{Z,w}}(\mathcal{O}_{Z,w})}{e_I(\mathcal{O}_{S,s})} = \mathrm{lg}_{\mathcal{O}_{Z,z}}(\mathcal{O}_{Z,z}) = m(Z_s, s^{\circledast}\alpha).$$

Assume now that s is not maximal. By remark 1.11 one may choose an extension $K/\kappa(s)$ such that the K-point

$$\mathrm{Spec}(K) \longrightarrow \mathrm{Spec}(\kappa(s)) \xrightarrow{s} S \quad (\text{composite } \underline{s})$$

is fat. Using remark 1.2, it is enough to show that $\underline{s}^{\circledast}\alpha = (s^{\circledast}_{\mathrm{mult}}\alpha)_K$. Fix a thickening $(\mathcal{O}, \tau, \sigma)$ of $\underline{s}$ and write α as

$$\alpha = \sum_{i=1}^{r} \alpha_i [Z_i]$$

where the Z_i's are closed integral subschemes of X equidimensional of dimension n over S. Up to renumbering the Z_i's, we may assume that the first k are those that dominate the irreductible component of S whose generic point is the image by τ of the generic point of $\mathcal{O}$. Let $\mathscr{Z}_i$ be the irreductible component of $\mathcal{O} \times_S Z_i$ that dominates $\mathcal{O}$. Since S is geometrically unibranch, the scheme $\mathcal{O} \times_S Z_i$ is equidimensional over $\mathcal{O}$. Lemmas 2.10 and 2.11 imply then that

$$\begin{aligned}(s^{\circledast}_{\mathrm{mult}}\alpha)_K &= \sum_{i=1}^{k} \alpha_i (s^{\circledast}_{\mathrm{mult}}[Z_i])_K = \sum_{i=1}^{k} \alpha_i \sigma^{\circledast}_{\mathrm{mult}}[\mathcal{O} \times_S Z_i] \\ &= \sum_{i=1}^{k} \alpha_i \sigma^{\circledast}_{\mathrm{mult}}[\mathscr{Z}_i].\end{aligned}$$

Since $\mathscr{Z}_i$ is flat over $\mathcal{O}$, lemma 2.9 proves that

$$(s^{\circledast}_{\mathrm{mult}}\alpha)_K = \sum_{i=1}^{k} \alpha_i [\mathrm{Spec}(K) \times_{\mathcal{O}} \mathscr{Z}_i] = \underline{s}^{\circledast}\alpha$$

and thus the proposition. □

With this explicit computation of the multiplicities appearing during a base change at hand, one can finally show that for regular base schemes, the base change defined by Suslin and Voevodsky coincide with the base change that may be defined using Serre's intersection multiplicities:

Corollary 2.13 *We assume S regular. Let X be an S-scheme and $\alpha \in \mathscr{Z}_{\text{equi}}(X/S, n)$. Then for all morphism of schemes $\theta : T \to S$ we have*

$$\theta^{\circledast}\alpha = \theta^{\circledast}_{\text{Tor}}\alpha. \tag{12.3}$$

Proof The two sides of (12.3) are relative cycles, it is therefore enough to show that those cycles coincide over the maximal points of T. Using remark 1.2, it is enough to show that $s^{\circledast}_{\text{Tor}}\alpha = s^{\circledast}\alpha$ for a point s in S image of a maximal point of T. The corollary is hence a consequence of proposition 2.12 and lemma 2.9. □

3 Operations on relative cycles

3.1 Let X and Y be two S-schemes and consider a S-morphism $p : X \to Y$ that is assumed to be a flat and equidimensional of dimension d. In that case we can check that, for any given cycle $\alpha \in \mathscr{Z}(Y/S, n)$, the flat pullback $p^*\alpha$ belongs to $\mathscr{Z}(X/S, n)$. Moreover the square

$$\begin{array}{ccc} \mathscr{Z}(Y/S,n)_{\mathbb{Q}} & \xrightarrow{p^*} & \mathscr{Z}(X/S,n)_{\mathbb{Q}} \\ \downarrow{\scriptstyle \theta^{\circledast}} & & \downarrow{\scriptstyle \theta^{\circledast}} \\ \mathscr{Z}(Y_T/T,n)_{\mathbb{Q}} & \xrightarrow{p_T^*} & \mathscr{Z}(X_T/T,n)_{\mathbb{Q}} \end{array}$$

is commutative for any morphism $\theta : T \to S$.

Proper pushforwards are also compatible with base change, however the proof is more involved than for the flat pullbacks. Let $p : X \to Y$ be a morphism of S-schemes. In the sequel we will say that the direct image by p of a relative cycle on X is well-defined if p is proper on the support of the cycle. We will denote by $Z(X/S, n; p_*)$ the set of all relative cycles on X of dimension n whose image by p is well-defined. Let α be such a cycle, written as

$$\alpha = \sum_{i=1} \alpha_i[Z_i]$$

where the Z_i's are closed integral subschemes of X that dominate an irreducible component of S and are of dimension n over S. Let z_i be the generic point of Z_i, W_i the closed irreductible subscheme of Y image of Z_i and w_i be its generic point. The pushforward is defined as usual by

$$p_*\alpha = \sum_{i=1}^{r} \alpha_i m_i [W_i] \tag{12.1}$$

where the integer m_i is given by

$$m_i = \begin{cases} [\kappa(z_i)/\kappa(w_i)] & \text{if } \kappa(z_i)/\kappa(w_i) \text{ is a finite extension} \\ 0 & \text{otherwise.} \end{cases} \tag{12.2}$$

It is easy to see that $p_*\alpha$ belongs to $Z(Y/S,n)$. The next lemma provides the compatibility of pushforwards with base change for flat cycles.

Lemma 3.2 *Let $p : X \to Y$ be a morphism of S-schemes, $\theta : T \to S$ a morphism of schemes and*

$$\alpha \in Z_{\mathrm{Hilb}}(X/S,n) \qquad \beta \in Z_{\mathrm{Hilb}}(Y/S,n)$$

two relative cycles. Assume that p is proper over the support of α and that $p_*\alpha = \beta$, then p_T is proper over the support of $\theta^{\circledast}\alpha$ and

$$p_{T*}\theta^{\circledast}\alpha = \theta^{\circledast}\beta. \tag{12.3}$$

Proof Write α and β as

$$\alpha = \sum_{i=1}^{r} \alpha_i [Z_i] \qquad \beta = \sum_{j=1}^{s} \beta_j [W_j]$$

where the Z_i's (resp. the W_j's) are are closed subschemes of X (resp. Y) flat over S and equidimensional of dimension n over S. We may assume that X is the union of the Z_i's and that Y is the union of the W_j's and of the images of the Z_i's. In particular we may assume that p is proper and that the fibres of Y over S are of dimension at most n.

Fix a closed integral subscheme W of $T \times_S Y$ and denote by w its generic point. It is enough to check the equality of multiplicities

$$m(W, p_{T*}\theta^{\circledast}\alpha) = m(W, \theta^{\circledast}\beta).$$

We may assume that W dominates an irreductible component of T and is of dimension n over T. Let z be the image of w in Y. Let V be an open neighborhood of z and consider the cartesian square

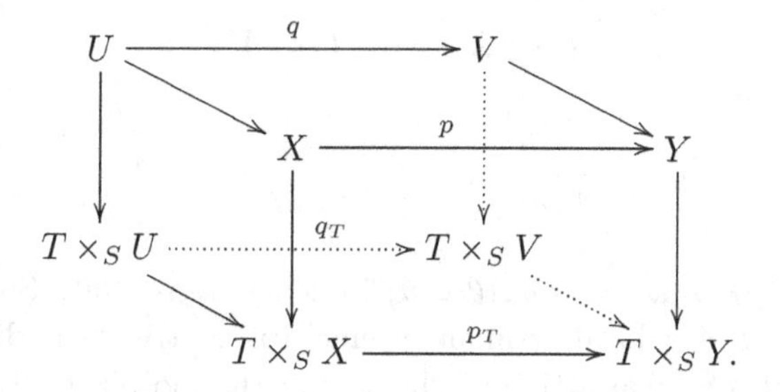

The equality (12.3) implies $q_*(\alpha|_U) = (p_*\alpha)|_V = \beta|_V$. Since

$$m(W, p_{T*}\theta^{\circledast}\alpha) = m\left(W \cap V, (p_{T*}\theta^{\circledast}\alpha)|_{T\times_S V}\right) = m(W \cap V, q_{T*}\theta^{\circledast}\alpha|_U)$$

and

$$m(W, \theta^{\circledast}\beta) = m\left(W \cap V, (\theta^{\circledast}\beta)|_{T\times_S V}\right) = m(W \cap V, \theta^{\circledast}\beta|_V),$$

we may consider only some small enough open neighborhood of z.

Assume that z belongs to an irreductible component C of Y which is not equidimensional over S. There exists an integer i such that C is contained in the image of Z_i. Since w is of dimension n in its fibre, this fibre is of dimension n and therefore $p(Z_i)$ has also a fibre of dimension n. Since p is proper, remark 5.14 implies that $p(Z_i)$ is equidimensional over S, and therefore that C also which is a contradiction. We may therefore assume that Y is equidimensional over S and via proposition 5.12 that the projection of Y on S is of the form $Y \xrightarrow{q} \mathbb{A}^n_S \to S$ where q is a maximal quasi-finite morphism. Let Z_i^0 (resp. W_i^0) be the closed subset of Z_i (resp. W_i) formed by the points of Z_i (resp. W_i) on a neighborhood of which $p \circ q$ (resp. q) is not flat. Since z is maximal in its fibre, proposition 5.12 implies that z does not belongs to W_i^0. Similarly a point x in Z_i having y for image by p is a maximal point in its fibre and therefore cannot be in Z_i^0 by proposition 5.12. We may therefore assume that the Z_i's and the W_j's are flat over $\mathbb{A}^n_S$, and so we may assume that $n = 0$.

Let ϑ be the morphism from T', the spectrum of the local ring $\mathscr{O}_{T,t}$, to T. To show that the multiplicities of W in the cycles $p_{T*}\theta^{\circledast}\alpha$ and $\theta^{\circledast}\beta$ are the same, it is enough to check it after the application of $\vartheta^{\circledast}$.

The cartesian square

$$\begin{array}{ccc} T\times_S X & \xrightarrow{p_T} & T\times_S Y \\ \uparrow{\scriptstyle \vartheta_X} & & \uparrow{\scriptstyle \vartheta_Y} \\ T'\times_S X & \xrightarrow{p_{T'}} & T'\times_S Y \end{array} \tag{12.4}$$

provides $\vartheta^{\circledast} p_{T*}\theta^{\circledast}\alpha = p_{T'*}(\theta\circ\vartheta)^{\circledast}\alpha$ since ϑ is plat. So we may assume that T is a local artinian scheme. Fix a strict henselization T' of T. It is also a local artinian scheme and the morphism $\vartheta : T' \to T$ is faithfully flat. By remark 1.2, it is enough to check that $p_{T*}\theta^{\circledast}\alpha$ and $\theta^{\circledast}\beta$ coincide after composition with $\vartheta^{\circledast}$. Using the base change formula of the square (12.4), we may assume that T is strictly local artinian. Let $\overline{s}$ be the geometric point of S defined by θ. There exists a commutative triangle

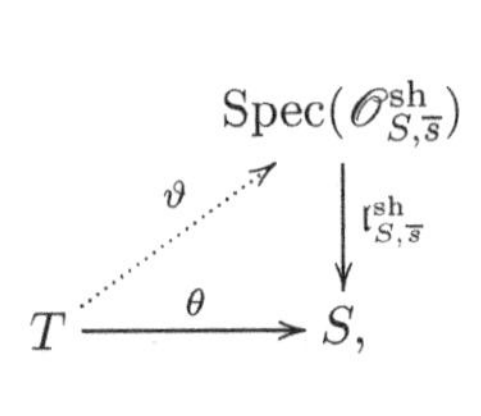

where $\mathrm{Spec}(\mathscr{O}^{\mathrm{sh}}_{S,\overline{s}})$ denotes the spectrum of the strict henselization of S at the geometric point $\overline{s}$. We may therefore assume that S is also a strictly henselian local scheme. In this case $T\times_S Y$ is also local. Being finite over the artinian scheme T, it is also local artinian, with sole point w. Let $x_1,\dots,x_m$ be the points of the scheme $T\times_S X$. Remark 1.6 implies

$$\begin{aligned} m(W, p_{T*}\theta^{\circledast}\alpha) &= \sum_{k=1}^{m}\sum_{i=1}^{r} n_i[\kappa(x_k):\kappa(w)]\,\mathrm{lg}(\mathscr{O}_{T\times_S Z_i,x_k}) \\ &= \sum_{i=1}^{r}\frac{n_i}{[\kappa(w):\kappa(t)]}\left(\sum_{k=1}^{m}[\kappa(x_k):\kappa(t)]\,\mathrm{lg}(\mathscr{O}_{T\times_S Z_i,x_k})\right) \\ &= \frac{\mathrm{lg}(\mathscr{O}_{T,t})}{[\kappa(w):\kappa(t)]}\left(\sum_{i=1}^{r} n_i \deg(T\times_S Z_i/T)\right) \\ &= \frac{\mathrm{lg}(\mathscr{O}_{T,t})}{[\kappa(w):\kappa(t)]}\left(\sum_{i=1}^{r} n_i \deg(Z_i/S)\right) \end{aligned}$$

and similarly

$$m(W, \theta^{\circledast}\beta) = \sum_{j=1}^{s} m_j \lg(\mathscr{O}_{T\times_S W_j, w})$$
$$= \frac{\lg(\mathscr{O}_{T,t})}{[\kappa(w):\kappa(t)]}\left(\sum_{j=1}^{s} m_j \deg(W_j/S)\right).$$

So we are reduced to check the equality

$$\sum_{i=1}^{r} n_i \deg(Z_i/S) = \sum_{j=1}^{s} m_j \deg(W_j/S). \tag{12.5}$$

Fiw a generic point s of S, the equality (12.3) implies that for any point y of Y above s

$$\sum_{i=1}^{r}\sum_{x\in p^{-1}(y)} n_i[\kappa(x):\kappa(y)] \lg_{\mathscr{O}_{Z_i,x}}(\mathscr{O}_{Z_i,x}) = \sum_{j=1}^{s} m_j \lg_{\mathscr{O}_{W_i,y}}(\mathscr{O}_{W_i,y}).$$

Multiplying each side by $[\kappa(y):\kappa(s)]$ and taking the sum over all points of Y above s, we get the equality

$$\sum_{i=1}^{r} n_i \left[\sum_{x\in\pi_{X/S}^{-1}(s)} [\kappa(x):\kappa(s)] \lg_{\mathscr{O}_{Z_i,x}}(\mathscr{O}_{Z_i,x})\right]$$
$$= \sum_{j=1}^{s} m_j \left[\sum_{y\in\pi_{Y/S}^{-1}(s)} [\kappa(y):\kappa(s)] \lg_{\mathscr{O}_{W_i,y}}(\mathscr{O}_{W_i,y})\right].$$

Applying remark 1.6 to it, we get (12.5). □

Proposition 3.3 *Let $p : X \to Y$ be a morphism of S-schemes and $\theta : T \to S$ be a morphism of schemes.*

1. The pushforward induces a commutative square

$$\begin{array}{ccc} Z(X/S, n; p_*) & \xrightarrow{p_*} & Z(Y/S, n) \\ \uparrow & & \uparrow \\ \mathscr{Z}(X/S, n; p_*) & \dashrightarrow & \mathscr{Z}(Y/S, n). \end{array}$$

2. *The following diagram is commutative*

$$\begin{array}{ccccc}
\mathscr{Z}(X/S,n)_{\mathbb{Q}} & \longleftarrow & \mathscr{Z}(X/S,n;p_*)_{\mathbb{Q}} & \xrightarrow{p_*} & \mathscr{Z}(Y/S,n)_{\mathbb{Q}} \\
\downarrow{\scriptstyle \theta^{\circledast}} & & \vdots & & \downarrow{\scriptstyle \theta^{\circledast}} \\
\mathscr{Z}(X_T/T,n)_{\mathbb{Q}} & \longleftarrow & \mathscr{Z}(X_T/T,n;p_{T*})_{\mathbb{Q}} & \xrightarrow{p_{T*}} & \mathscr{Z}(Y_T/T,n)_{\mathbb{Q}}.
\end{array}$$

Proof (1). Assume that α is a relative cycle in $\mathscr{Z}(X/S,n;p)$ written as

$$\alpha = \sum_{i=1}^{r} \alpha_i [Z_i]$$

where the Z_i's are closed integral subschemes of X of dimension n over S and that dominate an irreductible component of S. Replacing X by the union of the Z_i's, we may assume that p is proper. Let $\underline{s}$ be a fat K-point of S and $(\mathscr{O},\tau,\sigma)$ be a thickening of this fat point. Let z_i be the generic point of Z_i and W_i the image of Z_i by p. The proper pushforward of α by p is given by

$$p_*\alpha = \sum_{i=1}^{r} \alpha_i m_i [W_i]$$

where m_i is the integer defined by (12.2). We may assume that among the Z_i's, $Z_1,\dots,Z_k$, are the schemes that dominate the irreducible component of S whose generic point is the image by τ of the generic point of $\mathscr{O}$. In this case

$$(\mathscr{O},\tau,\sigma)^{\circledast}\alpha = \sum_{i=1}^{k} \alpha_i [\operatorname{Spec} K \times_{\mathscr{O}} \mathscr{Z}_i]$$

where $\mathscr{Z}_i$ is the unique closed subscheme of $\mathscr{O} \times_S Z_i$ flat over S and generically isomorphic to it. Similarly

$$(\mathscr{O},\tau,\sigma)^{\circledast} p_*\alpha = \sum_{i=1}^{k} \alpha_i m_i [\operatorname{Spec} K \times_{\mathscr{O}} \mathscr{W}_i]$$

where $\mathscr{W}_i$ is the unique closed subscheme of $\mathscr{O} \times_S W_i$ flat over S and generically isomorphic to it. Since the relative cycles

$$p_{\mathscr{O}*}\left(\sum_{i=1}^{r} \alpha_i [\mathscr{Z}_i]\right) \qquad \sum_i \alpha_i m_i [\mathscr{W}_i],$$

coincide generically, they are equal and lemma 3.2 implies that

$$p_{K*}(\mathscr{O},\tau,\sigma)^{\circledast}\alpha = (\mathscr{O},\tau,\sigma)^{\circledast} p_*\alpha.$$

This shows that the cycles $(\mathscr{O}, \tau, \sigma)^{\circledast} p_* \alpha$ are independent of the choice of the thickening and therefore that $p_* \alpha$ belong to $\mathscr{Z}(Y/S, n)$.

(2). Choose a relative cycle α in $\mathscr{Z}(X/S, n; p_*)_{\mathbb{Q}}$ and fix a maximal or fat K-point $\underline{t}$ of T such that its image $\underline{s}$ by θ is a maximal or fat K-point. The first part of the proof and proposition 1.23 imply that

$$\underline{t}^{\circledast} p_{T*} \theta^{\circledast} \alpha = p_{K*} \underline{t}^{\circledast} \theta^{\circledast} \alpha = p_{K*} \underline{s}^{\circledast} \alpha = \underline{s}^{\circledast} p_* \alpha$$

and it is enough to apply proposition 1.23 again to conclude that $p_{T*} \theta^{\circledast} \alpha = \theta^{\circledast} p_* \alpha$. This ends the proof of the proposition. □

3.4 One of the main operations on relative cycle is the *Cor* operation. In this section we recall its contruction and main properties. Let X be a S-scheme and Y a X-scheme. Assume to be given a relative cycle $\alpha \in \mathscr{Z}(Y/X, n)_{\mathbb{Q}}$ and a cycle on X with rational coefficients

$$\beta = \sum_Z \beta_Z [Z]$$

the sum being taken over the closed integral subschemes of X. For such a subscheme, we have a cartesian square

$$\begin{array}{ccc} Y \times_X Z & \xrightarrow{(\iota_Z)_Y} & Y \\ \downarrow & \square & \downarrow \\ Z & \xrightarrow{\iota_Z} & X \end{array}$$

and we may consider the cycle on Y with rational coefficients

$$Cor_{Y/X}(\alpha, \beta) = \sum_Z \beta_Z (\iota_Z)_{Y*} \iota_Z^{\circledast} \alpha.$$

We thus get a $\mathbb{Q}$-linear morphism

$$Cor_{Y/X} : \mathscr{Z}(Y/X, n)_{\mathbb{Q}} \otimes Z(X)_{\mathbb{Q}} \to Z(Y)_{\mathbb{Q}}. \tag{12.6}$$

Remark 3.5 If β is a relative cycle of dimension m over S, then $Cor(\alpha, \beta)$ is also a relative cycle of dimension $n + m$ over S. Indeed any closed integral subscheme of X with non zero mutiplicity in β, dominates an irreductible component of S and is of dimension m over S. For such a subscheme Z, we know that

$$\iota_Z^{\circledast} \in \mathscr{Z}(Y \times_X Z/Z, n)_{\mathbb{Q}}$$

therefore this cycle is a linear combination of closed integral subschemes $W_1, \ldots, W_r$ of $X \times_Y Z$ that dominate Z and are of dimension n over

Z. Thus each W_i dominates an irreductible component of S, and by definition of the relative dimension we get

$$\dim_S(W_i) = \dim_Z(W_i) + \dim_S(Z) = n + m.$$

This shows that $Cor(\alpha, \beta)$ is a relative cycle over S of dimension $n+m$.

The morphism (12.6) induces thus a morphism

$$Cor_{Y/X} : \mathscr{Z}(Y/X, n)_{\mathbb{Q}} \otimes Z(X/S, m)_{\mathbb{Q}} \to Z(Y/S, n+m)_{\mathbb{Q}}. \quad (12.7)$$

and a morphism

$$Cor_{Y/X} : z(Y/X, n) \otimes Z(X/S, m) \to Z(Y/S, n+m).$$

The behaviour of the Cor operation with respect to direct images is given by the following proposition:

Proposition 3.6 *Let X be a S-scheme and Y be a X-scheme.*

1. *Given a cartesian square of S-schemes*

$$\begin{array}{ccc} Y' & \xrightarrow{q} & Y \\ \downarrow & \square & \downarrow \\ X' & \xrightarrow{p} & X \end{array}$$

a cycle $\alpha \in \mathscr{Z}(Y/X, n)_{\mathbb{Q}}$ and a cycle $\beta \in Z(X'/S, m; p_*)_{\mathbb{Q}}$, then $Cor(p^{\circledast}\alpha, \beta) \in Z(Y'/S, n+m; q_*)_{\mathbb{Q}}$ and one has the equality

$$Cor(\alpha, p_*\beta) = q_*Cor(p^{\circledast}\alpha, \beta).$$

2. *Given a X-morphism $p : Y \to Y'$, a cycle $\alpha \in \mathscr{Z}(Y/X, n; p_*)_{\mathbb{Q}}$ and a cycle $\beta \in Z(X/S, m)_{\mathbb{Q}}$, then $Cor(\alpha, \beta) \in Z(Y/S, n+m; p_*)_{\mathbb{Q}}$ and*

$$Cor(p_*\alpha, \beta) = p_*Cor(\alpha, \beta).$$

Proof (1). By linearity we may assume $\beta = [Z']$ with Z' a closed integral subscheme of X'. Denote by Z the image of Z' by p and denote by ι and ι' the corresponding closed immersions. The faces of the

commutative cube

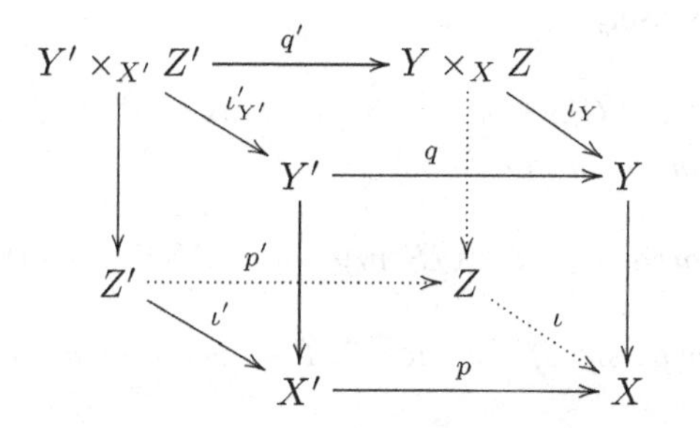

are cartesian except for the lower and upper faces. Let d be the integer equal to the degree of the field extension $F(Z)/F(Z')$ if this degree is finite and equal to zero otherwise. We have then

$$Cor(\alpha, p_*\beta) = d\iota_{Y*}\iota^{\circledast}\alpha$$

and also

$$q_*Cor(p^{\circledast}\alpha, \beta) = q_*\iota'_{Y'*}(\iota')^{\circledast}p^{\circledast} = \iota_{Y*}q'_*(p')^{\circledast}\iota^{\circledast}\alpha.$$

It is enough to check that $q'_*(p')^{\circledast}\gamma = d\gamma$ for a relative cycle γ in $\mathscr{Z}(Y \times_X Z/Z, n)_{\mathbb{Q}}$. We may replace the closed integral subscheme Z by its function field and therefore assume that Z is the spectrum of a field. In this case the morphisms p' and q' are flat and one has

$$q'_*(p')^{\circledast}\gamma = q'_*(q')^*\gamma = d\gamma.$$

This proves the assertion.

(2). By linearity we may assume $\beta = [Z]$ with Z a closed integral subscheme of X. Denote by ι the corresponding closed immersion, with the notation of the following commutative diagram

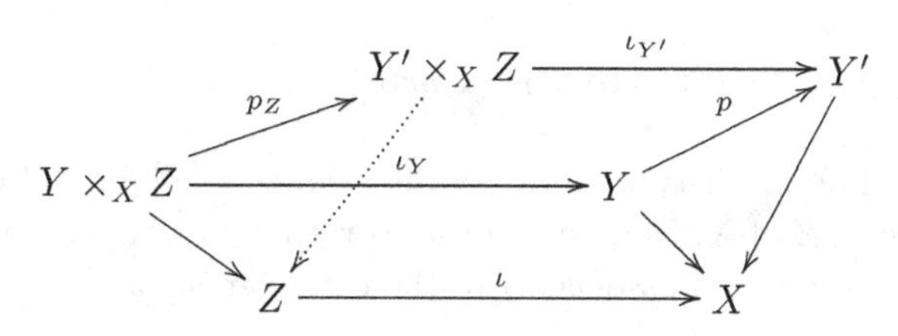

proposition 3.3 implies that

$$\begin{aligned} Cor(p_*\alpha, \beta) &= \iota_{Y'*}\iota^{\circledast}p_*\alpha = \iota_{Y'*}p_{Z*}\iota^{\circledast}\alpha \\ &= p_*\iota_{Y*}\iota^{\circledast}\alpha = p_*Cor(\alpha, \beta). \end{aligned}$$

□

The *Cor* operation is compatible with base change defined for relative cycles. More precisely:

Proposition 3.7 *Given a relative cycle $\alpha \in \mathscr{Z}(Y/X,n)_{\mathbb{Q}}$, the morphism (12.7) induces a morphism*

$$Cor(\alpha,-) : \mathscr{Z}(X/S,m)_{\mathbb{Q}} \to \mathscr{Z}(Y/S,n+m)_{\mathbb{Q}}$$

and for any morphism of schemes $\theta : T \to S$ the diagram

$$\begin{array}{ccc} \mathscr{Z}(X/S,m)_{\mathbb{Q}} & \xrightarrow{Cor(\alpha,-)} & \mathscr{Z}(Y/S,n+m)_{\mathbb{Q}} \\ \downarrow{\scriptstyle\theta^{\circledast}} & & \downarrow{\scriptstyle\theta^{\circledast}} \\ \mathscr{Z}(X\times_S T/T,m)_{\mathbb{Q}} & \xrightarrow{Cor(\theta^{\circledast}_X\alpha,-)} & \mathscr{Z}(Y\times_S T/T,n+m)_{\mathbb{Q}} \end{array}$$

is commutative.

Proof Fix a relative cycle β in $\mathscr{Z}(X/S,m)_{\mathbb{Q}}$. First remark that the statement holds if α belongs to $Z_{\mathrm{Hilb}}(Y/X,n)_{\mathbb{Q}}$ and β to $Z_{\mathrm{Hilb}}(X/S,m)_{\mathbb{Q}}$. Indeed by linearity we may assume that $\alpha = [Z]$ for a closed subscheme of Y flat and equidimensional over X of dimension n and that $\beta = [W]$ for a closed subscheme of X flat and equidimensional over S of dimension m. By definition $Cor(\alpha,\beta) = [Z\times_X W]$, which implies that $Cor(\alpha,\beta)$ belongs to $\mathscr{Z}(Y/S,n+m)_{\mathbb{Q}}$ and moreover that

$$\begin{aligned} \theta^{\circledast}Cor(\alpha,\beta) &= [T\times_S(Z\times_X W)] = [((X_T)\times_X Z)\times_{X_T}(W_T)] \\ &= Cor(\theta^{\circledast}_X\alpha,\theta^{\circledast}\beta). \end{aligned}$$

Similarly if we assume θ to be flat, then the relation

$$\theta^*_Y Cor(\alpha,\beta) = Cor(\theta^{\circledast}_X\alpha,\theta^{\circledast}\beta) \tag{12.8}$$

holds. Indeed by linearity we may assume that $\beta = [Z]$ for a closed integral subscheme Z of X. Denote by ι the corresponding closed immersion. Using the base formula associated to the cartesian square

$$\begin{array}{ccc} T\times_S(Y\times_X Z) & \xrightarrow{T\times_S\iota_Y} & T\times_S Y \\ {\scriptstyle\theta_{Y\times_X Z}}\downarrow & \square & \downarrow{\scriptstyle\theta_Y} \\ Y\times_X Z & \xrightarrow{\iota_Y} & Y \end{array}$$

and the funtoriality statement of proposition 1.23, we get the equality

$$\theta_Y^* Cor(\alpha,\beta) = \theta_Y^* \iota_{Y*} \iota^{\circledast}\alpha = (T\times_S \iota_Y)_* \theta_{Y\times_X Z}^* \iota^{\circledast}\alpha = (T\times_S \iota_Y)_* \theta_Z^{\circledast} \iota^{\circledast}\alpha$$
$$= (T\times_S \iota_Y)_* \iota_T^{\circledast} \theta_X^{\circledast} = Cor(\theta_X^{\circledast}\alpha, [T\times_S Z]) = Cor(\theta_X^{\circledast}\alpha, \theta^{\circledast}\beta)$$

and therefore the relation (12.8).

Consider now the general case and let $\underline{s}$ be a fat K-point of S and $(\mathscr{O},\tau,\sigma)$ be a thickening of it. To get the proposition, it is enough to check that

$$(\mathscr{O},\tau,\sigma)^{\circledast} Cor(\alpha,\beta) = Cor(\underline{s}_X^{\circledast}\alpha, \underline{s}^{\circledast}\beta). \tag{12.9}$$

Indeed assume this relation to be true. Then the left hand side is independent of the chosen thickening and thus $Cor(\alpha,\beta)$ belongs to the group $\mathscr{Z}(Y/S, n+m)_{\mathbb{Q}}$. On the other hand if $\underline{t}$ is a maximal or a fat K-point of T whose image $\underline{s}$ by θ is a maximal or fat K-point, then using relation (12.9) and the case of flat morphisms, we get

$$\underline{t}^{\circledast} Cor(\theta_X^{\circledast}\alpha, \theta^{\circledast}\beta) = Cor((\theta\circ\underline{s})_X^{\circledast}\alpha, \underline{t}^{\circledast}\theta^{\circledast}\beta) = Cor(\underline{s}_X^{\circledast}\alpha, \underline{s}^{\circledast}\beta)$$
$$= \underline{s}^{\circledast} Cor(\alpha,\beta)$$

which implies by proposition 1.23 that $\theta^{\circledast} Cor(\alpha,\beta) = Cor(\theta_X^{\circledast}\alpha, \theta^{\circledast}\beta)$.

Let us now check that equality (12.9) holds. For this we may assume that S is the spectrum of a DVR. Indeed to reduce to the case of a DVR, it is enough to check the relation

$$(\mathscr{O},\tau,\sigma)^{\circledast} Cor(\alpha,\beta) = (\mathscr{O},\mathrm{id}_{\mathscr{O}},\sigma)^{\circledast} Cor(\tau_X^{\circledast}, \tau^{\circledast}\beta). \tag{12.10}$$

For this denote by a the relative cycle $Cor(\alpha,\beta)$ and write this cycle as

$$a = \sum_{i=1}^{s} a_i[A_i]$$

where the A_i are some closed integral subschemes of Y that dominates an irreductible component of S. We may assume that $A_1,\dots,A_k$, are among the A_i's that dominate the irreductible component of S with generic point the image ρ of the generic point of $\mathscr{O}$. Denote by $\mathscr{A}_i$ the unique irreductible component of $\mathscr{O}\times_S A_i$ that dominates $\mathscr{O}$. To show (12.10), it is enough to check that the cycles

$$\sum_{i=1}^{k} a_i[\mathscr{A}_i] \qquad Cor(\tau_X^{\circledast}\alpha, \tau^{\circledast}\beta)$$

coincide generically. Let F be the function field of $\mathscr{O}$. Using the notation

of the diagram

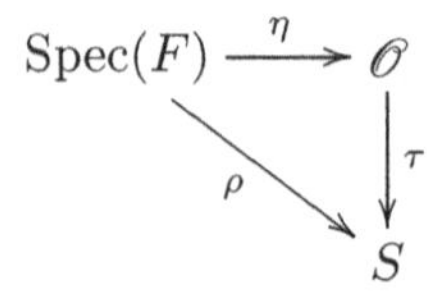

and the case of the flat morphisms, we get

$$\eta^{\circledast}\left(\sum_{i=1}^{k} a_i[\mathscr{A}_i]\right) = \rho^{\circledast}Cor(\alpha,\beta) = Cor(\rho_X^{\circledast}\alpha,\rho^{\circledast}\beta)$$
$$= Cor(\eta_X^{*}\tau_X^{\circledast}\alpha,\eta^{\circledast}\tau^{\circledast}\beta) = \eta^{\circledast}Cor(\tau_X^{\circledast}\alpha,\tau^{\circledast}\beta).$$

This proves (12.10) and we may therefore assume that S is the spectrum of a DVR. Moreover by (12.1) we may also assume that $\beta = [Z]$ with Z a closed integral subscheme of X that dominates $\mathscr{O}$ and is of dimension m over $\mathscr{O}$. Lemma 3.6 implies that

$$(\mathscr{O},\mathrm{id}_{\mathscr{O}},\sigma)^{\circledast}Cor(\alpha,\beta) = (\mathscr{O},\mathrm{id}_{\mathscr{O}},\sigma)^{\circledast}\iota_{Y*}\iota^{\circledast}\alpha = (\iota_Y)_{K*}(\mathscr{O},\mathrm{id}_{\mathscr{O}},\sigma)^{\circledast}\iota^{\circledast}\alpha$$
$$= (\iota_Y)_{K*}(\mathscr{O},\mathrm{id}_{\mathscr{O}},\sigma)^{\circledast}Cor(\iota^{\circledast}\alpha,[Z])$$

and yields also the equality

$$Cor(\sigma_X^{\circledast}\alpha,\sigma^{\circledast}\beta) = Cor(\sigma_X^{\circledast}\alpha,\sigma^{\circledast}\iota_*[Z]) = Cor(\sigma_X^{\circledast}\alpha,(\iota_K)_*\sigma^{\circledast}[Z])$$
$$= (\iota_Y)_{K*}Cor(\iota_K^{\circledast}\sigma_X^{\circledast}\alpha,\sigma^{\circledast}[Z])$$
$$= (\iota_Y)_{K*}Cor(\sigma_Z^{\circledast}\iota^{\circledast}\alpha,\sigma^{\circledast}[Z]).$$

It is enough to show the equality

$$(\mathscr{O},\mathrm{id}_{\mathscr{O}},\sigma)^{\circledast}Cor(\iota^{\circledast}\alpha,[Z]) = Cor(\sigma_Z^{\circledast}\iota^{\circledast}\alpha,\sigma^{\circledast}[Z])$$

and we may therefore assume that X is a closed integral scheme that dominates $\mathscr{O}$ and is of dimension m over $\mathscr{O}$ and moreover that $\beta = [X]$. Under those assumption X is flat over $\mathscr{O}$. Fix a generic abstract blow-up $\vartheta : X' \to X$ that flattens α. Using the first case, the functoriality statement of proposition 1.23, proposition 3.3, and lemma 3.6, we get

$$(\mathscr{O},\mathrm{id}_{\mathscr{O}},\sigma)^{\circledast}Cor(\alpha,[X]) = (\mathscr{O},\mathrm{id}_{\mathscr{O}},\sigma)^{\circledast}Cor(\alpha,\vartheta_*[X'])$$
$$= (\mathscr{O},\mathrm{id}_{\mathscr{O}},\sigma)^{\circledast}\vartheta_{Y*}Cor(\vartheta^{\circledast}\alpha,[X'])$$
$$= (\vartheta_Y)_{K*}(\mathscr{O},\mathrm{id}_{\mathscr{O}},\sigma)^{\circledast}Cor(\vartheta^{\circledast}\alpha,[X'])$$
$$= (\vartheta_Y)_{K*}Cor(\sigma_{X'}^{\circledast}\vartheta^{\circledast}\alpha,[X'_K])$$
$$= (\vartheta_Y)_{K*}Cor(\vartheta_K^{\circledast}\sigma_X^{\circledast}\alpha,[X'_K])$$
$$= Cor(\sigma_X^{\circledast}\alpha,\vartheta_{K*}[X'_K]) = Cor(\sigma_X^{\circledast}\alpha,[X_K]).$$

This proves the proposition. □

The *Cor* operation is associative, in other words:

Proposition 3.8 *Let X, Y, Z be S-schemes. Then*

$$Cor_{Z/Y}(\alpha, Cor_{Y/X}(\beta,\gamma)) = Cor_{Z/X}(Cor_{Z/Y}(\alpha,\beta),\gamma)$$

for any homogenous cycle $\alpha \in \mathscr{Z}(Z/Y,*)_{\mathbb{Q}}$, $\beta \in \mathscr{Z}(Y/X,*)_{\mathbb{Q}}$ and $\gamma \in \mathscr{Z}(X/S,*)_{\mathbb{Q}}$

Proof By linearity we may assume that $\gamma = [W]$ where W is a closed integral subscheme of X. Let ι be the corresponding closed immersion. Using lemma 3.6 and proposition 3.7 one gets

$$\begin{aligned} Cor(\alpha, Cor(\beta,\gamma)) &= Cor(\alpha, \iota_{Y*}\iota^{\circledast}\beta) = \iota_{Z*}Cor(\iota_Y^{\circledast}\alpha, \iota^{\circledast}\beta) \\ &= \iota_{Z*}\iota^{\circledast}Cor(\alpha,\beta) = Cor(Cor(\alpha,\beta),\gamma). \end{aligned}$$

□

3.9 The *Cor* operation allows to define an external product for relative cycle via the composition:

$$\begin{array}{ccccc} \mathscr{Z}(X/S,n)_{\mathbb{Q}} & \xrightarrow{\pi_{Y/S}^{\circledast}\otimes \mathrm{id}} & \mathscr{Z}(XY/Y,n)_{\mathbb{Q}} & & \\ \otimes & & \otimes & \xrightarrow{Cor} & \mathscr{Z}(XY/S,n+m)_{\mathbb{Q}}. \\ \mathscr{Z}(Y/S,m)_{\mathbb{Q}} & & \mathscr{Z}(Y/S,m)_{\mathbb{Q}} & & \end{array} \qquad (12.11)$$

(the composite is labelled $\times_S$)

where $XY = X \times_S Y$. This product is compatible with base change. Moreover it follows from the construction that this product is the usual one for relative cycles over fields.

Proposition 3.10 *The external product* (12.11) *enjoys the following properties:*

1. *it is associative and commutative ;*
2. *For any morphism of schemes $\theta : T \to S$ the square*

$$\begin{array}{ccc} \mathscr{Z}(X/S,*)_{\mathbb{Q}} \otimes \mathscr{Z}(Y/S,*)_{\mathbb{Q}} & \xrightarrow{\times_S} & \mathscr{Z}(X\times_S Y/S,*)_{\mathbb{Q}} \\ \downarrow{\scriptstyle \theta^{\circledast}\otimes\theta^{\circledast}} & & \downarrow{\scriptstyle \theta^{\circledast}} \\ \mathscr{Z}(X_T/T,*)_{\mathbb{Q}} \otimes \mathscr{Z}(Y_T/T,*)_{\mathbb{Q}} & \xrightarrow{\times_T} & \mathscr{Z}(X_T\times_T Y_T/T,*)_{\mathbb{Q}} \end{array}$$

is commutative.

Proof (2). Given two cycles $\alpha \in \mathscr{Z}(X/S,n)_{\mathbb{Q}}$ and $\beta \in \mathscr{Z}(Y/S,m)_{\mathbb{Q}}$, we get using proposition 3.7

$$\begin{aligned}\theta^{\circledast}(\alpha \times_S \beta) &= \theta^{\circledast} Cor(\pi_{Y/S}^{\circledast}\alpha, \beta) = Cor(\theta_Y^{\circledast}\pi_{Y/S}^{\circledast}\alpha, \theta^{\circledast}\beta) \\ &= Cor(\pi_{T\times_S Y/T}^{\circledast}, \theta^{\circledast}\beta) = (\theta^{\circledast}\alpha) \times_T (\theta^{\circledast}\beta)\end{aligned}$$

and the second assertion follows.

(1). Using the result above and remark 1.8, we may assume that S is the spectrum of a field. Writing α and β as

$$\alpha = \sum_{i=1}^{r} \alpha_i[Z_i] \qquad \beta = \sum_{j=1}^{s} \beta_j[W_j]$$

where the Z_i's are closed integral subschemes of X of dimension n and the W_j's closed integral subschemes of Y of dimension m, the definition gives

$$\begin{aligned}\alpha \times_S \beta &= \sum_{i=1}^{r}\sum_{j=1}^{s} \alpha_i\beta_j[(Z_i) \times_S (W_j)] = \sum_{i=1}^{r}\sum_{j=1}^{s} \alpha_i\beta_j[(W_j) \times_S (Z_i)] \\ &= \beta \times_S \alpha\end{aligned}$$

as wanted. □

This external product is also compatible with direct images:

Proposition 3.11 *Let $p : X \to X'$ and $q : Y \to Y'$ be morphisms of S-schemes. One has a commutative diagram*

$$\begin{array}{ccc}
\mathscr{Z}(X/S,n)_{\mathbb{Q}} \otimes \mathscr{Z}(Y/S,m)_{\mathbb{Q}} & \xrightarrow{\times_S} & \mathscr{Z}(XY/S,n+m)_{\mathbb{Q}} \\
\uparrow & & \uparrow \\
\mathscr{Z}(X/S,n;p_*)_{\mathbb{Q}} \otimes \mathscr{Z}(Y/S,m;q_*)_{\mathbb{Q}} & \rightarrowtail & \mathscr{Z}(XY/S,n+m;(p\times_S q)_*)_{\mathbb{Q}} \\
\downarrow{\scriptstyle p_*\otimes q_*} & & \downarrow{\scriptstyle (p\times_S q)_*} \\
\mathscr{Z}(X'/S,n)_{\mathbb{Q}} \otimes \mathscr{Z}(Y'/S,m)_{\mathbb{Q}} & \xrightarrow{\times_S} & \mathscr{Z}(X'Y'/S,n+m)_{\mathbb{Q}}
\end{array}$$

where $XY := X \times_S Y$ and similarly $X'Y' = X' \times_S Y'$.

Proof Using remark 1.8, proposition 3.3 and 3.10, we may assume that S is the spectrum of the field and are therefore reduced to the classical case. □

4 Finite correspondences and geometrical motives

4.1 Recall that a finite correspondence between two smooth schemes of finite type over a field k is linear combination of closed subschemes of the product $X \times_k Y$ which are finite and surjective over a connected components of X. For schemes of finite type (non necessarily smooth) over a more general base, the theory of relative cycles allows to general this definition. More precisely:

Definition 4.2 Let X, Y be schemes of finite type over S. The finite S-correspondences from X to Y are the elements of the abelian group

$$c_S(X,Y) := c_{\mathrm{equi}}(X \times_S Y/X, 0).$$

The composition of finite correspondences is given by the diagram

$$\begin{array}{ccc}
c_S(Y,Z) \otimes c_S(X,Y) & = & c_{\mathrm{equi}}(YZ/Y,0) \otimes c_{\mathrm{equi}}(XY/X,0) \\
\vdots & & \downarrow (p_Y^{XY})^{\circledast} \otimes \mathrm{id} \\
\vdots & & c_{\mathrm{equi}}(XYZ/XY,0) \otimes c_{\mathrm{equi}}(XY/X,0) \\
\vdots \circ & & \downarrow Cor \\
\vdots & & c_{\mathrm{equi}}(XYZ/X,0) \\
\vdots & & \downarrow (p_{XZ}^{XYZ})_* \\
c_S(X,Z) & = & c_{\mathrm{equi}}(XZ/X,0).
\end{array} \tag{12.1}$$

where $XY := X \times_S Y$ and similarly for other products. Given a morphism $f : X \to Y$ of S-schemes, we denote by Δ_f the closed immersion

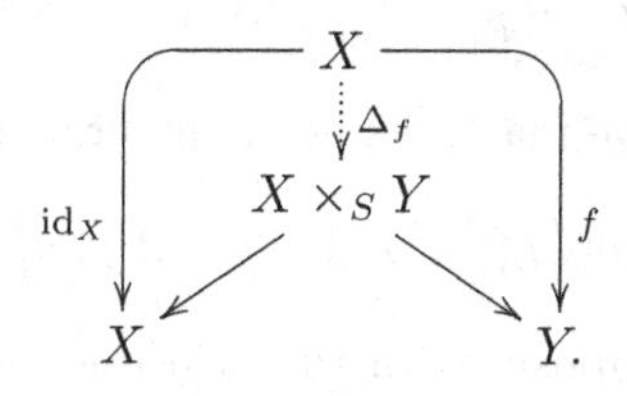

and by Γ_f the graph of f which defines a finite correspondence $[f] \in c_S(X,Y)$. The next lemma sums up the basic properties of the composition.

Lemma 4.3 *Let X, Y, Z and W be schemes of finite type over S.*

1. *Given any finite correspondences $\alpha \in c_S(X,Y)$, $\beta \in c_S(Y,Z)$ and $\gamma \in c_S(Z,W)$ we have*

$$\gamma \circ (\beta \circ \alpha) = (\gamma \circ \beta) \circ \alpha.$$

2. *For all morphisms of S-schemes $p : X \to Y$ and all finite correspondence $\alpha \in c_S(Y,Z)$ we have*

$$\alpha \circ [p] = p^{\circledast}\alpha.$$

3. *For all morphisms of S-schemes $p : Y \to Z$ and all finite correspondence $\alpha \in c_S(X,Y)$ we have*

$$[p] \circ \alpha = (\mathrm{id}_X \times_S p)_* \alpha.$$

4. *For all morphisms of S-schemes $p : X \to Y$ and $q : Y \to Z$ we have*

$$[q] \circ [p] = [q \circ p].$$

Proof (1). Using propositions 3.3, 3.7, 3.6 and 3.8 we get

$$\begin{aligned}
\gamma \circ (\beta \circ \alpha) &= \gamma \circ \left[p^{XYZ}_{XZ*} Cor(p^{XY\circledast}_{Y}\beta, \alpha)\right] \\
&= p^{XZW}_{XW*} Cor\left(p^{YZ\circledast}_{Z}\gamma, p^{XYZ}_{XZ*} Cor(p^{XY\circledast}_{Y}\beta, \alpha)\right) \\
&= p^{XZW}_{XW*} p^{XYZW}_{XZW*} Cor\left(p^{XYZ\circledast}_{XZ} p^{XZ\circledast}_{Z}\gamma, Cor(p^{XY\circledast}_{Y}\beta, \alpha)\right) \\
&= p^{XZW}_{XW*} p^{XYZW}_{XZW*} Cor\left(Cor(p^{XYZ\circledast}_{XZ} p^{XZ\circledast}_{Z}\gamma, p^{XY\circledast}_{Y}\beta), \alpha\right) \\
&= p^{XYW}_{XW*} p^{XYZW}_{XYW*} Cor\left(Cor(p^{XYZ\circledast}_{YZ} p^{YZ\circledast}_{Z}\gamma, p^{XY\circledast}_{Y}\beta), \alpha\right) \\
&= p^{XYW}_{XW*} Cor\left(p^{XYZW}_{XYW*} p^{XY\circledast}_{Y} Cor(p^{YZ\circledast}_{Z}\gamma, \beta), \alpha\right) \\
&= p^{XYW}_{XW*} Cor\left(p^{XY\circledast}_{Y} p^{YZW}_{YW*} Cor(p^{YZ\circledast}_{Z}\gamma, \beta), \alpha\right) \\
&= p^{XYW}_{XW*} Cor\left(p^{XY\circledast}_{Y}(\gamma \circ \beta), \alpha\right) = (\gamma \circ \beta) \circ \alpha.
\end{aligned}$$

(2). Using remark 1.8 we may assume X integral. We have then

$$\alpha \circ [p] = p^{XYZ}_{XZ*} Cor\left(p^{XY\circledast}_{Y}\alpha, [\Gamma_p]\right) = \Delta^{\circledast}_{p}\left(p^{XY\circledast}_{Y}\alpha\right) = p^{\circledast}\alpha.$$

(3). Let q be the composition of the projection of $X \times_S Y$ on Y and of the morphisms p. Write α as

$$\alpha = \sum_{i=1}^{r} \alpha_i [Z_i]$$

where Z_i is a closed integral subscheme of $X \times_S Y$. Let ι_{Z_i} be the closed immersion corresponding to Z_i and q_i be its composition with q. We have then

$$\begin{aligned}
[p] \circ \alpha &= p_{XZ*}^{XYZ} Cor\Big(p_Y^{XY\circledast}[\Gamma_p], \alpha\Big) \\
&= \sum_{i=1}^{r} \alpha_i \Big(p_{XZ*}^{XYZ} (\iota_{Z_i} \times_S \mathrm{id}_Z)_* \iota_{Z_i}^{\circledast} p_Y^{XY\circledast}[\Gamma_p]\Big) \\
&= \sum_{i=1}^{r} \alpha_i \Big(p_{XZ*}^{XYZ} (\iota_{Z_i} \times_S \mathrm{id}_Z)_* [\Gamma_{q_i}]\Big) = \sum_{i=1}^{r} \alpha_i \Big(p_{XZ*}^{XYZ} (\Delta_q)_*[Z_i]\Big) \\
&= p_{XZ*}^{XYZ}\Big((\Delta_q)_*\alpha\Big) = (\mathrm{id}_X \times_S p)_*\alpha.
\end{aligned}$$

(4). According to the second assertion, we know that $[q] \circ [p] = p^{\circledast}[\Gamma_q]$ and the result follows from the equality $p^{\circledast}[\Gamma_q] = [\Gamma_{q \circ p}]$. □

In particular lemma 4.3 assures that taking for objects the S-schemes of finite type and for morphisms the finite correspondences, we get a category VarCor_S with a fully faithful functor

$$[-] : \mathrm{Var}_S \to \mathrm{VarCor}_S \tag{12.2}$$

where Var_S is the category of schemes of finite type over S. Let SmCor_S be the subcategory formed by the smooth S-schemes of finite type with the finite correspondences as morphisms. The category VarCor_S has a natural tensor structure compatible via the functor (12.2) with the symmetric monoidal structure induced on Var_S by the fibre product. The tensor product of two finite correspondences is provided by the morphism

$$\begin{array}{ccc}
c_S(X,Y) \otimes c_S(X',Y') & = & c_{\mathrm{equi}}(X \times_S Y/X, 0) \otimes c_{\mathrm{equi}}(X' \times_S Y'/X', 0) \\
\vdots & & \Big\downarrow \Big(p_X^{X\times_S X'}\Big)^{\circledast} \otimes \Big(p_{X'}^{X\times_S X'}\Big)^{\circledast} \\
\vdots \otimes & & c_{\mathrm{equi}}(X \times_S X' \times_S Y/X \times_S X', 0) \otimes c_{\mathrm{equi}}(X \times_S X' \times_S Y'/X \times_S X') \\
\Big\downarrow & & \Big\downarrow \times_{X\times_S X'} \\
c_S(X \times_S X', Y \times_S Y') & = & c_{\mathrm{equi}}(X \times_S X' \times_S Y \times_S Y'/X \times_S X', 0)
\end{array}$$

Lemma 3.10 assures that this operation is symmetric. Moreover it is compabible with the fibre product. Indeed if $p : X \to Y$ and $q : Y \to Y'$ are two S-morphisms then $[p] \otimes [q] = [p \times_S q]$.

Remark 4.4 Let X, Y be two S-schemes and $\alpha \in c_S(X, Y)$ be a finite correspondence, the square

$$\begin{array}{ccc} X & \xrightarrow{[\Delta_X]} & X \times_S X \\ {\scriptstyle \alpha}\downarrow & & \downarrow{\scriptstyle \alpha\otimes\alpha} \\ Y & \xrightarrow{[\Delta_Y]} & Y \times_S Y \end{array}$$

is not necessarily commutative when α is not a finite correspondence associated to a morphism of S-schemes.

4.5 Over a field the composition of finite correspondences between smooth schemes of finite type is defined using the classical intersection theory. This is a special case of the general composition formula for finite correspondences. Indeed the next proposition shows that the composition of finite correspondences given by (12.1) coincides for smooth schemes of finite type over a regular scheme with the composition given by the classical intersection theory.

Proposition 4.6 *Let S be a regular scheme and X, Y, Z be smooth S-schemes of finite type. Then for any finite correspondence $\alpha \in c_S(X, Y)$ and $\beta \in c_S(Y, Z)$:*

1. *the intersection of the two cycles $p_{XY}^{XYZ*}\alpha$ and $p_{YZ}^{XYZ*}\beta$ is proper;*
2. *the composition of the finite correspondences α and β is given by*

$$\beta \circ \alpha = p_{XZ*}^{XYZ}\left(p_{XY}^{XYZ*}\alpha \frown p_{YZ}^{XYZ*}\beta\right).$$

Proof (1). Since X, Y, Z are smooth schemes of finite type over S, they are equidimensional over S. Since these schemes are regular, their irreductible components coincide with their connected components. Without loss of generality we may therefore assume that X, Y, Z and S are connected. In particular the schemes X, Y, Z are equidimensional over S with respective dimension n_X, n_Y and n_Z. Let d be the dimension of S let $|\alpha|$ be the support of α and $|\beta|$ the support of β. Since $X \times_S Y$ is equidimensional over S of dimension $n_X + n_Y$ and since the scheme $|\alpha|$ is finite over X, one sees that α is of codimension n_Y in $X \times_S Y$. Similarly β is of codimension n_Z in $Y \times_S Z$. Since all the schemes considered are regular, the notion of equidimensional morphism is stable by base change and therefore the closed subschemes $|\alpha| \times_S Z$ and $X \times_S |\beta|$ are respectively equidimensional of dimension 0 over $X \times_S Z$ and $X \times_S Y$. In particular they are of codimension n_Y and n_Z in $X \times_S Y \times_S Z$. It is enough to check that the codimension of

$$(|\alpha| \times_S Z) \times_{XYZ} (X \times_S |\beta|) = |\alpha| \times_Y |\beta|$$

in $X \times_S Y \times_S Z$ is $n_Y + n_Z$. This follows from the fact that this scheme is finite and equidimensional over X. Indeed one has a cartesian square

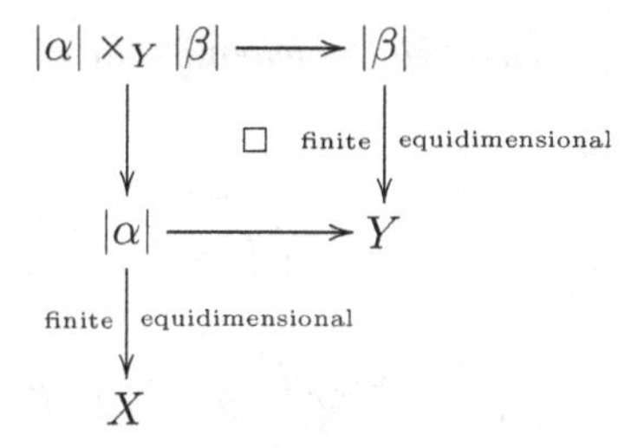

and finite equidimensional morphisms are stable under base change.

(2). By linearity we may assume $\alpha = [W]$ with W a closed integral subscheme of $X \times_S Y$ finite and equidimensional over X. Let ι be the corresponding closed immersion, one has

$$\beta \circ \alpha = p_{XZ*}^{XYZ} Cor\left(p_X^{XY\circledast}\beta, \alpha\right) = p_{XZ*}^{XYZ}\left(\iota_{Z*}\iota^{\circledast} p_X^{XY\circledast}\beta\right).$$

The first assertion and corollary 2.13 proves that

$$\iota_{Z*}\iota^{\circledast} p_X^{XY\circledast}\beta = \iota_{Z*}\iota_Z^* p_{YZ}^{XYZ*}\beta = [W \times_S Z] \frown p_{YZ}^{XYZ*}\beta$$

and this provides the desired formula:

$$\beta \circ \alpha = p_{XZ*}^{XYZ}\left(p_{XY}^{XYZ*}\alpha \frown p_{YZ}^{XYZ*}\beta\right).$$

□

4.7 The composition of finite correspondences is also well behaved with respect to base change. Indeed let $\theta : T \to S$ be a morphism of schemes let X and Y be two S-schemes of finite type. We may then consider the base change along θ_X, which provides a morphism:

$$-_T : c_S(X, Y) \to c_T(X_T, Y_T).$$

We will denote by α_T the finite T-correspondence obtained by base change from a finite S-correspondence α. The next lemma sums up the elementary properties of the base change.

Lemma 4.8 *Let $\theta : T \to S$ be a morphism of schemes, X, X', Y, Y', Z be some S-schemes.*

1. *Given two finite S-correspondences $\alpha \in c_S(X,Y)$ and $\beta \in c_S(Y,Z)$:*

$$\beta_T \circ \alpha_T = (\beta \circ \alpha)_T.$$

2. *Given two finite S-correspondences $\alpha \in c_S(X,Y)$ and $\beta \in c_S(Y,Y')$:*

$$\alpha_T \otimes \beta_T = (\alpha \otimes \beta)_T.$$

3. *If $p : X \to Y$ is a morphism of S-schemes, then $[p]_T = [p_T]$.*

Proof (1). Using proposition 3.7, we get

$$\begin{aligned}
\beta_T \circ \alpha_T &= \left(p^{X_T Y_T Z_T}_{X_T Z_T}\right)_* Cor\left(\left(p^{X_T Y_T}_{Y_T}\right)^{\circledast} \beta_T, \alpha_T\right) \\
&= \left(p^{X_T Y_T Z_T}_{X_T Z_T}\right)_* Cor\left(\left(p^{X_T Y_T}_{Y_T}\right)^{\circledast} \theta^{\circledast}_Y \beta, \theta^{\circledast}_X \alpha\right) \\
&= \left(p^{X_T Y_T Z_T}_{X_T Z_T}\right)_* Cor\left(\theta^{\circledast}_{XY} \left(p^{XY}_Y\right)^{\circledast} \beta, \theta^{\circledast}_X \alpha\right) \\
&= \left(p^{X_T Y_T Z_T}_{X_T Z_T}\right)_* \theta^{\circledast}_X Cor\left(\left(p^{XY}_Y\right)^{\circledast} \beta, \theta^{\circledast}_X \alpha\right).
\end{aligned}$$

Proposition 3.3 ensures finally that

$$\beta_T \circ \alpha_T = \theta^{\circledast}_X \left(p^{XYZ}_{XZ}\right)_* Cor\left(\left(p^{XY}_Y\right)^{\circledast} \beta, \theta^{\circledast}_X \alpha\right) = (\beta \circ \alpha)_T.$$

(2). Using lemma 3.10, we get

$$\begin{aligned}
\alpha_T \otimes \beta_T &= \left(p^{X_T X'_T \circledast}_{X_T} \theta^{\circledast}_X \alpha\right) \times_{X_T X'_T} \left(p^{X_T X'_T \circledast}_{X'_T} \theta^{\circledast}_{X'} \beta\right) \\
&= \left(\theta^{\circledast}_{XX'} p^{XX' \circledast}_X \alpha\right) \times_{X_T X'_T} \left(\theta^{\circledast}_{XX'} p^{XX' \circledast}_{X'} \beta\right) \\
&= \theta^{\circledast}_{XX'} (\alpha \otimes \beta) = (\alpha \otimes \beta)_T.
\end{aligned}$$

This shows the compatibility with the tensor product.

(3). It is enough to remark that $\theta^{\circledast}_X [\Gamma_p] = [\Gamma_{p_T}]$ to get the result. □

Lemma 4.8 shows that the base change morphism is a tensor functor and that moreover the square

$$\begin{array}{ccc}
\mathrm{Var}_S & \xrightarrow{T \times_S -} & \mathrm{Var}_T \\
\big\downarrow [-] & & \big\downarrow [-] \\
\mathrm{VarCor}_S & \xrightarrow{T \times_S -} & \mathrm{VarCor}_T.
\end{array}$$

is commutative. We refer to [7] for a finer study of correspondences, where in particular their local properties for the Nisnevich and étale topology is investigated[4].

4.9 With the notion of finite correspondences at hand, the definition of the triangulated category of geometrical motives over S is similar to the construction given in [19]. However over a general base one has to replace the Zariski topology by the Nisnevich topology. We refer to §5.1.1 and especially proposition 5.2 for the definition and the basic properties of the Nisnevich topology.

Definition 4.10 The category $\underline{DM}^{\mathrm{eff}}_{gm}(S)$ is the triangulated category

$$\underline{DM}^{\mathrm{eff}}_{gm}(S) = \mathrm{K}^{\mathrm{b}}(\mathrm{SmCor}_S)/E_{gm}$$

obtained by taking the quotient by the thick triangulated subcategory E_{gm} of $\mathrm{K}^{\mathrm{b}}(\mathrm{SmCor}_S)$ generated by the complexes:

- *Homotopy* :

$$[\mathbb{A}^1_X] \xrightarrow{[\pi]} \underset{\substack{\uparrow \\ \text{degree } 0}}{[X]} \tag{12.3}$$

  where X is a smooth S-scheme of finite type and π is the projection;
- *Mayer-Vietoris for the Nisnevich topology* :

$$[U \times_X V] \xrightarrow{[j_U]\oplus[j_V]} [U] \oplus [V] \xrightarrow{[i_U]\oplus(-[i_V])} \underset{\substack{\uparrow \\ \text{degree } 0}}{[X]} \tag{12.4}$$

  for any distinguished square for the Nisnevich topology of the form (12.1).

Remark 4.11 Since (12.3) is the mapping cone in $\mathrm{C}^{\mathrm{b}}(\mathrm{SmCor}_S)$ of the morphism $[\pi]$ and (12.4) is the mapping cone of the morphism between the mapping cone of $[U \times_X V] \to [U] \oplus [V]$ and $[X]$, a triangulated functor

$$F : \mathrm{K}^{\mathrm{b}}(\mathrm{SmCor}_S) \to \mathscr{T}$$

extends to the category $\underline{DM}^{\mathrm{eff}}_{gm}(S)$ if and only if the following two condition are satisfied:

[4] As shown in [7], finite correspondences possess a local decomposition for the Nisnevich and étale topology. However the construction uses in an essential way the fact that the local rings for these topologies are henselian: there is no such decomposition for the Zariski topology.

1. the morphism $F(\mathbb{A}^1_X) \to F(X)$ is an isomorphism;
2. the Mayer-Vietoris triangle

$$F(U \times_X V) \to F(U) \oplus F(V) \to F(X) \to F(U \times_X V)[1]$$

is distinguished.

Definition 4.12 The triangulated category of effective geometrical motives over S

$$DM^{\text{eff}}_{gm}(S)$$

is the pseudo-abelian hull of the triangulated category $\underline{DM}^{\text{eff}}_{gm}(S)$.

The category of definition 4.12 has a natural structure of triangulated category according to theorem 1.5 of [1]. We denote by M_{gm} the canonical functor

$$M_{gm} : \mathrm{Sm}_S \to DM^{\text{eff}}_{gm}(S)$$

which maps a smooth S-scheme of finite type X to its geometrical motives $M_{gm}(X)$ image of the object $[X]$ of SmCor_S in the triangulated category of effective geometrical motives.

5 The embedding theorem

We assume in this section that S has finite Krull dimension.

5.1 The Nisnevich topology introduced in [12] is the topology generated by the covering family $(X_i \xrightarrow{e_i} X)_i$ such that:

- the morphism e_i is étale;
- for any point x in X, there exists a point x_i in one of the X_i such that the morphism between the residue fields $\kappa(x)$ and $\kappa(x_i)$ is an isomorphism.

Since all the schemes are assumed to be noetherian, it is enough to consider only finite families. This defines a Grothendieck topology on the category Sm_S. The Nisnevich topology is finer than the Zariski topology but coarsest than étale topology. In particular any representable presheaf is a Nisnevich sheaf. A cartesian square in Sm_S

$$\begin{array}{ccc} W & \longrightarrow & V \\ \downarrow & \square & \downarrow p \\ U & \xrightarrow{e} & X \end{array}$$

is said to be excisive if e is an open immersion and the morphism p induces an homeomorphism between $X \setminus U$ and $V \setminus W$. The next proposition, fully proved in §2 of [20], may be seen as a compact reformulation of the classical properties of the Nisnevich topology. It implies in particular the theorem of Kato and Saito [8] on cohomological dimension. We refer to the appendix 5.7 for some recollections on cd-structures and their associated topologies.

Proposition 5.2 *The excisive cartesian squares*

$$\begin{array}{ccc} U \times_X V & \longrightarrow & V \\ \downarrow & \square & \downarrow p \\ U & \xrightarrow{e} & X \end{array} \tag{12.1}$$

such that p is an étale morphism, define on Sm_S *a regular complete and bounded cd-structure for the Krull density*[5] *and the associated topology is the Nisnevich topology.*

In particular a presheaf F on Sm_S is a Nisnevich sheaf if and only if the square

$$\begin{array}{ccc} F(X) & \longrightarrow & F(U) \\ \downarrow & \square & \downarrow \\ F(V) & \longrightarrow & F(U \times_X V) \end{array}$$

is cartesian for any distinguished square of the form (12.1). Let us recall the definition of a sheaf and a presheaf with transfers.

Definition 5.3 A presheaf with transfers on S is an additive presheaf of abelian groups on the category SmCor_S of smooth S-schemes of finite type with the finite correspondences as morphisms. A presheaf with transfer whose underlying preasheaf is a Nisnevich (resp. étale) sheaf is called a Nisnevich sheaf with transfers (resp. an étale sheaf with transfers).

In the sequel $\mathrm{PSh}^{\mathrm{tr}}(S)$ denotes the additive category of presheaves with transfers on S and by $\mathrm{Sh}^{\mathrm{tr}}_{\mathrm{Nis}}(S)$ the category of Nisnevich sheaves with transfers on S. The theory of relative algebraic cycles provide basic examples of presheaves with transfers:

[5] The category Sm_S has a natural density related to the Krull dimension. For any $X \in$: Sm_S the set $D_i(X)$ is formed by the open immersion $U \to X$ such that for any point $x_0 \in X \setminus U$ there exists points $x_1, \dots, x_i$ in X such that $x_r \neq x_{r+1}$ and $x_r \in \overline{\{x_{r+1}\}}$.

Lemma 5.4 *Let X be a S-scheme and n a nonnegative integer, the presheaves on* Sm_S

$$\underline{\mathscr{Z}}_{\mathrm{equi}}(X/S,n)_{\mathbb{Q}} \qquad \underline{z}_{\mathrm{equi}}(X/S,n) \qquad \underline{\mathscr{C}}_{\mathrm{equi}}(X/S,n)_{\mathbb{Q}} \qquad \underline{c}_{\mathrm{equi}}(X/S,n)$$

have a natural structure of presheaves with transfers.

Proof In order to shorten the notation, in the proof YZ will stand for the fiber product over S of the S-schemes Y, Z and similarly for other schemes. The construction of the transfers being the same for these four presheaves, we will denote by F_X one of them. Given two S-schemes X and Z, we denote by G_{XZ} the sub-presheaf of F_{XZ} whose sections on a S-scheme Y are the elements in $F_{XZ}(Y)$ with proper support over XY. Similarly given three S-schemes X, Z and W, we denote by H_{XZW} the sub-presheaf of F_{XZW} whose sections over a S-scheme Y are the elements in $F_{XZW}(Y)$ with proper support over XYZ.

Let $\alpha \in c_S(Y, Y')$ be a finite correspondence. Given an element $\beta \in F_X(YY')$, we may remark that the support of the cycle $Cor(\beta, \alpha)$ is proper over XY. Indeed α is a finite sum of the form

$$\alpha = \sum_{i=1}^{r} \alpha_i [Z_i]$$

where the Z_i's are closed integral subschemes of YY' finite and equidimensional over Y'. By definition of the cycle $Cor(\beta, \alpha)$, its support is contained is the union of $W_1, \ldots, W_r$ where W_i is the closed subscheme of XYY' given by the cartesian squares

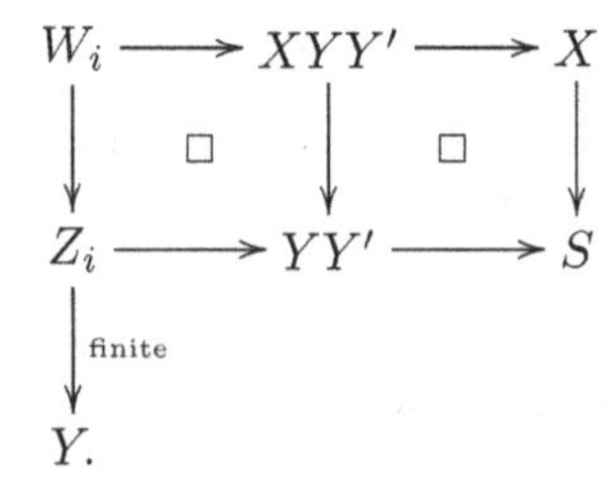

Since Z_i is finite over Y, one sees by base change that W_i is finite over XY and therefore that the support of $Cor(\beta, \alpha)$ id proper over XY. So we may consider the morphism $F_X(\alpha)$ definied by the diagram

$$\begin{array}{ccc} F_X(Y') & \xrightarrow{F_X(\alpha)} & F_X(Y) \\ {\scriptstyle p_{Y'}^{YY'\circledast}} \downarrow & & \uparrow {\scriptstyle p_{XY*}^{XYY'}} \\ F_X(YY') & \xrightarrow{Cor(-,\alpha)} & G_{XY'}(Y) \end{array}$$

this morphism provides the transfers on F_X.

One has now to check that it really endows F_X with a structure of presheaf with transfers. If $\alpha \in c_S(Y, Y')$ and $\alpha' \in c_S(Y', Y'')$ are two finite correspondences, the functoriality is a consequence of the commutative diagram

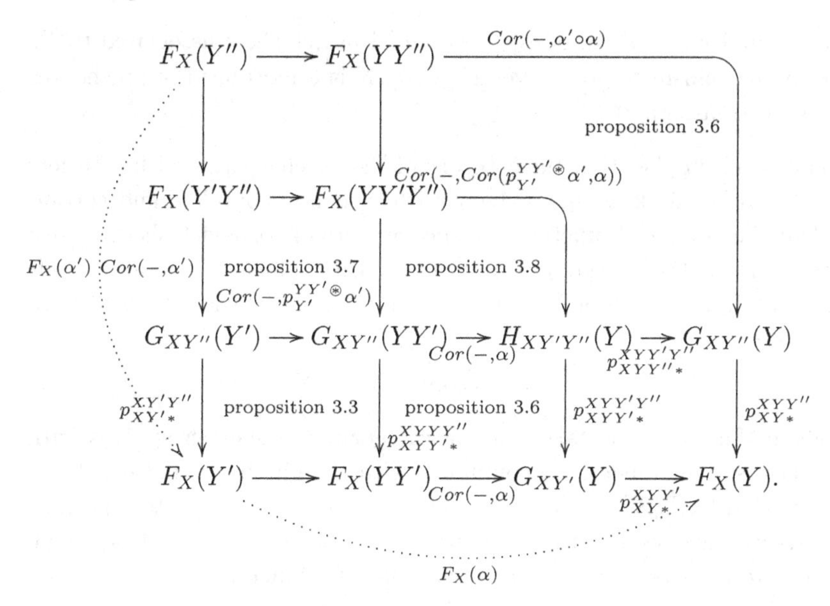

We have precised in the center of each square the result that ensures its commutativity. □

The presheaf with transfers on Sm_S represented by a S-scheme X is denoted by $\mathbb{Z}_{\mathrm{tr}}[X]$. Its sections over a smooth S-scheme of finite type Y are given by the finite correspondences

$$\mathbb{Z}_{\mathrm{tr}}[X](Y) = c_S(Y, X)$$

The Yoneda lemma provides an isomorphism

$$\mathrm{Hom}_{\mathrm{PSh}^{\mathrm{tr}}(S)}(\mathbb{Z}_{\mathrm{tr}}[X], F) = F(X)$$

for any preasheaf with transfers F. As in proposition 5.1.3 of [17], one can check that the presheaves in lemma 5.4 are sheaves for the étale topology. The proof of proposition 3.1.3 of [19][6] holds in greater generality, and shows that given a S-scheme X and a covering object U of X for the Nisnevich topology then the morphism of complexes of Nisnevich sheaves with transfers

$$\mathbb{Z}_{\mathrm{tr}}[\check{C}(U)] \to \mathbb{Z}_{\mathrm{tr}}[X]$$

[6] See also proposition 6.12 in [11].

is a quasi-isomorphism in the category of complexes of Nisnevich sheaves. As a corollary we get a generalization of proposition 3.1.6 in [19]: given a presheaf with transfers F, there exists a unique Nisnevich sheaves with transfers $a^{\mathrm{tr}}_{\mathrm{Nis}}F$ such that

- the underlying sheaf of $a^{\mathrm{tr}}_{\mathrm{Nis}}F$ is the Nisnevich sheaf associated to F,
- the morphism of presheaves $F \to a^{\mathrm{tr}}_{\mathrm{Nis}}F$ is a morphism of presheaves with transfers.

This result implies that the category of Nisnevich sheaves with transfers is a Grothendieck abelian category which is also complete and cocomplete. The forgetful functor from the category of Nisnevich sheaves with transfers to the category of presheaves with transfers has also a left adjoint L_S. It sends a Nisnevich sheaf F to the Nisnevich sheaf with transfers

$$\mathrm{L}_S(F) = \mathrm{colim}_{X/F}\, \mathbb{Z}_{\mathrm{tr}}[X]$$

where the colimit is taken in the category of Nisnevich sheaves with transfers. This functor L_S commutes with colimits but is not exact. By definition $\mathrm{L}_S(\mathbb{Z}[X]) = \mathbb{Z}_{\mathrm{tr}}[X]$ for a smooth S-scheme of finite type X.

As we have seen in the previous section, the category SmCor_S is an additive tensor category and the fully faithful functor

$$[-] : \mathrm{Sm}_S \to \mathrm{SmCor}_S$$

is a tensor functor. The tensor structure on SmCor_S extends to the category of presheaves with transfers. The tensor product of twopresheaves with transfers F and G, is defined by

$$F \otimes^{\mathrm{pr}}_{\mathrm{tr}} G = \mathrm{colim}_{X \in \mathrm{Sm}_S/F}\, \mathrm{colim}_{Y \in \mathrm{Sm}_S/G}\, \mathbb{Z}_{\mathrm{tr}}[X \times_S Y].$$

As in [15] we have a description of the sections by generators and relations. Indeed for any presheaf with transfers H, one has an exact sequence of presheaves with transfers

$$\bigoplus_{\alpha \in c_S(Y,Z)} H(Z) \otimes \mathbb{Z}_{\mathrm{tr}}[Y] \xrightarrow{\phi_H} \bigoplus_{Y \in \mathrm{Sm}_S} H(Y) \otimes \mathbb{Z}_{\mathrm{tr}}[Y] \xrightarrow{\psi_H} H \to 0$$

for which the morphisms ψ_H is given on the sections over a scheme $X \in \mathrm{Sm}_S$ by

$$\bigoplus_{Y \in \mathrm{Sm}_S} H(Y) \otimes c_S(X, Y) \to H(X)$$

$$s \otimes \beta \mapsto \beta^* s$$

and the morphism ϕ_H by

$$\bigoplus_{\alpha\in c_S(Y,Z)} H(Z)\otimes c_S(X,Y) \to \bigoplus_{Y\in \mathrm{Sm}_S} H(Y)\otimes c_S(X,Y)$$
$$s\otimes\beta \mapsto \alpha^* s\otimes\beta - s\otimes(\beta\circ\alpha).$$

By right exactness by taking the tensor product of the two sequences corresponding to F and G, we get the following description

$$(F\otimes_{\mathrm{tr}}^{\mathrm{pr}} G)(X) = \left(\bigoplus_{Y,Z\in\mathrm{Sm}_S} F(Y)\otimes G(Z)\otimes c_S(X, Y\times_S Z)\right)/R \quad (12.2)$$

with R the subgroup generated by the elements of the form

$$\alpha^* s_Y\otimes\beta^* s_Z\otimes\gamma - s_Y\otimes s_Z\otimes(\alpha\otimes\beta)\circ\gamma$$

with $\alpha\in c_S(Y',Y)$, $\beta\in c_S(Z',Z)$, $\gamma\in c_S(X,Y'\times_S Z')$, $s_Y\in F(Y)$ and $s_Z\in G(Z)$.

The internal Hom $\underline{\mathrm{Hom}}_{\mathrm{tr}}$ is constructed so as to get the usual adjunction

$$\mathrm{Hom}(F\otimes_{\mathrm{tr}}^{\mathrm{pr}} G, H)\simeq \mathrm{Hom}(F, \underline{\mathrm{Hom}}_{\mathrm{tr}}(G,H));$$

in other words it is defined by $\underline{\mathrm{Hom}}_{\mathrm{tr}}(F,G)(X) := \mathrm{Hom}(F\otimes_{\mathrm{tr}}^{\mathrm{pr}}\mathbb{Z}_{\mathrm{tr}}[X], G)$ and the category $\mathrm{PSh}^{\mathrm{tr}}(S)$ inherits a structure of closed tensor category.

The tensor product of two sheaves with transfers is then defined by

$$F\otimes_{\mathrm{tr}} G := a_{\mathrm{Nis}}^{\mathrm{tr}}(F\otimes_{\mathrm{tr}}^{\mathrm{pr}} G),$$

and we obtain this way a tensor structure on $\mathrm{Sh}_{\mathrm{Nis}}^{\mathrm{tr}}(S)$ and since if F are G are Nisnevich sheaves with transfers then $\underline{\mathrm{Hom}}_{\mathrm{tr}}(F,G)$ is also a Nisnevich with transfers, the category $\mathrm{Sh}_{\mathrm{Nis}}^{\mathrm{tr}}(S)$ is a closed tensor category.

5.5 The derived category of bounded above complexes of Nisnevich sheaves with transfers admits a nice and very compact description in terms of derived category of presheaves with transfers and distinguished squares.

Proposition 5.6 *For any distinguished square for the Nisnevich topology (12.1),*

$$\begin{array}{ccc} \mathbb{Z}_{\mathrm{tr}}[X] & \longrightarrow & \mathbb{Z}_{\mathrm{tr}}[V] \\ \downarrow & & \downarrow \\ \mathbb{Z}_{\mathrm{tr}}[U] & \longrightarrow & \mathbb{Z}_{\mathrm{tr}}[U\times_X V] \end{array}$$

is a cocartesian square in the category of Nisnevich sheaves with transfers. Moreover the functor $a^{\mathrm{tr}}_{\mathrm{Nis}}$ induces an equivalence of triangulated category

$$a^{\mathrm{tr}}_{\mathrm{Nis}} : \mathrm{D}^-(\mathrm{PSh}^{\mathrm{tr}}(S))/E_{MV} \xrightarrow{\sim} \mathrm{D}^-(\mathrm{Sh}^{\mathrm{tr}}_{\mathrm{Nis}}(S))$$

where E_{MV} is the thick triangulated subcategory of the left hand side generated by the complexes

$$\mathbb{Z}_{\mathrm{tr}}[U \times_X V] \to \mathbb{Z}_{\mathrm{tr}}[U] \oplus \mathbb{Z}_{\mathrm{tr}}[V] \to \mathbb{Z}_{\mathrm{tr}}[X]. \tag{12.3}$$

Proof Assume that a distinguished square of the form (12.1) is given. Since the functor L_S commutes with colimits, remark 5.15 implies that the square

$$\begin{array}{ccc} \mathbb{Z}_{\mathrm{tr}}[X] & \longrightarrow & \mathbb{Z}_{\mathrm{tr}}[U] \\ \downarrow & & \downarrow \\ \mathbb{Z}_{\mathrm{tr}}[V] & \longrightarrow & \mathbb{Z}_{\mathrm{tr}}[U \times_X V] \end{array}$$

is also cocartesian and that we have a Mayer-Vietoris exact sequence in the category of Nisnevich sheaves with transfers

$$0 \to \mathbb{Z}_{\mathrm{tr}}[U \times_X V] \to \mathbb{Z}_{\mathrm{tr}}[U] \oplus \mathbb{Z}_{\mathrm{tr}}[V] \to \mathbb{Z}_{\mathrm{tr}}[X] \to 0.$$

The exact functors $a^{\mathrm{tr}}_{\mathrm{Nis}}$ factorizes through the quotient by E_{MV}

$$a^{\mathrm{tr}}_{\mathrm{Nis}} : \mathrm{D}(\mathrm{PSh}^{\mathrm{tr}}(S))/E_{MV} \to \mathrm{D}(\mathrm{Sh}^{\mathrm{tr}}_{\mathrm{Nis}}(S)).$$

The right hand side is the localization of the left hand side by the thick triangulated subcategory kernel of the functor $a^{\mathrm{tr}}_{\mathrm{Nis}}$, it is therefore enough to show that an object F in this kernel belongs also to E_{MV}. In other words we have to show that for such an object F the presheaf H^i on Sm_S

$$H^i(X) := \mathrm{Hom}_{\mathrm{D}(\mathrm{PSh}^{\mathrm{tr}}(S))/E_{MV}}(\mathbb{Z}_{\mathrm{tr}}[X], F[i])$$

vanishes. For a distinguished square of the form (12.1), we have the distinguished triangle in $\mathrm{D}(\mathrm{PSh}^{\mathrm{tr}}(S))/E_{MV}$

$$\mathbb{Z}_{\mathrm{tr}}[U \times_X V] \to \mathbb{Z}_{\mathrm{tr}}[U] \oplus \mathbb{Z}_{\mathrm{tr}}[V] \to \mathbb{Z}_{\mathrm{tr}}[X] \to \mathbb{Z}_{\mathrm{tr}}[U \times_X V][1]$$

and thus a long exact sequence

$$\cdots \to H^i(U \times_X V) \to H^i(U) \oplus H^i(V) \to H^i(X) \overset{\partial_Q}{\to} H^{i+1}(U \times_X V) \to \cdots$$

Therefore the (H^i, ∂_Q) provide a Brown-Gersten functor for the cd-structure that defines the Nisnevich topology, according to theorem 3.2

of [18], it is enough to show that the Nisnevich sheaves associated to the H^i vanish, and this result is a consequence of the fact that F lies in the kernel of the triangulated functor $a^{\mathrm{tr}}_{\mathrm{Nis}}$. □

5.7 As in the case of a perfect, it is possible to define a triangulated category of effective motives via the derived category of Nisnevich sheaves with transfers:

Definition 5.8 The triangulated category $DM^{\mathrm{eff}}_{-}(S)$ of effective motives on S is the quotient

$$DM^{\mathrm{eff}}_{-}(S) = \mathrm{D}^{-}(\mathrm{Sh}^{\mathrm{tr}}_{\mathrm{Nis}}(S))/E_{\mathbb{A}^1}$$

where $E_{\mathbb{A}^1}$ is the thick triangulated subcategory generated by the objects $F \in \mathrm{D}^{-}(\mathrm{Sh}^{\mathrm{tr}}_{\mathrm{Nis}}(S))$ such that there exists a distinguished triangle

$$\mathbb{Z}_{\mathrm{tr}}[\mathbb{A}^1_X] \to \mathbb{Z}_{\mathrm{tr}}[X] \to F \xrightarrow{+1}$$

where X is a smooth scheme of finite type over S.

The next result, known as the embedding theorem, relates the triangulated category of geometrical motives to the version coming from the sheaf theoretic point of view. It also shows that the geometric category defined via the Nisnevich topology coincides over a perfect field with the geometric category defined via the Zariski topology as in [19].

Proposition 5.9 *There exists a fully faithful triangulated (tensor[7]) functor*

$$\iota : DM^{\mathrm{eff}}_{gm}(S) \to DM^{\mathrm{eff}}_{-}(S)$$

such that the diagram

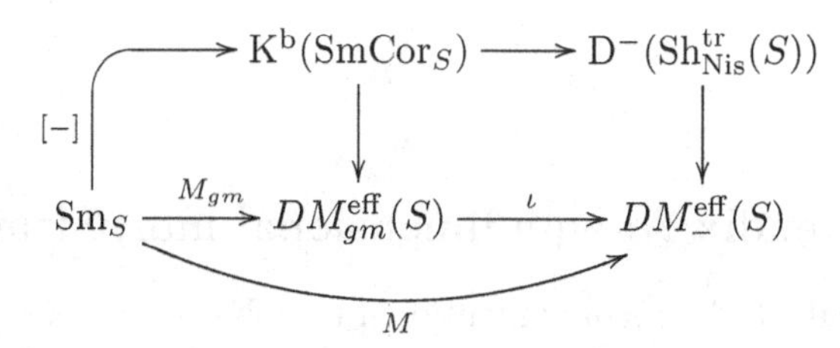

is commutative.

[7] It is possible to define a tensor structure on $DM^{\mathrm{eff}}_{-}(S)$ via the tensor structure on the category of Nisnevich sheaves with transfers. The first step is to derive the tensor product of sheaves with transfers, see [15, §2] for the construction over a field.

Proof Proposition 5.6 assures that the functor $a_{\mathrm{Nis}}^{\mathrm{tr}}$ induces an equivalence of triangulated categories

$$a_{\mathrm{Nis}}^{\mathrm{tr}} : \mathrm{D}(\mathrm{PSh}^{\mathrm{tr}}(S))/E_{MV}(S) \to \mathrm{D}(\mathrm{Sh}_{\mathrm{Nis}}^{\mathrm{tr}}(S)).$$

If we denote by $\mathscr{E}_{gm}$ the thick subcategory of the derived category of presheaves with transfers generated by

- $\mathbb{Z}_{\mathrm{tr}}[\mathbb{A}_X^1] \to \mathbb{Z}_{\mathrm{tr}}[X]$ where X is a smooth S-scheme of finite type,
- the Mayer-Vietoris complexes (12.3) for the Nisnevich topology,

we see that the functor $a_{\mathrm{Nis}}^{\mathrm{tr}}$ induces an equivalence of triangulated categories

$$a_{\mathrm{Nis}}^{\mathrm{tr}} : \mathrm{D}^-(\mathrm{PSh}^{\mathrm{tr}}(S))/\mathscr{E}_{gm} \to DM_-^{\mathrm{eff}}(S).$$

Moreover the functor $\mathbb{Z}_{\mathrm{tr}}[-] : \mathrm{SmCor}_S \to \mathrm{PSh}^{\mathrm{tr}}(S)$ induces a functor

$$\mathbb{Z}_{\mathrm{tr}}[-] : \mathrm{K}^{\mathrm{b}}(\mathrm{SmCor}_S) \to \mathrm{D}^-(\mathrm{PSh}^{\mathrm{tr}}(S)). \tag{12.4}$$

Since the objects $\mathbb{Z}_{\mathrm{tr}}[X]$ are projective in the category of presheaves with transfers, the functor (12.4) is fully faithful. Since the category E_{gm} of definition 4.10 satisfies

$$E_{gm} = \mathscr{E}_{gm} \cap \mathrm{K}^{\mathrm{b}}(\mathrm{SmCor}_S)$$

we get finally a fully faithful triangulated functor

$$\iota : \underline{DM}_{gm}^{\mathrm{eff}}(S) \to DM_-^{\mathrm{eff}}(S).$$

The result follows then from the fact that the right hand side is pseudo-abelian. □

Appendix A: equidimensional morphisms

For a thorough study of equidimensional morphisms we refer to [5, **IV**, §13]. Recall that we have defined a morphism $\theta : T \to S$ to be maximal if any irreductible component of T dominates an irreductible component of S. According to [5, **IV**, 2.3.5], open morphisms, hence flat morphisms, are maximal. Moreover the proof of the second assertion in [5, **IV**, 2.3.7] gives more generally the following result:

Lemma 5.10 *Consider a cartesian square*

$$\begin{array}{ccc} Y & \longrightarrow & X \\ \downarrow & \square & \downarrow \pi \\ T & \xrightarrow{\theta} & S \end{array} \tag{12.5}$$

and assume that the morphisms π and θ are maximal. Then any irreductible component of Y which dominates an irreductible component of T, dominates also an irreductible component of X.

The next lemma is also useful to keep track of the dimension when one works with relative cycles.

Lemma 5.11 *Let $\theta : T \to S$ be a maximal morphism and X be an integral S-scheme of dimension n over S. Then the irreductible components of X_T that dominate an irreductible component of T are of dimension n over T.*

Proof Fix an irreductible component Z of X_T that dominates a irreductible component of T and let z be its generic point. By lemma 5.10 we know that Z dominates X. In particular the generic point t of the irreductible component of T dominated by Z is sent to the image s of the generic point of X. Our assumptions imply that X_s is an integral scheme of dimension n and therefore that

$$(X_T)_t = \mathrm{Spec}(\kappa(t)) \times_{\mathrm{Spec}(\kappa(s))} X_s$$

is an equidimensional scheme de dimension n. The dimension of Z over T being equal to the dimension of the irreductible component of $(X_T)_t$ with generic point z, we finally have that Z is of dimension n over T. □

An equidimension is simply locally for the Zariski topology quasi-finite and maximal over an affine space (*i.e.* equidimensional of dimension zero over an affine space). More precisely, by [5, **IV**, 13.3.1], equidimensional morphisms can be described as follows:

Proposition 5.12 *Let X be a S-scheme of finite type. The following conditions are equivalent:*

1. *X is equidimensional of dimension n over S;*
2. *for any point x in X, there exists an open neighborhood U of x and*

a factorization

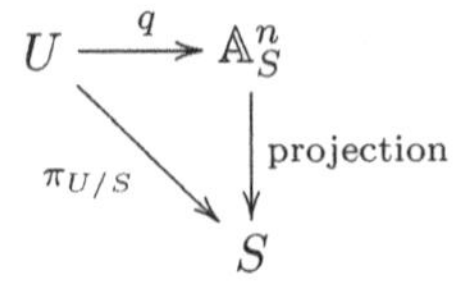

where q is a quasi-finite and maximal morphism.

3. X is maximal over S with equidimensional fibres of dimension n.

When the scheme is flat over S the previous proposition may be refined: the local factorization keeps track of the flatness.

Lemma 5.13 *Let X be a flat equidimensional scheme of dimension n over S. For any point w in X maximal in its fibre and any factorization*

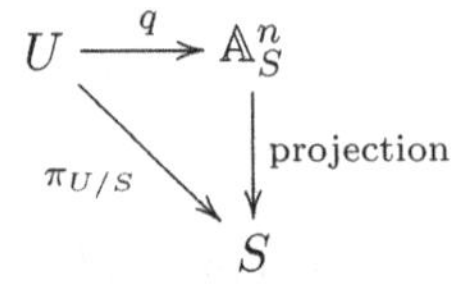

where q is a quasi-finite maximal and U is an open neighborhood of w, there exists an open neighborhood of w on which the morphism q is flat.

Proof Let v be the image of w in $\mathbb{A}^n_S$ and s be its projection on S. Let W be the closure of w in its fibre and V be the closure of v is its fibre. By hypothesis $\mathcal{O}_{X,w}$ is a flat $\mathcal{O}_{S,s}$-module, so by applying corollary 5.9 in [6, exposé IV] to the morphisms of local rings

$$\mathcal{O}_{S,s} \to \mathcal{O}_{\mathbb{A}^n_S,v} \to \mathcal{O}_{X,w}$$

it is enough to see that $\mathcal{O}_{X,w}/\mathfrak{m}_{S,s}\mathcal{O}_{X,w}$ is a flat $\mathcal{O}_{\mathbb{A}^n_S,v}/\mathfrak{m}_{S,s}\mathcal{O}_{\mathbb{A}^n_S,v}$-module. Since the quotient $\mathcal{O}_{\mathbb{A}^n_S,v}/\mathfrak{m}_{S,s}\mathcal{O}_{\mathbb{A}^n_S,v}$ is the local ring of the fibre of $\mathbb{A}^n_{\kappa(s)}$ at the point v, it is enough to check that v is the generic point of this fibre. Using [5, **IV**, 5.6.5] and the fact that q is quasi-finite, we get the inequality

$$\dim(W) \leqslant \dim(V).$$

Since X is equidimensional over S of dimension n over S and that w is maximal in its fibre, W is of dimension n. On the other hand the fibre $\mathbb{A}^n_{\kappa(s)}$ is equidimensional over S of dimension n, the last inequality assure that V is also of dimension n and therefore v is the generic point of the fibre. □

Remark 5.14 Let X be an equidimensional S-scheme of dimension n over S. Assume that the structural morphism of X has a factorization

$$X \xrightarrow{p} Y \xrightarrow{q} S$$

where p is a proper surjective morphism and q is a morphism having at least one fiber of dimension n. Then q is equidimensional of dimension n and the morphism p is finite at the generic point of Y.

Appendix B: cd-structures and topologies

The results and definitions given in this appendix are taken from [18]. The notion of cd-structure comes from an axiomatization of some of the properties of the Zariski topology. Among its main features, it allows to generalize the Brown-Gersten theorem [2, theorem 1'] to other topologies than the Zariski topology (in particular the Nisnevich topology). Recall that a cd-structure on a category $\mathscr{C}$ having an initial object $\varnothing$ is the data of a family P of squares

$$\begin{array}{ccc} B & \longrightarrow & Y \\ \downarrow & Q & \downarrow p \\ A & \xrightarrow{e} & X \end{array} \tag{12.6}$$

such that any square isomorphic to an element in P is also in P. A element in P is called a distinguished square. A cd-structure P defines a Grothendieck topology t_P on $\mathscr{C}$: it is the coarsest Grothentieck topology for which

- the sieve of X generated by the family of morphisms $\{p : Y \to X, e : A \to X\}$ is covering for any distinguished square (12.6);
- the empty sieve is a covering sieve for $\varnothing$.

In particular any distinguished square defines an elementary covering (p, e) of X. Let $\rho(X)$ be the t_P-sheaf associated to the presheaf represented by X and let $\mathbb{Z}[X]$ be the associated t_P-sheaf of abelian groups. Denote by $\mathrm{PSh}^{\mathbb{Z}}(\mathscr{C})$ the category of presheaf of abelian groups on $\mathscr{C}$ and by $\mathrm{Sh}^{\mathbb{Z}}(\mathscr{C}, t_P)$ the category of t_P-sheaves of abelian groups.

A cd-structure is said to be *complete* [18, definition 2.3] if any covering sieve of an object (not isomorphic to $\varnothing$) contains a sieve generated by a simple covering (*i.e.* a covering that can be obtained by iterating elementary coverings associated to distinguished squares). In that case any presheaf F such that

$$\begin{array}{ccc} F(X) & \longrightarrow & F(Y) \\ \downarrow & & \downarrow \\ F(A) & \longrightarrow & F(B) \end{array}$$

is cartesian for any elementary square and such that $F(\varnothing) = *$ is a t_P-sheaf. This condition becomes necessary for *regular* cd-structures (see [18, definition 2.10]).

Remark 5.15 In particular for a regular cd-structure, the squares

$$\begin{array}{ccc} \rho(X) & \longrightarrow & \rho(Y) \\ \downarrow & & \downarrow \\ \rho(A) & \longrightarrow & \rho(B) \end{array} \qquad \begin{array}{ccc} \mathbb{Z}[X] & \longrightarrow & \mathbb{Z}[Y] \\ \downarrow & & \downarrow \\ \mathbb{Z}[A] & \longrightarrow & \mathbb{Z}[B] \end{array}$$

are cocartesians and we have a Mayer-Vietoris exact sequence:

$$0 \to \mathbb{Z}[B] \to \mathbb{Z}[Y] \oplus \mathbb{Z}[A] \to \mathbb{Z}[X] \to 0$$

in the category of t_P-sheaves of abelian groups.

Recall now the definition 3.1 in [18]:

Definition 5.16 A Brown-Gersten functor for a cd-structure is the datum of a family of presheaves of abelian groups $T_q, q \geqslant 0$ and for any distinguished square of a morphism of abelian groups

$$\partial_Q : T_{q+1}(B) \to T_q(X)$$

such that:

1. the morphism ∂_Q are functorial with respect to morphisms of distinguished squares;
2. the sequences

$$T_{q+1}(B) \xrightarrow{\partial_Q} T_q(X) \to T_q(A) \oplus T_q(B)$$

   are exact.

A density D on $\mathscr{C}$ is the datum for any object X in $\mathscr{C}$ and any integer $i \geqslant 0$ of a family $D_i(X)$ of morphisms with target X such that :

- the morphism $\varnothing \to X$ belongs to $D_0(X)$ for any X;
- isomorphisms belongs to all the $D_i(X)$ for any i and any X;

- $D_{i+1}(X) \subset D_i(X)$ for any i and any X;
- if a morphism $a : Y \to X$ belongs to $D_i(X)$ and a morphism b belongs to $D_i(Y)$ then the composition $b \circ a$ is in $D_i(X)$.

The D-dimension of an object $X \in \mathscr{C}$ is the smallest integer n (maybe infinite) such that $D_n(X)$ contains only isomorphisms. A density is said to be *locally of finite dimension* if any object X is of finite D-dimension. When $\mathscr{C}$ is endowed with a cd-structure and a density D at the same time, a distinguished square is said to be *reducing* if for any $B_0 \in D_i(B), A_0 \in D_{i+1}(A), Y_0 \in D_{i+1}(Y)$, there exists a $X' \in D_{i+1}(X)$ and a morphism of squares

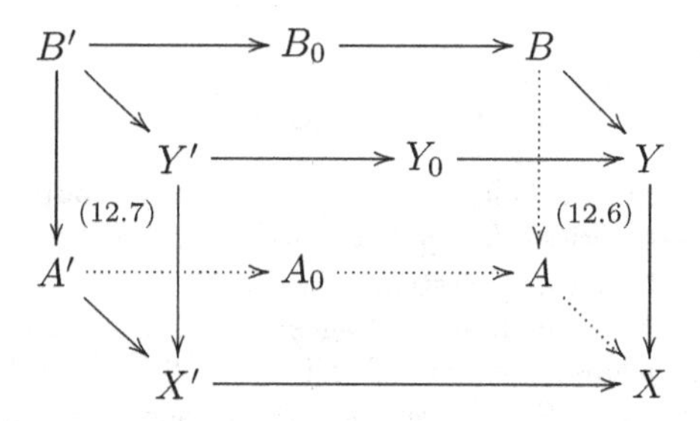

where (12.7) is a distinguished square. A density is said to be *reducing* when there exists for any distinguished square (12.6), a morphism of square

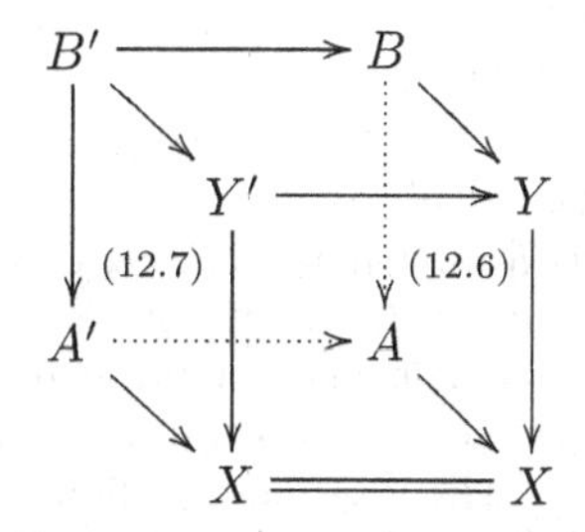

such that (12.7) is distinguished and reducing. A cd-structure is said to be bounded if there exists a reducing locally finite density. The regular complete and bounded cd-structures satisfy the two following essential properties:

- *Generalization of the Brown-Gersten theorem* [18, theorem 3.2]: one has $T_q = 0$ for any Brown-Gersten functor (T_q, ∂_Q) such that the t_P-sheaves associated to the T_q vanish and such that $T_q(\varnothing) = 0$.

- *Generalization of the Kato-Saito theorem* [18, theorem 2.26]: for any t_P-sheaf of abelian groups F, one has

$$H^n_{t_P}(X, F) = 0$$

for any integer n strictly greater than the D-dimension of X.

The second assertion is a generalization of the theorem of Kato-Saito [8] on cohomological dimension for the Nisnevich topology and of the analogous result for the Zariski topology due to Grothendieck [4, theorem 3.6.5]. From this result we can prove that the t_P-sheaves of abelian groups $\mathbb{Z}[X]$ are compact in the derived category $\mathrm{D}^-(\mathrm{Sh}^{\mathbb{Z}}(\mathscr{C}, t_P))$.

References

[1] Balmer, Paul, and Schlichting, Marco. 2001. Idempotent completion of triangulated categories. *J. Algebra*, **236**(2), 819–834.

[2] Brown, Kenneth S., and Gersten, Stephen M. 1973. Algebraic K-theory as generalized sheaf cohomology. Pages 266–292. Lecture Notes in Math., Vol. 341 of: *Algebraic K-theory, I: Higher K-theories (Proc. Conf., Battelle Memorial Inst., Seattle, Wash., 1972)*. Berlin: Springer.

[3] Fulton, William. 1998. *Intersection theory*. Second edn. Ergebnisse der Mathematik und ihrer Grenzgebiete. 3. Folge. A Series of Modern Surveys in Mathematics [Results in Mathematics and Related Areas. 3rd Series. A Series of Modern Surveys in Mathematics], vol. 2. Berlin: Springer-Verlag.

[4] Grothendieck, A. 1957. Sur quelques points d'algèbre homologique. *Tôhoku Math. J. (2)*, **9**, 119–221.

[5] Grothendieck, A. 1960-1967. Éléments de géométrie algébrique. (EGA). *Inst. Hautes Études Sci. Publ. Math.*

[6] Grothendieck, A. 1971. *Revêtements étales et groupe fondamental*. Berlin: Springer-Verlag. Séminaire de Géométrie Algébrique du Bois Marie 1960–1961 (SGA 1), Dirigé par Alexandre Grothendieck. Augmenté de deux exposés de M. Raynaud, Lecture Notes in Mathematics, Vol. 224.

[7] Ivorra, Florian. 2007. Réalisation ℓ-Adique des Motifs Triangulés Géométriques I. *Documenta. Math.*, **12**, 607–671.

[8] Kato, Kazuya, and Saito, Shuji. 1986. Global class field theory of arithmetic schemes. Pages 255–331 of: *Applications of algebraic K-theory to algebraic geometry and number theory, Part I, II (Boulder, Colo., 1983)*. Contemp. Math., vol. 55. Providence, RI: Amer. Math. Soc.

[9] Levine, Marc. 1998. *Mixed motives*. Mathematical Surveys and Monographs, vol. 57. Providence, RI: American Mathematical Society.

[10] Matsumura, Hideyuki. 1989. *Commutative ring theory*. Second edn. Cambridge Studies in Advanced Mathematics, vol. 8. Cambridge: Cambridge University Press. Translated from the Japanese by M. Reid.

[11] Mazza, Carlo, Voevodsky, Vladimir, and Weibel, Charles. 2006. *Lecture notes on motivic cohomology.* Clay Mathematics Monographs, vol. 2. Providence, RI: American Mathematical Society.

[12] Nisnevich, Ye. A. 1989. The completely decomposed topology on schemes and associated descent spectral sequences in algebraic K-theory. Pages 241–342 of: *Algebraic K-theory: connections with geometry and topology (Lake Louise, AB, 1987).* NATO Adv. Sci. Inst. Ser. C Math. Phys. Sci., vol. 279. Dordrecht: Kluwer Acad. Publ.

[13] Raynaud, Michel, and Gruson, Laurent. 1971. Critères de platitude et de projectivité. Techniques de "platification" d'un module. *Invent. Math.*, **13**, 1–89.

[14] Serre, Jean-Pierre. 1965. *Algèbre locale. Multiplicités.* Cours au Collège de France, 1957–1958, rédigé par Pierre Gabriel. Seconde édition, 1965. Lecture Notes in Mathematics, vol. 11. Berlin: Springer-Verlag.

[15] Suslin, Andrei, and Voevodsky, Vladimir. 2000a. Bloch-Kato conjecture and motivic cohomology with finite coefficients. Pages 117–189 of: *The arithmetic and geometry of algebraic cycles (Banff, AB, 1998).* NATO Sci. Ser. C Math. Phys. Sci., vol. 548. Dordrecht: Kluwer Acad. Publ.

[16] Suslin, Andrei, and Voevodsky, Vladimir. 2000b. Relative cycles and Chow sheaves. Pages 10–86 of: *Cycles, transfers, and motivic homology theories.* Ann. of Math. Stud., vol. 143. Princeton, NJ: Princeton Univ. Press.

[17] Voevodsky, Vladimir. 2000a. Cohomological theory of presheaves with transfers. Pages 87–137 of: *Cycles, transfers, and motivic homology theories.* Ann. of Math. Stud., vol. 143. Princeton, NJ: Princeton Univ. Press.

[18] Voevodsky, Vladimir. 2000b (September). *Homotopy theory of simplicial sheaves in completely decomposable topologies.* *www.math.uiuc.edu/K-theory/0443.*

[19] Voevodsky, Vladimir. 2000c. Triangulated categories of motives over a field. Pages 188–238 of: *Cycles, transfers, and motivic homology theories.* Ann. of Math. Stud., vol. 143. Princeton, NJ: Princeton Univ. Press.

[20] Voevodsky, Vladimir. 2000d (September). *Unstable motivic homotopy categories in Nisnevich and cdh-topology.* *www.math.uiuc.edu/K-theory/0444.*

[21] Voevodsky, Vladimir. 2002a (January). *Cancellation Theorem.* *www.math.uiuc.edu/K-theory/0541.*

[22] Voevodsky, Vladimir. 2002b. Motivic cohomology groups are isomorphic to higher Chow groups in any characteristic. *Int. Math. Res. Not.*, 351–355.

13

A survey of algebraic exponential sums and some applications

Emmanuel Kowalski

1 Introduction

This survey is a written and slightly expanded version of the talk given at the ICMS workshop on motivic integration and its interactions with model theory and non-archimedean geometry. Its presence may seem to require a few preliminary words of explanation: not only does the title apparently fail to reflect any of the three components of that of the conference itself, but also the author is far from being an expert in any of these. However, one must remember that there is but a small step from summation to integration. Moreover, as I will argue in the last section, there are some basic problems in the theory of exponential sums (and their applications) for which it seems not impossible that logical ideas could be useful, and hence presenting the context to model-theorists in particular could well be useful. In fact, the most direct connection between exponential sums and the topics of the workshop will be a survey of the extension to exponential sums of the beautiful counting results of Chatzidakis, van den Dries and Macintyre's [CDM]. These may also have some further applications.

Acknowledgements. I wish to thank warmly the organizers of the workshop for preparing a particularly rich and inspiring program, and for inviting me to participate.

Motivic Integration and its Interactions with Model Theory and Non-Archimedean Geometry (Volume II), ed. Raf Cluckers, Johannes Nicaise, and Julien Sebag. Published by Cambridge University Press.

2 Where exponential sums come from

Exponential sums, in the most general sense, are any type of finite sums of complex numbers

$$S = \sum_{1 \leqslant n \leqslant N} e(\theta_n)$$

where we write $e(z) = \exp(2i\pi z)$, as is customary in analytic number theory, and where the phases θ_n are real numbers. Such a sum is trivially bounded by the number of terms

$$|S| \leqslant N,$$

and of course if nothing more is known about θ_n, this can not be improved. However, in applications (whether arithmetic or otherwise), one knows something more, and the goal is very often to go from this to substantial improvements of the trivial bound: one typically wishes to prove

$$|S| \leqslant N\Sigma(N)^{-1}$$

with $\Sigma(N)$ increasing as $N \to +\infty$, as fast as possible. This is interpreted as being the result of substantial *oscillations* of the phases (in $\mathbf{R}/\mathbf{Z}$) which result in the sum being somewhat comparable with a random walk in the plane $\mathbf{C} = \mathbf{R}^2$.

In analytic number theory, exponential sums arise from many different sources. For the purpose of this survey, we will emphasize questions of *equidistribution* as leading naturally to exponential sums, as this will give a motivating framework for all the examples we want to consider here, and leads to problems and results which have obvious interest for all arithmeticians. So we will not speak about, e.g., the direct occurrence of exponential sums in the circle method, or in the distribution of primes (see, e.g., [IK, Ch. 20] for the former, and [IK, Ch. 5, 17–19] for the latter).

We therefore recall the definition of equidistribution, as well as the important Weyl criterion. Let (X, μ) be a compact topological space with a (Borel) probability measure μ. Given finite sequences $(x_n)_{n \leqslant N}$, where $x_n = x_n^{(N)}$ may depend on N, one says that they become μ-equidistributed in X if, for any open set U with $\mu(\partial U) = 0$, the sample counts

$$\frac{1}{N}|\{n \leqslant N \mid x_n \in U\}|$$

converge to the measure $\mu(U)$ of U as N gets large. Equivalently, for any continuous function f on X, the sample average

$$\frac{1}{N}\sum_{n\leqslant N} f(n)$$

is close to the integral of f over X. In the important special case where X is the torus $\mathbf{R}/\mathbf{Z}$ and μ the (unit) Lebesgue measure, this means that

$$\frac{1}{N}|\{n\leqslant N \mid a < x_n < b\}| \to (b-a)$$

for any a, b with $0 \leqslant a < b \leqslant 1$. The basic criterion of H. Weyl states that there is μ-equidistribution if and only if, for some orthonormal basis (f_h) of $L^2(X,\mu)$, elements of which are continuous functions with $f_0 = 1$, we have

$$\lim_{N\to+\infty}\frac{1}{N}\sum_{n\leqslant N} f_h(x_n) = 0, \qquad \text{for any fixed } h \neq 0.$$

If we consider again $X = \mathbf{R}/\mathbf{Z}$, we can take $f_h(x) = e(hx)$, and the criterion becomes

$$\lim_{N\to+\infty}\frac{1}{N}S_h(N) = 0, \qquad \text{where} \qquad S_h(N) = \sum_{n\leqslant N} e(hx_n),$$

which are clearly some sort of exponential sum, and the goal is to exhibit *some* cancellation, for every non-zero *frequency* h.

There are quite a few techniques available to deal with sums of the type

$$\sum_{1\leqslant n\leqslant N} e(f(n))$$

if f is some smooth real-valued function defined on $\mathbf{R}$, leading to many equidistribution statements. However, currently, the sums which are best understood are those of "algebraic nature", where extremely deep techniques of algebraic geometry are available to analyze the sums ("dissect" wouldn't be too strong a word!).

As an example of such algebraic exponential sum, which we will also carry along this survey, we take one of the most important one in analytic number theory, the *Kloosterman sums*. They are also among the first examples to have been considered historically in a non-trivial way.[1]

[1] Though the very first case is probably to be found in Gauss sums.

In keeping with our emphasis, we introduce those sums by means of an equidistribution statement which depends on estimates for Kloosterman sums.[2]

Theorem 2.1 *For $n \geqslant 1$, let $R_4(n)$ denote the set of 4-tuples $(n_1, \ldots, n_4) \in \mathbf{Z}^4$ such that*

$$n = n_1^2 + \cdots + n_4^2.$$

(1) *We have*

$$|R_4(n)| = 8n \prod_{\substack{p|n \\ p \geqslant 3}} \Big(1 + \frac{1}{p}\Big)$$

for all $n \geqslant 1$, in particular $r_4(n) = |R_4(n)|$ tends to infinity as n does.

(2) *Consider the set of points*

$$\tilde{R}_4(n) = \{x \in \mathbf{R}^4 \mid \|x\| = 1, \text{ and } n^{1/2}x \in R_4(n)\}$$

in the unit 3-sphere $\mathbf{S}^3 \subset \mathbf{R}^4$. Then, as $n \to +\infty$, $\tilde{R}_4(n)$ becomes equidistributed with respect to the Lebesgue measure on $\mathbf{S}^3$, i.e., for any continuous function f on the sphere, we have

$$\lim_{n \to +\infty} \frac{1}{r_4(n)} \sum_{x \in R_4(n)} f\Big(\frac{x}{\|x\|}\Big) = \int_{\mathbf{S}^3} f(x)dx.$$

It is the proof of Part (2) which requires exponential sums, while (1) is a result going back at least to Jacobi, which can be proved in many different ways (arithmetic of quaternions, theta functions, etc; see, e.g., [HW, §20.12] for an elementary approach).

What is the link with exponential sums? Kloosterman's approach was based on a refinement of the circle method of Hardy-Littlewood-Ramanujan (which remains of great importance today), but this particular problem is probably better understood by an appeal to modular forms. More precisely, Weyl's criterion shows that it is enough to prove that the limit exists and is equal to zero for non-constant spherical (harmonic) polynomials P on the three-sphere, which are the eigenvalues of the (Riemannian) Laplace operator on $\mathbf{S}^3$ (those functions, including the constant 1, are well-known to form an orthonormal basis of $L^2(\mathbf{S}^3)$).

[2] This is not quite what Kloosterman himself did, which involved counting solutions to diagonal quadratic equations in four variables $a_1n_1^2 + \cdots + a_4n_4^2 = N$; however, the analysis which is required is very similar.

Fix one P; then one can form the generating function

$$\theta(z;f)=\sum_{n\geqslant 1} n^{d/2}\Big(\sum_{x\in R_4(n)} P(x/\|x\|)\Big)e(nz),$$

which exists for $z \in \mathbf{H} = \{z \mid \operatorname{Im}(z) > 0\}$, where d is the degree of f. Using modularity properties, it is possible to express the coefficient of $e(nz)$ as a finite linear combination of those of other functions called Poincaré series. These last coefficients, as shown already by Poincaré himself, are roughly of the form

$$\sum_{c\geqslant 1}\frac{1}{c}S(m,n;c)J(4\pi\sqrt{mn}/c)$$

where m is a non-zero integer, the function J is a Bessel function, and $S(m,n;c)$ is a *Kloosterman sum*:

$$S(m,n;c)=\sum_{\substack{x \ (\mathrm{mod}\ c)\\ (x,c)=1}} e\Big(\frac{mx+nx^{-1}}{c}\Big),$$

where the inverse of x is of course the inverse in the unit group $(\mathbf{Z}/c\mathbf{Z})^{\times}$. One then uses facts about Bessel functions to verify that the theorem follows from any non-trivial estimate for Kloosterman sums of the type

$$S(m,n;c) \ll c^{1-\delta}$$

for a fixed $\delta > 0$, and for all m, n coprime with c. This is what Kloosterman already proved:

Proposition 2.2 *Let $c \geqslant 1$ be an integer, let m, $n \in \mathbf{Z}$ be such that the gcd $(m,n,c) = 1$. Then*

$$|S(m,n;c)| < 2c^{3/4}.$$

The main difference between sums like Kloosterman's and general exponential sums is that the range of summation is the set of points of an algebraic variety over a finite ring, and the phases are obtained by evaluating rational functions defined on this variety (in his case, the variety is the multiplicative group, and the rational function is $x \mapsto mx + n/x$). This crucial algebraicity suggests that the theory of more general sums like

$$S(f,V,c)=\sum_{x\in V(\mathbf{Z}/c\mathbf{Z})} e(f(x)/c)$$

should be accessible for general algebraic varieties V (defined over $\mathbf{Z}/c\mathbf{Z}$ or over $\mathbf{Z}$ more simply), and for f an algebraic function on V.

There is indeed such a theory, which in fact splits into two fairly distinct questions. This can be seen from the original example of Kloosterman: his argument started by using the Chinese Remainder Theorem to relate $S(m,n;c_1c_2)$ to $S(m_1,n_1;c_1)$ and $S(m_2,n_2;c_2)$ when c_1 and c_2 are coprime (with m_1, n_1, m_2, n_2 simple rational functions of m, n, c_1, c_2). So Proposition 2.2 was reduced to the case where $c = p^k$, with $k \geqslant 1$, is the power of a prime number. Now it turns out (and this is a general feature) that the case $c = p$ is very different from $c = p^k$ with $k \geqslant 2$.

Precisely, in the case of Kloosterman sums, one gets *exact formulas* for $k \geqslant 2$ (due to Salié, see, e.g., [I, 4.3]); as an example, if $p \geqslant 3$ is $\equiv 1 \pmod 4$, we have

$$S(1,1;p^k) = 2p^{k/2} \cos\Big(\frac{4\pi n}{p^k}\Big)$$

if $k \geqslant 2$, and those formulas lead immediately to the desired estimate (with the better exponent $1/2$) without any more work.

Such a drastic simplification is not rare in concrete examples arising in analytic number theory, but it is not to be expected in all cases. Indeed, there is a whole theory of trying to understand the behavior of exponential sums of algebraic origin over $\mathbf{Z}/p^k\mathbf{Z}$ as $k \to +\infty$, to which the name of Igusa is most commonly attached; for recent results along these lines, see for instance the paper [C] of R. Cluckers.

We will now concentrate (almost) exclusively on the case $k = 1$, i.e., on sums over finite fields. For many concrete applications, this turns out to be the most important, and this justifies somewhat this special attention, but we repeat that one should not immediately consider a problem solved when the relevant sums are understood in that case: sometimes, it is higher powers of primes which are most difficult. An example of this is given in the work of Belabas and Fouvry [BF, 3.c].

3 Weil's interpretation of exponential sums over finite fields

Kloosterman's argument to prove Proposition 2.2 for $c = p$ a prime number was elegant, but not immediately generalizable to more situations (it can be seen in [I, 4.4]). Around 1940, A. Weil, following hints of Davenport and Hasse in particular, realized that the *Riemann Hypothesis* for curves over finite fields led to a stronger result and a better understanding of the underlying problem: namely, using results on the number of points over finite fields of the curves with equations

$$y^p - y = mx + \frac{n}{x}, \tag{13.1}$$

Weil showed that

$$|S(m,n;p)| \leqslant 2\sqrt{p}$$

for all primes p and m, n coprime with p. In fact, this was based on combining almost trivially two remarkable properties. The first one, already showing the importance of viewing the problem algebraically, is the *rationality* of the generating function

$$Z_{m,n}(T) = \exp\Bigl(\sum_{\nu\geqslant 1} S_\nu(m,n)\frac{T^\nu}{\nu}\Bigr) \in \mathbf{C}[[T]]$$

formed with the analogues of the Kloosterman sum *over extension fields* $\mathbf{F}_{p^\nu}/\mathbf{F}_p$ – not to be mistaken with the finite rings $\mathbf{Z}/p^k\mathbf{Z}$ – defined by

$$S_\nu(m,n) = \sum_{x\in\mathbf{F}_{p^\nu}^\times} e\Bigl(\frac{\mathrm{Tr}_\nu(xm+n/x)}{p}\Bigr)$$

in terms of the trace map $\mathbf{F}_{p^\nu} \xrightarrow{\mathrm{Tr}_\nu} \mathbf{F}_p$. These sums do not (as far as the author knows!) occur naturally in problems of analytic number theory,[3] but the whole family $(S_\nu(m,n))$ really contains the key to the general theory. Indeed, the rationality property already mentioned takes the form of the formula

$$Z_{m,n}(T) = \frac{1}{1 - S_1(m,n)T + pT^2}$$

with (of course) $S_1(m,n) = S(m,n;p)$, the original Kloosterman sum. What this means, in particular, is that the size of $S_1(m,n)$ is dictated precisely by the poles of the generating function, which is not surprisingly called the zeta function associated with the Kloosterman sum. Precisely, factor the quadratic term as

$$1 - S_1(m,n)T + pT^2 = (1-\alpha_{m,n}T)(1-\beta_{m,n}T),$$

so that

$$|S_1(m,n)| = |\alpha_{m,n} + \beta_{m,n}| \leqslant |\alpha_{m,n}| + |\beta_{m,n}|. \tag{13.2}$$

[3] Of course, for a fixed ν, they may arise for problems over a number field with a finite residue field of order p^ν, but then extensions of the latter would not occur.

The next result, due to Weil, then explains his bound for Kloosterman sums: the two (inverse) roots $\alpha_{m,n}$, $\beta_{m,n}$, for m, n coprime with p, satisfy

$$|\alpha_{m,n}| = |\beta_{m,n}| = \sqrt{p}.$$

This is called the Riemann Hypothesis (for Kloosterman sums), because of the following interpretation: if we introduce a complex variable s and take $T = p^{-s}$, the resulting complex function $\zeta_{m,n}(s) = Z_{m,n}(p^{-s})$ is meromorphic and its only poles are situated on the line $\mathrm{Re}(s) = 1/2$.

Weil showed that this type of interpretation could be extended naturally (and beautifully) to all sums of algebraic origin in *one* variable, i.e., ranging over points of a curve $C/\mathbf{F}_p$, using suitable Artin-Shreier coverings of C and the corresponding zeta functions, and most importantly appealing to the Riemann Hypothesis for all curves over finite fields, which he proved during the 1940s. This leads, for instance, to the following estimate, which has been used in many contexts in analytic number theory:

Theorem 3.1 (Weil) *Let $P \in \mathbf{Z}[X]$ be a monic polynomial of degree d which no repeated factor. Let p be a prime such that P has no repeated root modulo p, i.e., such that p does not divide the discriminant of P. Then we have*

$$\Big| \sum_{x \,(\mathrm{mod}\, p)} e\Big(\frac{P(x)}{p}\Big)\Big| \leqslant (d-1)\sqrt{p}.$$

4 The yoga of Grothendieck-Deligne

Weil's theory, beautiful and revolutionary as it was, did not extend well to situations involving sums in more than one variable, or in other words, beyond the case where the underlying set of summation is (the set of $\mathbf{F}_p$-points of) a curve. Or, rather, it doesn't provide a way to analyze those sums except by "fibering by curves", which is simply to say writing

$$\Big|\sum_{x,y} e(f(x,y)/p)\Big| = \Big|\sum_{x}\Big(\sum_{y} e(f(x,y)/p)\Big)\Big| \leqslant \sum_{x}\Big|\sum_{y} e(f(x,y)/p)\Big|$$

and trying to understand individually the inner sums before putting them all together. This can often be done using Theorem 3.1, but note that if x, y range over $\mathbf{F}_p$, this means that – in terms of p – the best possible estimate one can then expect is of order $p^{3/2}$, as no cancellation can be obtained from the sum over x.

Such estimates, although non-trivial, are often not enough for the purposes of applications. Here is an example, with an equidistribution problem. Recall that, for a prime p and a multiplicative character $\chi : \mathbf{F}_p^\times \to \mathbf{C}^\times$ of the field $\mathbf{F}_p$, the associated Gauss sum (itself a type of exponential sum over finite field) is given by

$$\tau(\chi) = \sum_{x \in \mathbf{F}_p^\times} \chi(x) e\Big(\frac{x}{p}\Big).$$

For a non-trivial character $\chi \neq 1$, it is well known that $|\tau(\chi)| = \sqrt{p}$, which means one can write

$$\tau(\chi) = p^{1/2} e(\theta(\chi)), \qquad \text{with} \qquad \theta(\chi) \in [0,1].$$

Is is then an interesting question to understand how the angles of the Gauss sums $\theta(\chi)$ vary. Indeed, Gauss considered the case where $\chi = (\cdot/p)$ is a real-valued character (the Legendre symbol modulo p), and succeeded in proving after much effort that, for $p \geqslant 3$, we have

$$\tau((\cdot/p)) = 1, \text{ i.e., } \theta((\cdot/p)) = 0, \text{ if } p \equiv 1 \pmod 4,$$
$$\tau((\cdot/p)) = i, \text{ i.e., } \theta((\cdot/p)) = \frac{1}{4}, \text{ if } p \equiv 3 \pmod 4.$$

However, the angles seem intractable for most other characters,[4] and one may suspect that they are spread all over the unit circle. To test this, equidistribution theory suggests strongly to try to estimate the Weyl-type sums, which are

$$W_n = \frac{1}{p-2} \sum_{\substack{\chi \pmod p \\ \chi \neq 1}} e(n\theta(\chi))$$

for $n \in \mathbf{Z} - \{0\}$, where the sum is over non-trivial multiplicative characters modulo p (the number of which is $p-2$). For $n \geqslant 1$, this can be

[4] A famous conjecture of Kummer considered what happens when χ is of order 3; for this story, see, e.g., [HP].

expressed as

$$\begin{aligned} W_n &= \frac{1}{p-2} \sum_{\substack{\chi \,(\mathrm{mod}\, p) \\ \chi \neq 1}} \Big(\frac{\tau(\chi)}{\sqrt{p}}\Big)^n \\ &= \frac{1}{p^{n/2}(p-2)} \sum_{x_1,\ldots,x_n} e\Big(\frac{x_1+\cdots+x_n}{p}\Big) \sum_{\chi \neq 1} \chi(x_1 \cdots x_n) \\ &= \frac{p-1}{p-2} \frac{K_n(p)}{p^{n/2}}, \end{aligned}$$

where

$$K_n(p) = \sum_{\substack{x_1,\ldots,x_n \,(\mathrm{mod}\, p) \\ x_1 \cdots x_n = 1}} e\Big(\frac{x_1+\cdots+x_n}{p}\Big)$$

is an exponential sum with $n-1$ variables, called an *hyper-Kloosterman sum* (because, in the case $n=2$, we recover $K_2(p) = S(1,1;p)$).

If we sum over one variable using Weil's bound, we can easily show that $K_n(p) \ll p^{n-3/2}$ for all p, but note that such a bound does not even allow us to recover the trivial fact that $|W_n| \leqslant 1$ if $n \geqslant 4$! On the other hand, standard probabilistic considerations ("square-root cancellation philosophy", taking root in the fact that a random walk of length N with random, uniformly distributed phases, has modulus of size about $\sqrt{N}$ with overwhelmingly large probability) suggest that one should have

$$K_n(p) \ll p^{(n-1)/2},$$

or in other words, each of the $n-1$ variables should, independently of the others, gain a factor $\sqrt{p}$. This would lead to

$$W_n \ll \frac{1}{\sqrt{p}}$$

for $p \geqslant 3$ (and n fixed), and therefore[5] we would conclude that the angles of the Gauss sums $\tau(\chi)$, where χ ranges over all non-trivial characters, become equidistributed, as $p \to +\infty$, on the unit circle.

This fact could only be proved (by Deligne [D2]) after Grothendieck and his school of algebraic geometry had developed a new framework to study algebraic exponential sums, which goes much further than the one of Weil in many respects: not only does it encompass sums in arbitrarily many variables, but also it is formally much more flexible, and makes it

[5] For the application of the Weyl criterion to $n < 0$, one can use an obvious symmetry.

possible to study and exploit variations of exponential sums in families, and indeed to analyze certain types of sums which do not look like the standard ones $S(f,V,p)$ we have mentioned earlier. However, because of considerations of space, we will only sketch how this formalism applies to exponential sums of this type, pointing (as an introduction) to the book of Katz [K2] for particularly striking illustrations of the more general sums that can naturally be considered, in the setting of the distribution of angles related to Kloosterman sums (see also another survey of the author [Ko2] for some motivation and discussion of Deligne's Equidistribution Theorem, which is an important part of this type of issues).

To present a fairly general case, we consider an affine scheme of finite type $V/\mathbf{Z}$, and two functions f, g on V, with g invertible. Then, for any prime number p, any additive character $\psi : \mathbf{F}_p \to \mathbf{C}^\times$ and any multiplicative character $\chi : \mathbf{F}_p^\times \to \mathbf{C}^\times$, we look at the sum

$$S = S(V,f,g,\chi,\psi;p) = \sum_{x\in V(\mathbf{F}_p)} \psi(f(x))\chi(g(x)).$$

The case of hyper-Kloosterman sums corresponds to the affine subvariety $V \subset \mathbf{A}^n$ defined by the equation $x_1\cdots x_n = 1$, to the additive character $x \mapsto e(x/p)$ (note that all additive characters are of the form $x \mapsto e(ax/p)$ for some a modulo p) and to $g = 1$, $\chi = 1$.

The analysis of such sums in the Grothendieck framework (see [D1] for a more detailed presentation, [IK, 11.11] for another survey tailored to analytic number theorists) starts by choosing, for a given prime p, another prime $\ell \neq p$. Then the formalism of the so-called *Lang torsor* (which is, concretely, a fairly systematic analysis of Artin-Schreier coverings (13.1) and of Kummer coverings for the multiplicative part) gives an object, called an ℓ-adic (lisse) sheaf of rank 1, $\mathscr{L} = \mathscr{L}_{\psi(f)} \otimes \mathscr{L}_{\chi(g)}$, depending on all the data, which can be seen as associating (in particular) to every rational point $x \in V(\mathbf{F}_p)$ a one-dimensional $\bar{\mathbf{Q}}_\ell$-vector space (the "stalk" $\mathscr{L}_x$ of $\mathscr{L}$ at x) together with an action of the Frobenius automorphism, which is the natural generator of the Galois group of $\mathbf{F}_p$, in such a way that the basic formula

$$\mathrm{Tr}(\mathrm{Fr}_x \mid \mathscr{L}_x) = \psi(f(x))\chi(g(x))$$

holds, for Fr_x the inverse of $x \mapsto x^p$ (the geometric Frobenius; up to changing f by $-f$ and g by g^{-1}, we could use the standard Frobenius as well). Of course, since the stalk is one-dimensional, speaking of the trace is somewhat pedantic, but the generalizations briefly mentioned earlier will involve similar sheaves such that $\mathscr{L}_x$ is a vector space of higher dimension.

Hence the exponential sums take the form

$$S = \sum_{x \in V(\mathbf{F}_p)} \operatorname{Tr}(\mathrm{Fr}_x \mid \mathscr{L}_x),$$

and the next basic steps will work for sums defined in this way for an arbitrary lisse ℓ-adic sheaf on $V/\mathbf{F}_p$.

The first transformation which is done is the analogue of the application of the rationality of the zeta function for Kloosterman sums (and it can be interpreted in this manner, although this would be anachronistic, since the rationality, in general, is proved exactly in this way): the *trace formula* of Grothendieck, a deep analogue of a formula of Lefschetz in classical algebraic topology, states that

$$\sum_{x \in V(\mathbf{F}_p)} \operatorname{Tr}(\mathrm{Fr}_x \mid \mathscr{L}_x) = \sum_{i=0}^{2d} (-1)^i \operatorname{Tr}(F \mid H^i_c(V_{\bar{\mathbf{F}}_p}, \mathscr{L})),$$

where the sum now runs only over integers up to $2d$, with d the dimension of $V/\mathbf{F}_p$, and F denotes the global action of the geometric Frobenius automorphism on the various ℓ-adic cohomology spaces with compact support of V, base-changed to an algebraic closure of $\mathbf{F}_p$.

Example 4.1 The trace formula is already interesting for the "trivial" sheaf $\bar{\mathbf{Q}}_\ell$ itself, for which all local traces are equal to 1, so that the associated sum is

$$\sum_{x \in V(\mathbf{F}_p)} \operatorname{Tr}(\mathrm{Fr}_x \mid \bar{\mathbf{Q}}_\ell) = |V(\mathbf{F}_p)|,$$

which is the number of points on V over $\mathbf{F}_p$. Indeed, the original Weil conjectures, which motivated the general theory (and in fact, much of the development of modern algebraic geometry, see the summary in [Ha, Appendix C]) concerned precisely this case. The trace formula states that

$$|V(\mathbf{F}_p)| = \sum_{i=0}^{2d} (-1)^i \operatorname{Tr}(F \mid H^i_c(V_{\bar{\mathbf{F}}_p}, \mathbf{Q}_\ell))$$

which is highly non-trivial in all but the simplest case.

One can see these formulas as black boxes, but of course this may become somewhat unsatisfactory. As a baby step towards enlightenment, let us consider one of the very few elementary situations where the formula becomes transparent. Consider the case where V is 0-dimensional, given by the equation $f(x) = 0$, for some monic polynomial $f \in \mathbf{Z}[X]$ of degree $\deg(f) \geqslant 1$. Then, for any prime p, $V/\mathbf{F}_p$ is zero-dimensional and $V(\mathbf{F}_p)$ is the number of zeros of f in the base field $\mathbf{F}_p$. Since $d = 0$, the trace formula gives

$$|\{x \pmod p \mid f(x) = 0\}| = \mathrm{Tr}(F \mid H_c^0(V_{\bar{\mathbf{F}}_p}, \mathbf{Q}_\ell)).$$

But what is the 0-th cohomology space? Since $V_{\bar{\mathbf{F}}_p}$ is simply the finite collection of the zeros of f in $\bar{\mathbf{F}}_p$, the intuition from topology (which can be confirmed by the barest introduction to the definition of étale cohomology) states that $H_c^0(V_{\bar{\mathbf{F}}_p}, \bar{\mathbf{Q}}_\ell)$ should be isomorphic to $\bar{\mathbf{Q}}_\ell^\delta$, where δ is the number of distinct zeros of f (which was not assumed to be squarefree, so repeated roots are possible). And how should the global Frobenius act? It seems obvious (and again is confirmed easily) that its action on $\bar{\mathbf{Q}}_\ell^\delta$ is simply obtained from the permutation action of $x \mapsto x^p$ (or rather its inverse) on the zeros of f. In other words, the matrix representing this action is the permutation matrix associated with this permutation of the zeros of f. What is its trace? It is, as is well-known, the number of fixed points of the permutation, and this is precisely the number of zeros in $\mathbf{F}_p$.

This example illustrates also one property which is important and may seem doubtful at first: the trace formula works equally well for non-reduced schemes (e.g., if there are repeated roots) as for reduced ones. This is useful for applications, where checking that $V/\mathbf{F}_p$ is reduced might be quite bothersome.

As we leave this example, note that – even in this very simple case – the variation of $|V(\mathbf{F}_p)|$ with p is by no means an easy question!

Coming back to the general application of the trace formula, we note that it does not yet lead to any non-trivial estimate, because it might be that the traces on the various cohomology groups are enormous. In fact, it is not even clear (and it is open in general!) that the various terms on the right-hand side are independent of the choice of the auxiliary prime $\ell \neq p$. The eigenvalues may also conceivably be elements of $\bar{\mathbf{Q}}_\ell$ which are not algebraic over $\mathbf{Q}$.

However, the extraordinary general Riemann Hypothesis for varieties and sheaves over finite fields, proved by Deligne [D3], leads to quite precise information concerning those eigenvalues, from which non-trivial estimates may often be deduced.

In the context we consider, the result is the following: (1) any eigenvalue α of the Frobenius acting on a cohomology space $H_c^i(V_{\bar{\mathbf{F}}_p}, \mathscr{L})$, for

$\mathscr{L}$ of the type described, is an algebraic integer; (2) any such α has the property that if $\beta \in \mathbf{C}$ is an arbitrary Galois conjugate of α (e.g., $\beta = \alpha$ if $\bar{\mathbf{Q}}$ is identified with a subfield of $\mathbf{C}$), we have

$$|\beta| = p^{j/2}$$

where the *weight* $j = j(\alpha)$ depends only on α and satisfies $0 \leqslant j \leqslant i$.

It is this last estimate of the weight which "is" the Riemann Hypothesis. To see why, consider the example of Kloosterman sums $S(1,1;p)$: since the variety V is the multiplicative group, of dimension 1, the yoga leads to

$$S(1,1;p) = \sum_{i=0}^{2} (-1)^i \operatorname{Tr}(F \mid H_c^i(\mathbf{G}_{m,\bar{\mathbf{F}}_p}, \mathscr{K})),$$

for some suitable sheaf, and the result of Deligne says that the eigenvalues of F on H_c^0 are of modulus $\leqslant 1$, those on H_c^1 are of modulus $\leqslant \sqrt{p}$, and those on H_c^2 of modulus $\leqslant p$. Comparing with Weil's expression, one can guess (and it is true) that in fact $H_c^0 = H_c^2 = 0$ here, and H_c^1 is 2-dimensional with eigenvalues of Frobenius given by the algebraic integers $\alpha_{1,1}$ and $\beta_{1,1}$ occurring in (13.2).

Here is a more general explanation of the relevance of Deligne's result. We expect that the number of points in the sum over $x \in V(\mathbf{F}_p)$ is roughly of size p^d (in simple cases, such as hyper-Kloosterman sums $K_n(p)$, V is so simple that this is obvious; there, $\dim V = n - 1$ and $|V(\mathbf{F}_p)| = (p-1)^{n-1}$). Deligne's bound shows that only the topmost cohomology group H_c^{2d} may have eigenvalues as large as $p^d = p^{2d/2}$. If we let $\beta_i(\mathscr{L}) = \dim H_c^i$, we get

$$|S| = |S(V, f, g, \chi, \psi; p)| \leqslant \beta_{2d} p^d + R, \qquad |R| \leqslant q^{\kappa/2} B$$

(after choosing a fixed embedding of $\bar{\mathbf{Q}}$ in $\mathbf{C}$), where

$$\kappa = \max\{j \leqslant 2d - 1 \mid \beta_j \neq 0\} \leqslant 2d - 1,$$
$$B = B(p) = \sum_{0 \leqslant i \leqslant \kappa} \dim H_c^i(V_{\bar{\mathbf{F}}_p}, \mathscr{L}).$$

Thus *getting some non-trivial estimate* depends on showing that the topmost group has dimension $\beta_{2d} = 0$. This, it turns out, is often the case, because Poincaré duality can be used to relate H_c^{2d} to some H^0, which can be analyzed fairly simply. Here is a special case: for a fixed $V/\mathbf{Z}$, and for any prime p large enough, if $V/\mathbf{F}_p$ is geometrically connected, we have $\beta_{2d} = 0$ *unless* the function f is constant on $V(\mathbf{F}_p)$. In that case (if $\chi = 1$ at least), it is clear that we can not expect cancellation, since we are just counting the number of points of summation. In particular, this condition is trivially true for hyper-Kloosterman sums, and thus one gets "for free" that

$$|K_n(p)| \leqslant Bp^{n-3/2}. \tag{13.1}$$

Strangely enough, in terms of exponential sums, this remains a trivial estimate, because B has not been estimated explicitly,[6] and it depends on p because the sheaves $\mathscr{L}$ that were introduced to represent the exponential sums themselves depend on p (through the additive and multiplicative characters). Thus the following theorem is crucial (see [K1] for a very general statement):

Theorem 4.2 (Dwork, Bombieri, Adolphson-Sperber) *Let V, p, χ, ψ, $f \in \mathbf{Z}[V]$, $g \in \mathbf{Z}[V]^\times$ be as above. Then there exists a constant B_0, independent of p, such that*

$$\sum_{0 \leqslant i \leqslant 2d} \dim H^i_c(V_{\bar{\mathbf{F}}_p}, \mathscr{L}) \leqslant B_0$$

for all p.

In fact one can often write down a concrete value for B_0.

Let us come back, to conclude this section, to the special case of hyper-Kloosterman sums. The bound (13.1), reinforced by the bound $B(p) \leqslant B_0$, is no better than the one coming from Weil's bound for one-dimensional sums and "fibering". This is because it is based on assuming (a worst case scenario) that $\kappa = 2(n-1) - 1$, i.e., that the cohomology space "topmost minus one" is non-zero, namely that

$$H_c^{2(n-1)-1} \neq 0.$$

This *would be the case* if the hyper-Kloosterman sums were replaced, indeed, by sums where only one variable is involved non-trivially, such as

$$\sum_{\substack{x_1,\ldots,x_n \\ x_1\cdots x_n=1}} e\Big(\frac{x_1^2}{p}\Big)$$

where nothing may be gained from the contribution of the $n-2$ variables $x_2, \ldots, x_n$. Of course, the actual hyper-Kloosterman sums do not look like that, but it is quite a bit more delicate to prove that, in fact, the

[6] The corresponding estimate for the sums ranging over finite extensions $\mathbf{F}_{p^\nu}$ would be non-trivial.

only possible contribution to the cohomology comes from the "middle" dimension $i-1$. This is a result of Deligne:

Theorem 4.3 (Deligne) *For the variety $V \subset \mathbf{A}^n$ with equation $X_1 \cdots X_n = 1$ and the sheaf $\mathscr{K}_n$ corresponding to hyper-Kloosterman sums, we have*

$$H_c^i(V_{\bar{\mathbf{F}}_p}, \mathscr{K}_n) = 0 \text{ if } i \neq n-1,$$
$$\dim H_c^{n-1}(V_{\bar{\mathbf{F}}_p}, \mathscr{K}_n) = n.$$

As a corollary, the very precise bound

$$|K_n(p)| \leqslant np^{(n-1)/2}$$

follows from Deligne's analysis, hence also the equidistribution of angles of Gauss sums!

The proof of this theorem is quite intricate, but note that, because of the base change, this is a *geometric* statement, not so much a theorem of arithmetic anymore. This, in fact, explains partly why the method is so successful: it isolates *geometric* reasons for the smallness of the exponential sums, and these reasons may be accessible to arguments (e.g., deformation, "continuity") which are not available or visible from the purely arithmetic viewpoint.

A more direct illustration of this is found in the next theorem, also due to Deligne:

Theorem 4.4 (Deligne) *Let $\mathbf{F}_q$ be a finite field, $P \in \mathbf{F}_q[X_1, \ldots, X_n]$ a homogeneous polynomial of degree d. Assume that the homogeneous part of degree d of P has the property that its zero set in the projective space of dimension $d-1$ (over $\bar{\mathbf{F}}_q$) is a smooth hypersurface. Then we have*

$$\Big| \sum_{x \in \mathbf{F}_q^n} e\Big(\frac{P(x)}{p}\Big) \Big| \leqslant (d-1)^n q^{n/2}.$$

Notice that this generalizes in a powerful way Weil's Theorem 3.1 to arbitrarily many dimensions (the condition on P, though natural, is not necessary for the estimate to hold). Deligne's proof, which is quite instructive, can be sketched very roughly as follows: the goal is to prove that the relevant cohomology groups (say of $\mathscr{L}_P$) satisfy

$$H_c^i(V_{\bar{\mathbf{F}}_p}, \mathscr{L}_P) = 0 \text{ if } i \neq n, \ \dim H_c^n(V_{\bar{\mathbf{F}}_p}, \mathscr{L}_P) = (d-1)^n.$$

However, when P varies among all the "Deligne" polynomials of degree d (those satisfying the condition concerning the zero set of the part of degree d), it turns out that the dimensions of the cohomology groups remain constant (this is a type of "smoothness" having to do with the fact that the Deligne polynomials are themselves parameterized by a nice affine algebraic variety), and hence it is enough to check the desired property for a *single* well-chosen P. This can be done for instance for

$$P = X_1^d + \cdots + X_n^d,$$

for which the exponential sums factors

$$\sum_{x\in\mathbf{F}_q^n} e\Big(\frac{P(x)}{p}\Big) = \Big(\sum_{x\in\mathbf{F}_q} e\Big(\frac{x^d}{p}\Big)\Big)^n$$

and (reverting the link between estimates and dimensions, as can be done) one obtains the required statement, e.g. from Weil's estimate (although these special one-variable sums, which are variants of Gauss sums, can also be estimated more directly).

Remark 4.5 The richness of the formalism of Grothendieck-Deligne is such that, quite often, it is possible to recover comparably precise estimates without requiring geometric analysis as detailed as that of Theorem 4.3 – in other words, by combining some extra arithmetic information that may well be available, for a given problem, with the geometric structure. For hyper-Kloosterman sums, this was done very cleverly by Bombieri; see [IK, 11.11, Example 2] for his argument based on mean-square averages of a family of hyper-Kloosterman sums and Galois-conjugacy.

5 Exponential sums over definable sets

We come back to the motivations from analytic number theory. Quite frequently, a natural problem is to estimate sums which are obtained from an exponential sum of algebraic origin by *shortening the range of summation*: for instance, estimates of

$$\sum_{n\leqslant p^\delta} \chi(p), \tag{13.1}$$

where χ is a multiplicative character modulo p and $0 < \delta < 1$ are of considerable importance in the theory of Dirichlet L-functions. This set is not the set of points of an algebraic variety,[7] but one may wonder

[7] In a reasonable way, at least: of course, any finite set is the set of zeros of suitable polynomials.

if it is possible to extend the language used to define algebraic sets to allow for a richer spectrum of possibilities that might contain these "short intervals". This, and the encounter with the paper [CDM], led the author (see [Kol]) to try to look at exponential sums over *definable sets* over finite fields. These are probably the simplest generalizations of algebraic varieties: they amount to replacing the set of points $V(\mathbf{F}_p)$ with a definable set $\varphi(\mathbf{F}_p)$ associated with a formula in the *first order language of rings*, i.e., the set of $x \in \mathbf{F}_p$ for which the formula is satisfied.

The case of algebraic varieties corresponds to positive formulas without quantifiers: if $V/\mathbf{F}_p$ is affine, embedded in $\mathbf{A}^n$, we have

$$x = (x_1, \ldots, x_n) \in V(A) \Leftrightarrow f_1(x) = \ldots = f_m(x) = 0$$

for any $\mathbf{F}_p$-algebra A, where $(f_1, \ldots, f_m)$ are generators of the ideal in $\mathbf{F}_p[X_1, \ldots, X_n]$ defined by V (i.e., those are possible equations defining V).

But, over finite fields at least, definable sets are more general. As a very simple example, consider the formula φ given by

$$\varphi(x) \, : \, \exists \, y, \; x = y^2,$$

so that $\varphi(A)$, for any ring A, is the set of squares in A (recall that all quantifiers are implicitly running over the ring A for which the formula is "evaluated"). In particular, note that

$$|\varphi(\mathbf{F}_p)| = \frac{p+1}{2}, \qquad \text{if } p \text{ is odd,}$$

and there is no subvariety $V/\mathbf{Z} \subset \mathbf{A}^1/\mathbf{Z}$ with this number of $\mathbf{F}_p$-points for infinitely many primes.

Example 5.1 There are other examples of definable sets which are quite a bit more refined. Here are two types of important examples.

(1) Let $f(X, Y) \in \mathbf{Z}[X, Y]$ be a non-constant irreducible polynomial in two variables. There is then a formula $\varphi_f(x)$ such that, for any A, we have

$$\varphi_f(A) = \{x \in A \mid \text{the polynomial } f(x, Y) \in A[Y] \text{ is irreducible}\}$$

(these sets occur in the context of the Hilbert Irreducibility Theorem, and are often called *Hilbert sets*). Even more is true here: if we let $(a_{i,j})_{i+j \leqslant d}$ denote variables representing the coefficients of a polynomial $f \in \mathbf{Z}[X, Y]$ of degree $\leqslant d$, there is a single formula $\varphi_d(x, a_{i,j})$ with parameters $(a_{i,j})$ such that

$$\varphi_d(A, a_{i,j}) = \{x \in A \mid \text{the polynomial } \sum_{i,j} a_{i,j} x^i Y^j \in A[Y] \text{ is irreducible}\}$$

for any choice of parameters $(a_{i,j})$. In other words, even the variation of the Hilbert sets $\varphi_f(A)$ with f is "definable", and this is an important property which can be crucial in applications. In algebraic geometry, the analogue variation would be that of the fibers $V_y = \pi^{-1}(y)$ for a morphism $V \xrightarrow{\pi} W$, where W is seen as the space of parameters.

Concretely, assume $d = 2$ (the simplest case). Then the parameters are $(a, b, c, d, e, f) = (a_{0,0}, a_{1,0}, a_{0,1}, a_{1,1}, a_{2,0}, a_{0,2})$ for the polynomials

$$a + bX + cY + dXY + eX^2 + fY^2 \in A[X, Y],$$

and since a polynomial of degree 2 is irreducible if and only it has no zero, we have

$$\varphi_2(x, a, b, c, d, e, f) \, : \, \forall \; y, \; a + bx + ex^2 + (c + dx)y + fy^2 \neq 0.$$

The nature of the sets $\varphi_2(\mathbf{F}_p, a, b, c, d, e, f)$ is not clear a priori, and even less so when the number of variables grows.

(2) Let $\mathbf{G}/\mathbf{F}_p$ be a linear algebraic group defined over $\mathbf{F}_p$, for instance $\mathbf{G} = GL(n)$ or $SL(n)$, or a symplectic group. After embedding $\mathbf{G}$ in a suitable affine space $\mathbf{A}^m$, we have coordinates $x = (x_1, \ldots, x_m)$ for $\mathbf{G}$, and $\mathbf{G}$ is defined by finitely many equations in terms of these, while the product and inverse maps of $\mathbf{G}$ are also given by polynomials with coefficients in $\mathbf{F}_p$. Then, we can see for instance that the conjugacy classes C_g are definable: if $\psi_{\mathbf{G}}(x_1, \ldots, x_m)$ is a formula defining $\mathbf{G}$ in the affine space, then

$$\varphi_{conj}(x, g) \, : \, \psi_{\mathbf{G}}(g) \wedge \psi_{\mathbf{G}}(x) \wedge (\exists \; z, \; \psi_{\mathbf{G}}(z) \wedge x = zgz^{-1})$$

with variables $x = (x_1, \ldots, x_m)$ and parameters $g = (g_1, \ldots, g_m)$ is such that $\varphi_{conj}(A, g)$ is naturally identified with the conjugacy class in $\mathbf{G}(A)$ of $g \in \mathbf{G}(A)$. In general, of course, such conjugacy classes are not the set of points of an algebraic variety.

Chatzidakis, van den Dries and Macintyre studied such definable sets (with parameters) in [CDM], in the case of finite fields, and established a beautiful result concerning the possible number of points. (See also [FHJ]).

Theorem 5.2 (Chatzidakis, van den Dries, Macintyre) *Let $\varphi(x, y)$ be a formula in the language of rings with n variables $x = (x_1, \ldots, x_n)$ and*

m parameters $y = (y_1, \ldots, y_m)$. There exist a set D of finitely many pairs $(\delta, \mu) \in \mathbf{Q}^+ \times \mathbf{N}$ and a constant C depending only on $\varphi(x, y)$, with the following properties:

(1) *For any finite field $\mathbf{F}_q$ with q elements, and $y \in \mathbf{F}_q^m$ such that $\varphi(\mathbf{F}_q, y) \neq \emptyset$, there exist $(\delta, \mu) \in D$ for which*

$$\Big| |\varphi(\mathbf{F}_q, y)| - \delta q^{\mu} \Big| \leqslant C q^{\mu - 1/2}.$$

(2) *For any $(\delta, \mu) \in D$, there exists a formula $\mathscr{C}_{d,\mu}(y)$ with m parameters in the language of rings such that* (1) *holds for $\mathbf{F}_q$ and $y \in \mathbf{F}_q^m$ if and only if $y \in \mathscr{C}_{d,\mu}(\mathbf{F}_q)$.*

So, intuitively, the number of points $|\varphi(\mathbf{F}_p, y)|$ always looks like δq^{μ} for some rational "density" δ and some integral "dimension" μ, and although those may vary with y (and thus with p), there are only finitely many possibilities for a given φ – a type of "tameness" property of the variation of definable sets –, and moreover, the sets of parameters for which the density and dimension are fixed are themselves definable.

Some remarkable applications of this result in group theory where found by Hrushovski and Pillay in [HP], including a new proof of a difficult theorem of Mathews-Vaserstein-Weisfeiler on Strong Approximation for algebraic groups over $\mathbf{Z}$, which has many important applications in analytic number theory, in particular in applications of sieve methods.

It was fairly natural to try to extend this to results concerning exponential sums over definable sets, with the hope that these could have number-theoretic applications. More precisely (restricting to additive sums only for simplicity), given a formula $\varphi(x, y)$, a polynomial $f \in \mathbf{Z}[X]$, we can define a family (indexed by y) of exponential sums

$$S_{\varphi}(f, y) = S_{\varphi}(\mathbf{F}_q, f, \psi, y) = \sum_{x \in \varphi(\mathbf{F}_q, y)} \psi(f(x))$$

where ψ is an additive character of $\mathbf{F}_q$. These sums, to an analytic number theorist, are even worth investigating independently of possible applications, and this study was begun in [Ko1].

It turns out that the proof of Theorem 5.2 can be combined with the procedure of Section 4, with some care. This leads to the following statement (see [Ko1, Th. 13], which is a bit more general and more precise).

Theorem 5.3 *Let $\varphi(x, y)$ be a formula in the language of rings with n variables $x = (x_1, \ldots, x_n)$ and m parameters $y = (y_1, \ldots, y_m)$, and let $f \in \mathbf{Z}[X]$. Let D be the set given by Theorem 5.2 for $\varphi(x, y)$. There* *exists $p_0 \geqslant 1$, a constant $\eta > 0$, and $B \geqslant 0$, depending only on φ and the degree of f, such that for $p \geqslant p_0$ and $y \in \mathbf{F}_q^m$, we have*

$$|S_\varphi(f,y)| \leqslant \frac{BS_\varphi(1,y)}{\sqrt{p}} = \frac{B|\varphi(\mathbf{F}_p,y)|}{\sqrt{p}},$$

unless there exists $c \in \mathbf{F}_p$ with

$$|\{x \in \varphi(\mathbf{F}_p,y) \mid f(x) = c\}| \geqslant \eta S_\varphi(1,y) = \eta|\varphi(\mathbf{F}_p,y)|.$$

This statement is not entirely satisfactory, in that by itself it does not provide a way to understand when (or if) there is more cancellation than the $p^{-1/2}$ gain stated, which is comparable with the outcome of the Weil and fibering method. The condition which is excluded to obtain cancellation is easily understood: it means that f is constant for a positive proportion of the points of summation. If that is the case, any compensation in the sum (there may not be any, e.g., if $f = 1$) will have entirely different origins than what we expect generically. (This happens already for sums over varieties V if $V_{\bar{\mathbf{F}}_p}$ is not geometrically connected, where f might be constant on the geometric components, and the sum of these values might lead to cancellation).

From the proof of this theorem, one can extract some geometric information to help investigate further cancellation, which could, in principle, sometimes lead to stronger bounds. It would be highly interesting to find cases where this can be done, beyond cases involving only algebraic varieties, but the geometry is not as easily understood as that case (which may already be quite involved!). In particular, note that there may be little relation between $\varphi(\bar{\mathbf{F}}_p)$ and $\varphi(\mathbf{F}_p)$, i.e., no analogue of the fact that $\mathbf{F}_p$-rational points on V are fixed point of the Frobenius among the $\bar{\mathbf{F}}_p$-points. This is already clear for the set of squares in $\mathbf{F}_p$, which is non-trivial, but which of course is the whole line geometrically.

On the positive side, at least in the case of one-variable sums, the result is essentially optimal. So although we have not yet been able to find convincing purely arithmetic applications, we can prove the following statement (see [Ko1, Remark 19]), where equidistribution reappears:

Corollary 5.4 *Let $\varphi(x)$ be a formula in one variable in the first order language of rings. Let p run over any increasing sequence of primes, if it exists, such that $|\varphi(\mathbf{F}_p)| \to +\infty$ as p grows. Then the finite families $(\frac{x}{p})_{x\in\varphi(\mathbf{F}_p)}$, where $\{\frac{x}{p}\}$ denotes the fractional part, become equidistributed in $\mathbf{R}/\mathbf{Z}$, with respect to Lebesgue measure, as $p \to +\infty$.*

Indeed, we simply apply the Weyl Criterion: for $h \neq 0$, we need to show that

$$\frac{1}{|\varphi(\mathbf{F}_p)|} \sum_{x \in \varphi(\mathbf{F}_p)} e\Big(\frac{hx}{p}\Big)$$

converges to 0 as p grows. This is an exponential sum over the definable set $\varphi(\mathbf{F}_p)$, hence by Theorem 5.3, we can find constants $B \geqslant 0$ and $\eta > 0$ for which

$$\sum_{x \in \varphi(\mathbf{F}_p)} e\Big(\frac{hx}{p}\Big) \leqslant B|\varphi(\mathbf{F}_p)|p^{-1/2},$$

unless the function $x \mapsto hx$ is constant for at least $\eta|\varphi(\mathbf{F}_p)|$ values of $x \in \varphi(\mathbf{F}_p)$. However, the latter is clearly impossible if $|\varphi(\mathbf{F}_p)| \to +\infty$.

In particular, this corollary means that it is not possible to find a formula $\varphi(x)$ such that, for infinitely many primes p, we have

$$\varphi(\mathbf{F}_p) = \{x \pmod p \mid p/2 \leqslant x \leqslant 3p/2\}.$$

Note that this could not be derived directly from the point counting of Theorem 5.2, since the right-hand side forms a set which always has a density (namely, 1/2) and a dimension (namely, 1) in the sense of that theorem.

6 Questions

We conclude with a few questions which, possibly, could be the occasion of useful meetings of the minds between analytic number theory, algebraic geometry and model-theoretic ideas.

– We have seen that for applications to number theory, when p varies, Theorem 4.2 is crucial to the efficiency of the Grothendieck-Deligne approach to algebraic exponential sums. Except in the case where the additive character is trivial, its proof is not entirely satisfactory: one would hope to derive it as a type of "uniformity in parameters", flowing naturally from the fact that the sums over finite fields come from an object "defined over $\mathbf{Z}$". However, algebraic geometry does not provide such a global object for additive character sums, because the Artin-Schreier coverings (13.1) depend on p. Can one use model theory, possibly using a richer language than that of rings, to develop such a theory of "exponential sums over $\mathbf{Z}$"? (See the paper [K3] of Katz for some speculations on this hypothetical theory).

– In some cases, "elementary" methods give better results than the algebraic methods (either Weil's, or those of Grothendieck-Deligne): consider for instance sums like

$$S_{m,p} = \sum_{x \,(\mathrm{mod}\, p)} e\Big(\frac{x^m + x}{p}\Big)$$

where now m is not bounded, but increases with p. By Theorem 3.1, we get

$$|S_{m,p}| \leqslant m\sqrt{p}$$

(this is a one-variable sum, and in the cohomological framework, we have $H_c^0 = H_c^2 = 0$ and $\dim H_c^1 = d - 1$, so the method of Section 4 give the same result). Thus, if $m > \sqrt{p}$, this result is *worse* than the trivial bound $|S_{m,p}| \leqslant p$. Analytic heuristics, however, still suggest that $S_{m,p}$ should be "small" for larger values of m, provided certain conditions hold. Such results are indeed known (see, e.g., [B]), but the methods are completely different (based on additive combinatorics). Can progress be made on understanding this type of examples using algebraic geometry or model theory?

– Are there applications of Theorem 5.3 to model theory or algebraic geometry, maybe in the spirit of the work of Hrushovski and Pillay [HP]? This is also suggested by the work of Tomašić [T], but the author lacks competence to suggest any plausible track to follow...

– And the most important, maybe, for analytic number theory: is there an approach to more general exponential sums (such as those over short intervals (13.1)) involving a formalism as efficient (or comparable!) to the Grothendieck-Deligne approach?

References

[BF] K. Belabas and É. Fouvry: *Sur le 3-rang des corps quadratiques de discriminant premier ou presque premier*, Duke Math. J. 98 (1999), no. 2, 217–268.

[B] J. Bourgain: *Mordell's exponential sum estimate revisited*, Journal A.M.S 18 (2005), 477–499.

[CDM] Z. Chatzidakis, L. van den Dries and A. Macintyre: *Definable sets over finite fields*, J. reine angew. Math. 427 (1992), 107–135

[C] R. Cluckers: *Igusa and Denef-Sperber conjectures on nondegenerate p-adic exponential sums*, Duke Math. J. 141 (2008), no. 1, 205–216.

[D1] P. Deligne: *Cohomologie Étale*, S.G.A $4\frac{1}{2}$, L.N.M 569, Springer Verlag (1977).

[D2] P. Deligne: *La conjecture de Weil : I*, Publ. Math. IHÉS 43 (1974), 273–307

[D3] P. Deligne: *La conjecture de Weil, II*, Publ. Math. IHÉS 52 (1980), 137–252.

[FHJ] M. Fried, D. Haran and M. Jarden: *Effective counting of the points of definable sets over finite fields*, Israel J. of Math. 85 (1994), 103–133.

[HW] G.H. Hardy and E.M. Wright: *An introduction to the theory of numbers*, 5th ed., Oxford Univ. Press, 1979.

[Ha] R. Hartshorne: *Algebraic geometry*, Grad. Texts in Math. 52, Springer-Verlag (1977).

[HP] D.R. Heath-Brown and S.J. Patterson: *The distribution of Kummer sums at prime arguments*, J. reine angew. Math. 310 (1979), 111–130.

[HP] E. Hrushovski and A. Pillay: *Definable subgroups of algebraic groups over finite fields*, J. reine angew. Math 462 (1995), 69–91.

[I] H. Iwaniec: *Topics in classical automorphic forms*, Grad. Studies in Math. 17, A.M.S (1997).

[IK] H. Iwaniec and E. Kowalski: *Analytic Number Theory*, A.M.S Colloq. Publ. 53, A.M.S (2004).

[K1] N. Katz: *Sums of Betti numbers in arbitrary characteristic*, Finite Fields Appl. 7 (2001), no. 1, 29–44.

[K2] N. Katz: *Gauss sums, Kloosterman sums and monodromy*, Annals of Math. Studies, 116, Princeton Univ. Press, 1988.

[K3] N. Katz: *Exponential sums over finite fields and differential equations over the complex numbers: some interactions*, Bull. A.M.S 23 (1990), 269–309.

[Ko1] E. Kowalski: *Exponential sums over definable subsets of finite fields*, Israel J. Math. 160 (2007), 219–251.

[Ko2] E. Kowalski: *Some aspects and applications of the Riemann Hypothesis over finite fields*, Milan J. of Mathematics 78(2010), 179–220.

[T] I. Tomašić: *Exponential sums in pseudofinite fields and applications*, Illinois J. Math. 48 (2004), no. 4, 1235–1257.

14

A motivic version of p-adic integration

Karl Rökaeus

1 Motivic measures

I will discuss how one can specialize motivic integration to p-adic integration via point counting. We use $\mathscr{M}_k$ to denote the Grothendieck ring of k-varieties localized at $\mathbb{L}$, the class of the affine line.

Let $\mathscr{X}$ be a scheme of finite type, defined over a complete discrete valuation ring with perfect residue field k. In the theory of geometric motivic integration, as developed by Denef and Loeser [DL02] and Looijenga [Loo02] in the case when the discrete valuation ring is of equal characteristic zero, and by Sebag [Seb04] in the mixed and positive characteristic cases, one defines the space of arcs on $\mathscr{X}$, which we will denote $\mathscr{X}_\infty$. For our purposes, this is just a set, and in the case when the discrete valuation ring is absolutely unramified of mixed characteristic, which is the case we are mainly interested in, it equals $\mathscr{X}(\mathbf{W}(\overline{k}))$, where $\mathbf{W}$ is the ring scheme of Witt vectors and $\overline{k}$ is an algebraic closure of k. For every positive integer n there is a k-variety $\mathscr{X}_n$, the *Greenberg variety* [Gre61], with the property that, in the mixed characteristic case, $\mathscr{X}_n(K) = \mathscr{X}(\mathbf{W}_n(K))$ for every k-algebra K. The arc space is the projective limit of the set of $\overline{k}$-points of these.

The subsets of $\mathscr{X}_\infty$ that (in a certain sense) are defined by a finite truncation of arcs are called *stable*. The collection of stable subsets is a Boolean algebra; on this algebra one defines an additive measure, $\tilde{\mu}_{\mathscr{X}}$, taking valuses in $\mathscr{M}_k$. When we work over $\mathbb{Z}_p$, so that $k = \mathbb{F}_p$, there is a ring homomorphism $\mathrm{C}_p\colon \mathscr{M}_k \to \mathbb{Q}$, induced by counting $\mathbb{F}_p$-points on

Motivic Integration and its Interactions with Model Theory and Non-Archimedean Geometry (Volume II), ed. Raf Cluckers, Johannes Nicaise, and Julien Sebag.
Published by Cambridge University Press.

varieties. If $\mathscr{X}$ is smooth, and $A \subset \mathscr{X}_\infty$ is stable, then we may specialize $\tilde{\mu}_{\mathscr{X}}(A)$ to the ordinary p-adic measure by applying C_p. For example, if $\mathscr{X} = \mathbb{A}^d_{\mathbb{Z}_p}$, then $\mathscr{X}_\infty = \mathbf{W}(\overline{\mathbb{F}}_p)^d$, and $\mathrm{C}_p\, \tilde{\mu}_{\mathscr{X}}(A) = \mu_{\mathrm{Haar}}(A \cap \mathbb{Z}_p^n)$, *cf.* [LS03], Lemma 4.6.2.

When defining the geometric motivic measure theory, one uses these stable sets as building blocks, using coverings of them to define general measurable sets. This procedure involves taking limits, and so the measure of A, $\mu_{\mathscr{X}}(A)$, now lies in a completion of $\mathscr{M}_k$. The standard choice is to use $\widehat{\mathscr{M}_k}$, the completion with respect to the dimension filtration. However, for our purpose of specializing to p-adic integration, this is not a suitable choice. The reason is that C_p is not continuous with respect to the dimension filtration. (To see this, note for example that the sequence $x_n = p^n/\mathbb{L}^n \in \mathscr{M}_{\mathbb{F}_p}$ tends to 0 with respect to dimension, but $\mathrm{C}_p\, x_n = 1$ for every n.) For this reason, C_p does not have a natural extension to $\widehat{\mathscr{M}_k}$ (which would be by continuity).

In fact, this has been the main problem so far in adapting the theory of geometric motivic integration to fit my purposes. The main part of this talk will be devoted to a discussion of which topology to use. Before that, let me just show with an example how the specialization works.

Let $\overline{\mathrm{K}_0}(\mathsf{Var}_k)$ be the "correct" completion of $\mathscr{M}_k$, one on which the topology is strong enough for C_p to be defined and continuous when $k = \mathbb{F}_p$, while at the same time coarse enough for the motivic measure to be well defined. We define the measurable subsets of $\mathscr{X}_\infty$ in the standard way, see e.g., [DL02] [Seb04]. If $A \subset \mathscr{X}_\infty$ is measurable, then its measure, $\mu_{\mathscr{X}}(A)$, is an element of $\overline{\mathrm{K}_0}(\mathsf{Var}_k)$. Now, suppose for example that $\mathscr{X} = \mathbb{A}^d_{\mathbb{Z}_p}$, so that $k = \mathbb{F}_p$. With the definition of the arc space that I use, we then have $\mathscr{X}_\infty = \mathbf{W}(\overline{\mathbb{F}}_p)^d$ where $\mathbf{W}$ is the ring scheme of Witt vectors. $\mathbf{W}(\overline{\mathbb{F}}_p)$ may be identified with the completion of the integers in an unramified closure of $\mathbb{Q}_p$, and it contains $\mathbb{Z}_p = \mathbf{W}(\mathbb{F}_p)$ as a subring. Suppose now that $A \subset \mathbf{W}(\overline{\mathbb{F}}_p)^d$ is measurable, so that $\mu_{\mathscr{X}}(A)$ lies in $\overline{\mathrm{K}_0}(\mathsf{Var}_k)$. Then, similarly as in the case of a stable set, we have

$$\mathrm{C}_p\, \mu_{\mathscr{X}}(A) = \mu_{\mathrm{Haar}}(A \cap \mathbb{Z}_p^d).$$

More generally, if q is a power of p, we will have

$$\mathrm{C}_q\, \mu_{\mathscr{X}}(A) = \mu_{\mathrm{Haar}}(A \cap \mathbf{W}(\mathbb{F}_q)^d).$$

Example 1.1 Let $\mathscr{X} = \mathbb{A}^1_{\mathbb{Z}_p}$. Define the function $|\cdot|\colon \mathscr{X}_\infty \to \overline{\mathrm{K}_0}(\mathsf{Var}_k)$ as $x \mapsto \mathbb{L}^{-\operatorname{ord} x}$. Within the framework that I discuss in this talk one proves that

$$\int_{\mathscr{X}_\infty} |X^2+1| d\mu_{\mathscr{X}} = 1 - \left[\mathrm{Spec}\big(\mathbb{F}_p[X]/(X^2+1)\big)\right]\frac{1}{\mathbb{L}+1} \in \overline{\mathrm{K}_0}(\mathsf{Var}_{\mathbb{F}_p}).$$

This integral has the property that for every power of p, $\mathrm{C}_q \int_{\mathscr{X}_\infty} |X^2 + 1| d\mu_{\mathscr{X}} = \int_{\mathbf{W}(\mathbb{F}_q)} |X^2+1|_p dX$. So by computing the motivic integral we have simultaneously computed the corresponding integral over $\mathbf{W}(\mathbb{F}_q)$, for every power of p.

If $p \equiv 3 \pmod 4$ then -1 is a non-square in $\mathbb{F}_q$ and $\mathbb{F}_p[X]/(X^2+1) = \mathbb{F}_{p^2}$. So $\int_{\mathscr{X}_\infty} |X^2+1| d\mu_{\mathscr{X}} = 1 - [\mathbf{Spec}(\mathbb{F}_{p^2})]\frac{1}{\mathbb{L}+1}$, showing, e.g., that $\int_{\mathbb{Z}_p} |X^2+1|_p dX = 1$ and $\int_{\mathbf{W}(\mathbb{F}_{p^2})} |X^2+1|_p dX = 1 - 2/(p^2+1)$.

2 Euler characteristics

The simplest solution to the problem that the counting homomorphism is not continuous with respect to the dimension filtration would perhaps be to just introduce the weakest topology on $\mathscr{M}_k$ such that it is. However, we want a definition that is independent of p, and works over any field k.

We will use the ℓ-adic cohomology of varieties to define this topology. In fact, we will follow the lead of Ekedahl, [Eke07], who develops a topology for which taking the trace of Frobenius on the ℓ-adic cohomology is continuous, and then uses this to extend C_p.

It will turn out that, in the end, Ekedahl's topology is too strong for our purposes. However, we will still use it to define the topology that we use. So we begin by a discussion of the topology from [Eke07], and for this we will first need to discuss the Euler characteristic with values in a ring of Galois representations.

In this section, we let $\mathscr{G}$ be the absolute Galois group of k. For X a separated k-scheme of finite type we use $\mathrm{H}^i_c(X)$ to denote the ℓ-adic cohomology of the extension of X to a separable closure of k, with its natural $\mathscr{G}$-action.

2.1 Euler characteristic taking values in a ring of Galois representation

Let k be a field with absolute Galois group $\mathscr{G}$, and let $\mathrm{K}_0(\mathsf{Rep}_{\mathscr{G}}\mathbb{Q}_\ell)$ be the Grothendieck ring of continuous $\mathbb{Q}_\ell$-representations of $\mathscr{G}$. Let X be a k-variety. The assignment

$$X \mapsto \sum_i (-1)^i [\mathrm{H}_c^i(X)]$$

defines a compactly supported Euler characteristic, $\chi_c\colon \mathrm{K}_0(\mathsf{Var}_k) \to \mathrm{K}_0(\mathsf{Rep}_{\mathscr{G}}\mathbb{Q}_\ell)$. Since $\chi_c(\mathbb{L}) = [\mathbb{Q}_\ell(-1)]$, the class of the dual to the cyclotomic representation, is invertible, it follows that χ_c factors through $\mathscr{M}_k$. Therefore it can be used to distinguish elements in $\mathscr{M}_k$ as well as in $\mathrm{K}_0(\mathsf{Var}_k)$. To illustrate this we consider the following example by Naumann, [Nau07]:

Example 2.1 (Naumann) Let L/k be a Galois extension of degree d. Then

$$[\mathbf{Spec}(L)]^2 = [\mathbf{Spec}(L \otimes_k L)] = [\mathbf{Spec}(L^d)] = d[\mathbf{Spec}(L)],$$

consequently $[\mathbf{Spec}(L)]$ is a zero divisor if it is not equal to 0 or d. Naumann proves this for finitely generated k by reducing to the case when $k = \mathbb{F}_q$ and then using the point counting homomorphism. Using χ_c one can generalize this proof: Let $S = \mathrm{Gal}(L/k)$, so that $\chi_c(\mathbf{Spec}(L)) = [\mathbb{Q}_\ell[S]]$. To every $\sigma \in \mathscr{G}$ there exists an invariant $\mathrm{C}_\sigma\colon \mathrm{K}_0(\mathsf{Rep}_{\mathscr{G}}\mathbb{Q}_\ell) \to \overline{\mathbb{Q}}_\ell$, induced by mapping V to the character of V evaluated in σ. (It has the property that when $k = \mathbb{F}_q$ and σ is the Frobenius automorphism, $\mathrm{C}_q = \mathrm{C}_\sigma \circ \chi_c$, see Section 3.) Using C_σ for different σ it is easy to show that $[\mathbb{Q}_\ell[S]] \notin \mathbb{Z}$, in particular it is not equal to zero or d. Hence $[\mathbf{Spec}(L)]$ is a zero divisor in both $\mathrm{K}_0(\mathsf{Var}_k)$ and $\mathscr{M}_k$. (The details of this argument are given in [Rök07].)

Remark 2.2 This example constructs zero divisors in $\mathscr{M}_k$ when k is not separably closed. On the other hand, the classical example of zero divisors in $\mathrm{K}_0(\mathsf{Var}_k)$ is by Poonen, and holds over any field of characteristic zero, including the algebraically closed ones. Recently, it was shown ([Eke07], Corollary 3.5) that these remain non-zero also in $\mathscr{M}_k$. For a discussion of the status of the problem of zero divisors, see [Nic08], Section 2.6. For a discussion about the effects of inverting $\mathbb{L}$, see [DL04].

2.2 Euler characteristic taking values in a ring of mixed Galois representations

For our purpose, it is not optimal to let χ_c take values in $\mathrm{K}_0(\mathsf{Rep}_{\mathscr{G}}\mathbb{Q}_\ell)$. In this section we instead define the ring $\mathrm{K}_0(\mathsf{Coh}_k)$, which has the advantage of being graded. The topology corresponding to this grading is compatible with the dimension filtration on $\mathrm{K}_0(\mathsf{Var}_k)$, in that χ_c is continuous. Moreover, another disadvantage of $\mathrm{K}_0(\mathsf{Rep}_{\mathscr{G}}\mathbb{Q}_\ell)$ is that it is too small when k is not finitely generated (for example, if $k = \overline{k}$ then $\mathrm{K}_0(\mathsf{Rep}_{\mathscr{G}}\mathbb{Q}_\ell) = \mathbb{Z}$). On the other hand, the structure of $\mathrm{K}_0(\mathsf{Coh}_k)$ is always sufficiently rich.

We now define $\mathrm{K}_0(\mathsf{Coh}_k)$. Firstly, for k finitely generated, we use Coh_k to denote the category of mixed Galois $\mathbb{Q}_\ell$-representations (sometimes written $\mathsf{WRep}_{\mathscr{G}}\mathbb{Q}_\ell$). This is the full subcategory of $\mathsf{Rep}_{\mathscr{G}}\mathbb{Q}_\ell$ consisting of those representations that have a weight filtration. We then have an injection $\mathrm{K}_0(\mathsf{Coh}_k) \hookrightarrow \mathrm{K}_0(\mathsf{Rep}_{\mathscr{G}}\mathbb{Q}_\ell)$, the image being generated by modules of pure weight. (When $k = \mathbb{F}_q$ a representation is of pure weight n if all the archimedeian absolute values of all eigenvalues of the geometric Frobenius are in $\overline{\mathbb{Q}}$ and have absolute value $q^{n/2}$.) The image of χ_c is contained in this subring, i.e., we have $\chi_c\colon \mathscr{M}_k \to \mathrm{K}_0(\mathsf{Coh}_k)$.

For an arbitrary field k the construction of the category Coh_k is due to Ekedahl, [Eke07], Section 2. (See this reference for the construction). It has the property that if $\{k_\alpha\}$ is the collection of finitely generated subfields of k then we have an isomorphism $\lim_{\to\alpha} \mathrm{K}_0(\mathsf{Coh}_{k_\alpha}) \to \mathrm{K}_0(\mathsf{Coh}_k)$ (actually, one does not have to define Coh_k for fields that are not finitely generated, one could instead use this as the definition of $\mathrm{K}_0(\mathsf{Coh}_k)$ in that case).

Now, if X is a k-scheme of finite type then it is defined over some finitely generated subfield k_{α_0}. The Euler characteristic of X can therefore be defined in $\mathrm{K}_0(\mathsf{Coh}_{k_{\alpha_0}})$, we define $\chi_c(X)$ to be the image of this in $\mathrm{K}_0(\mathsf{Coh}_k)$. This is well defined, so we have a ring homomorphism $\chi_c\colon \mathscr{M}_k \to \mathrm{K}_0(\mathsf{Coh}_k)$ for any field k.

When $k = \mathbb{C}$, we could also use Coh_k to denote the category of mixed polarisable rational Hodge structures. Everything in this section would then hold also with this alternative definition of Coh_k.

One reason for us to work in $\mathrm{K}_0(\mathsf{Coh}_k)$ rather than in $\mathrm{K}_0(\mathsf{Rep}_{\mathscr{G}}\mathbb{Q}_\ell)$ is that we have some control of the structure of $\mathrm{K}_0(\mathsf{Coh}_k)$: If V is any mixed representation then there is a Jordan-Hölder sequence $0 = V_0 \subset V_1 \subset \cdots \subset V_n = V$. Here V_{i+1}/V_i is simple, hence of pure weight. Since $[V] = \sum_i [V_{i+1}/V_i]$, and since this sum is independent of the chosen sequence, it follows that any $v \in \mathrm{K}_0(\mathsf{Coh}_k)$ may be written uniquely as

$$v = \sum_n v_n, \text{ where } v_n = \sum_i c_{ni}[V_{ni}] \tag{14.1}$$

and the V_{ni} are irreducible, pairwise non-isomorphic representations of pure weight n. Since this also respects the multiplication, it follows that $\mathrm{K}_0(\mathsf{Coh}_k)$ is graded by weight. A important fact is that the image of $\mathrm{K}_0(\mathsf{Var}_k)$ under χ_c is contained in the part of non-negative degree.

Example 2.3 This gives an easy way to see that $\mathbb{L}$ is not invertible in $\mathrm{K}_0(\mathsf{Var}_k)$. For if $x\mathbb{L} = 1$ then $\chi_c(x) = [\mathbb{Q}_\ell(1)]$ in $\mathrm{K}_0(\mathsf{Coh}_k)$. Since the right hand side is pure of weight -2, this is impossible.

More generally, it is showed in [Eke07], Proposition 1.3, that if P and Q are integer polynomials such that $P(\mathbb{L})$ divides $Q(\mathbb{L})$ in $\mathrm{K}_0(\mathsf{Var}_k)$, then P divides Q in $\mathbb{Z}[T]$.

Before we continue, we illustrate with two examples the usefulness of this grading. The first example is a result of Nicaise (Corollary 3.13 in [Nic08]). Our proof is essentially the same, we just use a different setting.

Example 2.4 Let X and Y be k-varieties. If $[X] = [Y]$ in $\mathscr{M}_k$ then X and Y have the same dimension n, and the same number of geometric components of dimension n.

We use the following results of Deligne: $\mathrm{H}^i_c(X)$ is of mixed weight $\leq i$, and if $i > 2\dim X$ then $\mathrm{H}^i_c(X) = 0$. Furthermore, if $\dim X = n$ then $\mathrm{H}^{2n}_c(X)$ is of pure weight $2n$, and its dimension equals the number of geometric components of X of dimension n.

Now, let n be the dimension of X and m the dimension of Y. We have that

$$\chi_c(X) = [\mathrm{H}^{2n}_c(X)] + \big(\text{terms of weight} < 2n\big).$$

Since the weight $2n$-part of this is a non-zero, non-virtual representation, it follows that the same must hold for the weight $2n$-part of $\chi_c(Y)$, hence that $m = n$, and also that these parts have the same dimension.

The following example is a variation of the preceding one, which we will need when defining our topology.

Example 2.5 Let X_i be non-empty k-varieties and let n_i be non-negative integers. Let $x := \sum_{i=1}^N n_i[X_i]$. If $x = 0$ then $n_i = 0$ for every i. Moreover, if $x \in \mathscr{F}^m \mathscr{M}_k$ then $[X_i] \in \mathscr{F}^m \mathscr{M}_k$ for every i. (We use $\mathscr{F}$ to denote the dimension filtration.)

The first part is a special case of the preceding example. For the second part, let m' be the maximal dimension of the X_i and suppose that $m' > m$. Since $x \in \mathscr{F}^m \mathscr{M}_k$, the weight $2m'$ part of $\chi_c(x)$ is 0. On the other hand

$$\chi_c(x) = \sum_{\dim X_i = m'} n_i[\mathrm{H}_c^{2m'}(X_i)] + (\text{terms of weight} < 2m'),$$

and the weight $2m'$-part of this is non-zero, a contradiction.

3 Topologies

3.1 The dimension filtration and its drawbacks

The grading of $\mathrm{K}_0(\mathsf{Coh}_k)$ imposes a filtration, the *weight filtration*: For $v \in \mathrm{K}_0(\mathsf{Coh}_k)$, let v_n be its component of degree n, as in (14.1). Define the filtration in the negative direction as $\mathscr{F}^{\leq N}\mathrm{K}_0(\mathsf{Coh}_k) = \{v = \sum_{n \leq N} v_n : \text{finite sums}\} \subset \mathrm{K}_0(\mathsf{Coh}_k)$. If we complete with respect to the weight filtration, we get $\mathrm{K}_0\widehat{(\mathsf{Coh}_k)}$, a ring that behaves nicely in that elements can be represented uniquely as sums, infinite in the negative direction:

$$\mathrm{K}_0\widehat{(\mathsf{Coh}_k)} = \Big\{\sum_{n \leq N} v_n : \text{infinite sums}\Big\}.$$

The Euler characteristic $\chi_c\colon \mathscr{M}_k \to \mathrm{K}_0(\mathsf{Coh}_k)$ is continuous with respect to this topology and the dimension filtration. For suppose that $\dim x_i = d$, where $x_i \in \mathscr{M}_k$. Then, using the above mentioned results of Deligne, one shows that $\chi_c(x_i) \in \mathscr{F}^{\leq 2d}\mathrm{K}_0(\mathsf{Coh}_k)$. Hence, if $x_i \to 0$, then $\chi_c(x_i) \to 0$. Therefore χ_c extends by continuity to a continuous homomorphism $\chi_c\colon \widehat{\mathscr{M}_k} \to \mathrm{K}_0\widehat{(\mathsf{Coh}_k)}$.

In [Eke07] the author wants to extend C_q to a completion of $\mathscr{M}_k$ in such a way that it still is compatible with taking the trace of Frobenius. In other words, when $k = \mathbb{F}_q$ we have, by the Lefschetz trace formula, a commutative diagram

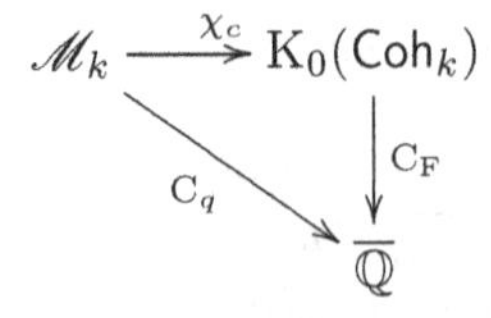

where C_F is defined by taking the trace of Frobenius, i.e., by mapping the representation V to the character of V evaluated at the Frobenius automorphism F. We then first need to complete $\mathrm{K}_0(\mathsf{Coh}_k)$ in such a way that it is possible to extend C_F to the completion in case k is finite, and for this we need a stronger topology on $\mathrm{K}_0(\mathsf{Coh}_k)$ than the above mentioned. For define, for $v = \sum_n v_n$ as in (14.1),

$$\overline{\mathrm{w}}_n(v) := \sum_i |c_{ni}| \dim V_{ni}.$$

Then, if $k = \mathbb{F}_q$, the only possible extension of C_F to $\mathrm{K}_0(\mathsf{Coh}_k)$ would be by continuity, i.e., to define $\mathrm{C}_\mathrm{F}\, v = \sum_n \mathrm{C}_\mathrm{F}\, v_n$. However, $\sum_n |\mathrm{C}_\mathrm{F}\, v_n| \approx \sum_n q^{n/2}\, \overline{\mathrm{w}}_n(v)$. So in case $\overline{\mathrm{w}}_n$ grows too fast, $\mathrm{C}_\mathrm{F}\, v$ is not convergent. (For an example, let $v = \sum_{0 \leq n} q^n [\mathbb{Q}_\ell(n)]$. This is convergent, but $\mathrm{C}_\mathrm{F}\, q^n [\mathbb{Q}_\ell(n)] = 1$ for all n.) In [Eke07] this is resolved in the following way:

3.2 The topology of uniform polynomial growth

Let $\{v_i\}_{i \in \mathbb{N}}$ be a sequence in $\mathrm{K}_0(\mathsf{Coh}_k)$. We say that the sequence is of *uniform polynomial growth* is there exist constants D and d such that for every i and n we have

$$\overline{\mathrm{w}}_n(v_i) \leq |n|^d + D. \tag{14.2}$$

We now define the sequence to be convergent if it is of uniform polynomial growth, and convergent with respect to the weight filtration. This topology does not come from a filtration anymore, neither from a metric. However, we may define the completion as the ring of all Cauchy sequences (meaning that $v_i - v_j \to 0$ as $i, j \to \infty$) modulo those converging to zero. It has the properties of a completion: $\mathrm{K}_0(\mathsf{Coh}_k)$ is dense in it, and every Cauchy sequence is convergent. Denote this completion by $\overline{\mathrm{K}_0}^{\mathrm{pol}}(\mathsf{Coh}_k)$. It has a very simple description as

$$\overline{\mathrm{K}_0}^{\mathrm{pol}}(\mathsf{Coh}_k) = \Big\{ v = \sum_{n \leq N} v_n \,:\, v \text{ of uniform polynomial growth} \Big\}.$$

If $k = \mathbb{F}_q$ we may now define C_F, the trace of Frobenius, on $\overline{\mathrm{K}_0}^{\mathrm{pol}}(\mathsf{Coh}_k)$. For if $v = \sum_{n \leq N} v_n \in \overline{\mathrm{K}_0}^{\mathrm{pol}}(\mathsf{Coh}_k)$ then

$$|\mathrm{C}_\mathrm{F}\, v| \leq \sum_{n \leq N} |\mathrm{C}_\mathrm{F}\, v_n| \leq \sum_{n \leq N} q^{n/2}\, \overline{\mathrm{w}}_n(v) \leq \sum_{n \leq N} q^{n/2}(|n|^d + D)$$

which is convergent. Similarly one proves that C_F is continuous.

So this has the property that we want. However, with this stronger topology, χ_c is no longer continuous. We therefore need a stronger topology on $\mathscr{M}_k$ as well. This definition also uses the weight concept. Basically, one defines, for X a k-variety, $\overline{\mathrm{w}}_n(X) := \sum_i \overline{\mathrm{w}}_n(\mathrm{H}^i_c(X))$. (To make this well defined on $\mathscr{M}_k$ there are some technical details to take care of; however, this is the basic idea.) Note that $\overline{\mathrm{w}}_n(X) \geq \overline{\mathrm{w}}_n(\chi_c(X))$. (An alternative choice would be to just define $\overline{\mathrm{w}}_n(X) := \overline{\mathrm{w}}_n(\chi_c(X))$, however, this would make the definition less intrinsic; $\overline{\mathrm{w}}_n(X)$ could be small due to cancellations in the cohomology that do not correspond to cancellations in $\mathscr{M}_k$.)

We now define a sequence in $\mathscr{M}_k$ to be of uniform polynomial growth in the same way as for sequences in $\mathrm{K}_0(\mathsf{Coh}_k)$, i.e., if it satisfies (14.2). Then, a sequence in $\mathscr{M}_k$ converges to 0 if it does so with respect to the dimension filtration, and is of uniform polynomial growth. Again we define $\overline{\mathrm{K}_0}^{\mathrm{pol}}(\mathsf{Var}_k)$ as the ring of Cauchy sequences, modulo those converging to zero. The dimension and weight functions extends to this ring, and it has the properties that every Cauchy sequence converges, and that the image of $\mathscr{M}_k$ is dense in it.

Now, for $k = \mathbb{F}_q$, we get a commutative diagram of continuous maps, where the extension of C_q is defined by continuity, or equivalently as just the composition $\mathrm{C}_{\mathrm{F}} \circ \chi_c$.

$$\begin{array}{ccc} \overline{\mathrm{K}_0}^{\mathrm{pol}}(\mathsf{Var}_k) & \xrightarrow{\chi_c} & \overline{\mathrm{K}_0}^{\mathrm{pol}}(\mathsf{Coh}_k) \\ & \searrow^{\mathrm{C}_q} & \downarrow \mathrm{C}_{\mathrm{F}} \\ & & \mathbb{C} \end{array}$$

3.3 Our topology

At first sight, the topology of uniform polynomial growth seems to be perfect for our purposes of motivic integration. The problem is that it is too strong. To define a satisfying theory of motivic measures (in particular, to copy the existing theory) one needs to have some control of the measure of $U \subset V$, if one knows the measure of V. For example, suppose that $U_n \subset V_n$. Then it is reasonable to demand that if $\mu_{\mathscr{X}}(V_n) \to 0$ then $\mu_{\mathscr{X}}(U_n) \to 0$. However, we may have arbitrary complicated subvarieties of affine spaces, and for an affine space we have $\overline{\mathrm{w}}_n(\mathbb{A}_k^m) = 0$ if $n \neq 2m$ and $\overline{\mathrm{w}}_{2m}(\mathbb{A}_k^m) = 1$. Therefore there is no way to give an upper bound for $\overline{\mathrm{w}}_n(U)$ just by knowing $\overline{\mathrm{w}}_n(V)$ for $U \subset V$.

To overcome this problem we introduce a partial ordering on $\mathrm{K}_0(\mathsf{Var}_k)$ by saying that $x \leq y$ if there exists a variety X such that $x + [X] = y$. (That this really is a partial ordering is proved using Example 2.5.) One then extends this partial ordering to $\mathscr{M}_k$. We say that a sequence x_i in $\mathscr{M}_k$ is *strongly convergent* to zero, if it is convergent in the topology of uniform polynomial growth. We then define the sequence to be *convergent* to zero if there are two sequences a_n and b_n, strongly convergent to zero, such that $a_n \leq x_n \leq b_n$. Using Example 2.5 one shows that $\dim x_n \leq \max\{\dim a_n, \dim b_n\}$, hence that if a sequence is convergent, then it is convergent with respect to the dimension filtration. It is now easy to show that C_q is continuous when $k = \mathbb{F}_q$. We construct the completion of $\mathscr{M}_k$ with respect to this topology in the standard way, as a quotient of the ring of Cauchy sequences. We denote it $\overline{\mathrm{K}_0}(\mathsf{Var}_k)$.

This topology has one major drawback: The Euler characteristic is no longer continuous. However, it has the property that we are after: When $k = \mathbb{F}_q$, the counting homomorphism is continuous and hence extends to $\overline{\mathrm{K}_0}(\mathsf{Var}_k)$.

$$\overline{\mathrm{K}_0}(\mathsf{Var}_k) \xrightarrow{\mathrm{C}_q} \mathbb{R}$$

Using $\overline{\mathrm{K}_0}(\mathsf{Var}_k)$, instead of $\widehat{\mathscr{M}_k}$, we may now define the motivic measure theory in the standard way, everything is well defined also with respect to this stronger topology. Moreover, we may specialize to p-adic integration via point counting (in particular, we may use this theory to make sense of Example 1.1). The details of this construction are given in [Rök08].

References

[DL02] Jan Denef and François Loeser, *Motivic integration, quotient singularities and the McKay correspondence*, Compositio Math. **131** (2002), no. 3, 267–290. MR MR1905024 (2004e:14010)

[DL04] ———, *On some rational generating series occurring in arithmetic geometry*, Geometric aspects of Dwork theory. Vol. I, II, Walter de Gruyter GmbH & Co. KG, Berlin, 2004, pp. 509–526. MR MR2099079 (2005h:11267)

[Eke07] Torsten Ekedahl, *On the class of an algebraic stack*, www.mittag-leffler.se/preprints/0607/info.php?id=66 (2007).

[Gre61] M. Greenberg, *Schemata over local rings*, Ann. Math. (1961), no. 73, 624–648.

[Loo02] Eduard Looijenga, *Motivic measures*, Astérisque (2002), no. 276, 267–297, Séminaire Bourbaki, Vol. 1999/2000. MR MR1886763 (2003k:14010)

[LS03] François Loeser and Julien Sebag, *Motivic integration on smooth rigid varieties and invariants of degenerations*, Duke Math. J. **119** (2003), no. 2, 315–344. MR MR1997948 (2004g:14026)

[Nau07] N. Naumann, *Algebraic independence in the Grothendieck ring of varieties*, Trans. Amer. Math. Soc. **359** (2007), no. 4, 1653–1683 (electronic). MR MR2272145 (2007j:14012)

[Nic08] Johannes Nicaise, *A trace formula for varieties over a discretely valued field.* To appear in *J. Reine Angew. Math.*, arXiv:0805.1323v2 [math.AG] (2008).

[Rök07] Karl Rökaeus, *The computation of the classes of some tori in the grothendieck ring of varieties*, arXiv:0708.4396v3 [math.AG] (2007).

[Rök08] ———, *A version of geometric motivic integration that specializes to p-adic integration via point counting*, arXiv:0810.4496v1 [math.AG] (2008).

[Seb04] Julien Sebag, *Intégration motivique sur les schémas formels*, Bull. Soc. Math. France **132** (2004), no. 1, 1–54. MR MR2075915 (2005e:14017)

15

Absolute desingularization in characteristic zero

Michael Temkin

1 Introduction

1.1 Preamble

This paper is an expository lecture notes originally based on a lecture on the results of [Tem1] given by the author at the workshop on Motivic Integration in May 2008, at ICMS, Edinburgh. Since a substantial progress was done since May 2008, it seemed natural to include the new results of [BMT], [Tem2] and [Tem3] in this exposition. We will mainly concentrate on the functorial non-embedded desingularization constructed in [Tem2] because it seems that the results of [Tem3] on the embedded case can be improved further. We pursue expository goals, so we will concentrate on explaining the results and the main ideas of our method and we will refer to the cited papers for proofs and technical details. Also, we try to include more examples and general remarks than in a pure research paper. Thus, this survey can serve as a companion to or a light version of [Tem1] and [Tem2]. I would like to warn the reader that the current situation described in the paper can change soon (similarly

I am grateful to E. Bierstone and P. Milman for useful discussions and correcting an example in a preliminary version of the paper. I would like to thank the ICMS and the organizers of the workshop on Motivic Integration and its Interactions with Model Theory and Non-Archimedean Geometry held in May 2008 for the hospitality and excellent working conditions during the workshop. This work was finished during my visit at the Institute for Advanced Study at Princeton when I was supported by the NFS grant DMS-0635607.

Motivic Integration and its Interactions with Model Theory and Non-Archimedean Geometry (Volume II), ed. Raf Cluckers, Johannes Nicaise, and Julien Sebag. Published by Cambridge University Press.

to the change since 2008), but this is always a danger with a survey on an active research area.

1.2 The history

In 1964 Hironaka proved many fundamental desingularization results including strong desingularization of algebraic varieties in characteristic zero. The latter means that any reduced variety of characteristic zero can be modified to a smooth one by successive blow ups along nowhere dense smooth centers. Hironaka's method was extremely difficult for understanding (due to a complicated inductive structure of the proof), and perhaps the main reason for this was that his method was not constructive, canonical or functorial, unlike many new proofs. In particular, unlike the new proofs, Hironaka could not work within the category of varieties since some arguments with formal completions were involved. Probably for this reason, Hironaka proved his desingularization for all schemes of finite type over local rings R with regular completion homomorphism $R \to \widehat{R}$.

A year later, Grothendieck introduced quasi-excellent (or *qe*) schemes in [EGA, IV_2, §7.9] in order to provide the most general framework for desingularization. Grothendieck observed that the schemes studied by Hironaka were schemes of finite type over a local qe scheme k, and proved that if any integral scheme of finite type over a base scheme k admits a desingularization in the weakest possible sense then k is qe. Grothendieck conjectured that the converse is probably true (i.e. any integral qe scheme admits a desingularization) and claimed without proof that the conjecture holds true for noetherian qc schemes over $\mathbf{Q}$ as can be proved by Hironaka's method. The latter claim was never checked in published literature, and its status is unclear until now. Nevertheless, this fact was occasionally used by other mathematicians, for example for desingularizing affinoid spaces over $\mathbf{Q}_p$.

In [Tem1] the author proved that indeed, any noetherian integral qe scheme over $\mathbf{Q}$ admits a desingularization. Moreover, the regular locus of the scheme can be preserved by the desingularization and one can also resolve a closed subscheme to a normal crossings divisor. The construction of [Tem1] uses any desingularization of varieties as a black box input, but modifies it significantly. In particular, many good properties were lost in the resulting algorithm, including functoriality and regularity of the centers. Very recently the method was strengthened in [Tem2] and [Tem3] in order to preserve the two above properties as well. In particular, a desingularization $\mathcal{F}$ (resp. embedded desingularization $\mathcal{E}$) of all generically reduced qe schemes over $\mathbf{Q}$ (resp. closed subschemes in regular qe schemes) is now available and $\mathcal{F}$ and $\mathcal{E}$ go by blowing up regular centers and are functorial with respect to all regular morphisms. The functoriality property is a serious achievement since it rigorously implies desingularization in many other categories in characteristic zero, including Artin stacks, schemes acted on by regular group schemes, qe formal schemes and complex or non-archimedean analytic spaces.

1.3 Motivation and applications

Non-functorial desingularization of qe schemes in [Tem1] only allowed to desingularize affine formal schemes. In order to obtain a global desingularization result for formal schemes one had to construct a desingularization which is functorial at least with respect to formal localizations. These are regular morphisms of not finite type and it seems that the most natural way to ensure such functoriality is to achieve functoriality with respect to all regular morphisms, as was done in [Tem2] and [Tem3]. Already desingularization of formal varieties over $\mathbf{C}[[T]]$ is a new result, and currently it seems that desingularization of formal varieties will be most useful for applications. Actually, it were few requests about formal desingularization that convinced me to continue the research of [Tem1]. In particular, it seems that desingularization of formal varieties may have applications to motivic integration (see [Nic]), log canonical thresholds (see [FEM]), desingularization of meromorphic connections (see [Ked]) and motivic Donaldson-Thomas invariants studied by Kontsevich-Soibelman. Finally, it seems that the desingularization of rigid spaces and Berkovich analytic spaces (not necessarily good) is also new.

1.4 Overview

We introduce all necessary terminology (e.g. qe schemes, blow ups, regular locus, etc.) in §2. The reader can look through this section and return to it when needed. In §3 we formulate our main results, explain how our method works in general and divide it to two stages. Then, both stages are studied in details in §4 and §5. In addition, we consider in §5.2 few examples that illustrate our algorithm. Finally, in §6 we deduce similar results for other categories including stacks, formal scheme and various analytic spaces both in compact and non-compact settings.

2 Setup

Throughout this paper all schemes and formal schemes are assumed to be locally noetherian.

2.1 Varieties

Variety or *algebraic variety* in this paper always means a scheme X which admits a finite type morphism $X \to \mathrm{Spec}(k)$ to the spectrum of a field. If such a morphism is fixed then we say that X is a *k-variety* and k is the *ground field* of X. It is an easy fact (see [BMT]) that any reduced connected variety X possesses a maximal (and hence canonical) ground field $k \subset \mathcal{O}_X(X)$. Unfortunately, this is not true for non-reduced varieties as the following example shows.

Example 2.1 Let k be a field of characteristic zero with an irreducible curve C which is not reduced at its generic point η. Then $\mathrm{Spec}(\mathcal{O}_{C,\eta})$ possesses various structures of a zero-dimensional $k(\eta)$-variety, but none of them is "better" than another.

Remark 2.2 (i) The above example extends to formal varieties. Moreover, even smooth formal varieties do not have to have a canonical ground field. For example, already for $k = \mathbf{Q}(x)$ there exist many embeddings $k \hookrightarrow k[[t]]$, which are as "good" as the obvious embedding, and, more generally, the field of coefficients in Cohen's theorem is not unique.

(ii) As we will see later, the above observation is responsible for the main obstacle to proving functorial desingularization by our method. In addition, it indicates that even for varieties it is more natural to study absolute algorithms rather than the algorithms that take k into account (for example by working with k-derivations). We will return to this discussion in §2.5.9.

2.2 Regularity

2.2.1 Regular schemes

There are many equivalent ways to say that a local ring A is regular and here are two possibilities: (a) the associated graded ring $\oplus_{n=0}^{\infty} m^n/m^{n+1}$ is isomorphic to $k[T_1, \dots, T_d]$, where m is the maximal ideal and $k = A/m$ is the residue field, (b) the dimension of A (i.e. the maximal length of a chain of prime ideals decreased by one) equals to the dimension of the cotangent k-vector space m/m^2. We define the *regular locus* X_{reg}

of a scheme X as the set of points $x \in X$ with regular $\mathcal{O}_{X,x}$ and say that X is *regular* at each $x \in X_{\rm reg}$. The *singular locus* $X_{\rm sing}$ is defined as the complement of $X_{\rm reg}$. Although regularity is an analog of smoothness, it is an absolute property, while smoothness is a relative property. For example, a variety of positive characteristic can be smooth and not smooth over different fields of definition.

2.2.2 Monomial divisors

A divisor Z in a regular scheme X is called *snc* (or strictly normal crossings) if its irreducible components are regular and transversal, i.e. each $Z_{i_1} \cap Z_{i_2} \cap \cdots \cap Z_{i_n}$ is regular of codimension n in X (or empty). This is equivalent to the condition that Zariski locally at each point x the divisor Z is given by an equation $\prod_{i=1}^{l} t_i = 0$ where $t_1, \ldots, t_n \in \mathcal{O}_{X,x}$ is a regular family of parameters. If Z is given by an equation $\prod_{i=1}^{n} t_i^{m_i} = 0$ then we say that it is *strictly monomial.* Finally, if the above conditions hold only étale-locally (i.e. the parameters can be chosen in the strict henselization $\mathcal{O}_{X,x}^{\rm sh}$) then we say that Z is *normal crossings* or *monomial*, respectively. Note that a closed subscheme $Z \hookrightarrow X$ is a (strictly) monomial divisor if and only if it is a Cartier divisor whose reduction is (strictly) normal crossing.

2.2.3 Regular morphisms

A morphism $f : Y \to X$ is called *regular* if it is flat and has geometrically regular fibers. Since a finite type morphism is regular iff it is smooth, this can be viewed as a generalization of smoothness to "large" morphisms. A homomorphism of algebras $f : A \to B$ is *regular* if Spec(f) is regular (Hironaka calls regular morphisms "universally regular", but our terminology is the standard one). It is well known that regular/singular locus is compatible with regular morphisms, i.e. for a regular morphism $f : Y \to X$ we have that $Y_{\rm sing} = f^{-1}(X_{\rm sing})$ and $Y_{\rm reg} = f^{-1}(X_{\rm reg})$. Similarly, one shows that the monomiality locus of a divisor is compatible with regular morphisms. We warn the reader that the same is not true for strictly monomial locus, since the preimage of a not strictly monomial divisor under an étale morphism can be strictly monomial.

2.2.4 Equisingularity

We say that a scheme X is *equisingular* at a point x if its reduction X_0 is regular at x and $X_{\rm red}$ is normally flat along $X_{\rm red}$ at x. Recall that the latter means that the $\mathcal{O}_{X_{\rm red}}$-sheaves $\mathcal{N}_X^i/\mathcal{N}_X^{i+1}$ are locally free at x, where $\mathcal{N}_X \subset \mathcal{O}_X$ is the radical. The set of all points $x \in X$ at which X

is equisingular will be called the *equisingular locus* of X. Equisingular loci behave similarly to regular loci. In particular, they are compatible with regular morphisms, etc.

Remark 2.3 We prefer the notion "equisingular" since it is much shorter than other alternatives. Also, it reflects the geometric meaning pretty well because in some sense the singularity of an equisingular scheme along any irreducible component is constant; that is, the singularity on the entire component is as bad as at its generic point. Note also that an equisingular scheme is regular if and only if it is generically reduced.

2.3 Quasi-excellent schemes

2.3.1 The definition

For shortness, we will abbreviate the word quasi-excellent as qe. Quasi-excellent schemes were introduced by Grothendieck in [EGA, IV_2, §7.9] though the word "quasi-excellent" was invented later. These are locally noetherian schemes X satisfying the following conditions N and G (after Nagata and Grothendieck): (N) for any Y of finite type over X the regular locus Y_{reg} is open, (G) for any point $x \in X$ the completion homomorphism $\mathcal{O}_{X,x} \to \widehat{\mathcal{O}}_{X,x}$ is regular. A qe scheme which is universally catenary (see §2.3.4) is called excellent.

2.3.2 Connection to desingularization

Obviously, the condition (N) is necessary in order to have a universal desingularization theory over X (i.e. in order to be able to desingularize schemes of finite type over X). Grothendieck proved the same for the condition (G) in [EGA, IV_2, 7.9.5]: if any integral scheme of finite type over X admits a regular modification then X is qe. It was suggested by Grothendieck and is believed by many mathematicians that the converse is also true. Moreover, it is a common belief (or at least hope) that qe schemes admit much stronger variants of desingularization discussed in §2.5.

2.3.3 Basic properties

Main properties of quasi-excellence and excellence are as follows:

(1) They are invariant under many operations including passing to a scheme of finite type, localization and henselization along a closed subscheme.

(2) If a ring A is qe (i.e. $\mathrm{Spec}(A)$ is qe) and $\widehat{A}$ is its completion along any ideal then the completion homomorphism $A \to \widehat{A}$ is regular.

(3) It is a very difficult result recently proved by Gabber that a noetherian I-adic ring A is qe iff A/I is qe. In particular, quasi-excellence is preserved under formal completions. See Remark 2.16 below for more details.

(4) Basic examples of excellent rings are $\mathbf{Z}$, fields, noetherian convergent power series rings in complex and non-archimedean analytic geometries, and schemes obtained from those by use of operations (1) and (2).

Remark 2.4 (i) Intuitively, general qe schemes have no "floor" unlike the algebraic varieties. For example, one cannot fiber them by curves and many pathologies can occur with the dimension, as we will see below. Usually, in the study of qe schemes one uses that they are in a "good relation with their roof" by the G-condition; that is, their formal completions are regular over them in the affine case (i.e. the homomorphism $A \to \widehat{A}$ is regular when $X = \mathrm{Spec}(A)$). For example, the completion of a qe scheme X along a subvariety (e.g. a closed point) is a formal variety $\mathfrak{X}$, and the desingularization theory for X is closely related to that of $\mathfrak{X}$ because of the G-condition.

(ii) Formal varieties, in their turn, can be studied by various methods. In particular, one can fiber them by formal curves (Gabber's adoption of de Jong's approach), one can algebraize them in the rig-regular case (our adoption of Elkik's theory), and, very probably, one can generalize for them the algorithms for varieties by switching to the sheaves of continuous derivations.

2.3.4 Caveats with the dimension theory

A scheme X is called *catenary* if for any point $x \in X$ with a specialization y all maximal chains of specializations between x and y have the same length. A scheme X is *universally catenary* if any scheme of finite type over X is catenary. Actually, it is the catenarity condition which makes dimension theory reasonable. The following simple example from [EGA, IV$_2$] shows that non-catenary schemes can be not as horrible as one might expect.

Example 2.5 Let k be a field with an isomorphism $\phi : k \widetilde{\to} k(t)$ (so, k is of infinite absolute transcendence degree, and one can take $k = F(t_1, t_2 \dots)$, where F is any field). Let z be a closed k-point in $\mathbf{A}^n_k = \mathrm{Spec}(k[x_1, \dots, x_n])$ and y be the generic point of an affine line not containing z. Let $\widetilde{X}_n$ be a localization of $\mathbf{A}^n_k$ with $n \geq 2$ on which both y and z are closed points and let X_n be obtained from $\widetilde{X}_n$ by gluing y and

z via ϕ (i.e. we consider only functions $f \in \mathcal{O}_{\widetilde{X}_n}$ with $f(y) = \phi(f(z))$). Note that X_n is a "nice" qe scheme; for example, its normalization is the localized variety $\widetilde{X}_n$. However, our operation obviously destroys the dimension theory on X_n, and indeed one can easily show that X_n is not catenary for $n \geq 3$ and is catenary but not universally catenary for $n = 2$.

Remark 2.6 (i) The above example is in a sense the most general one. Namely, it follows from §2.3.3(2) that a local qe ring A is normal if its completion $\widehat{A}$ is normal, and one can use this to show that any normal qe scheme is catenary. Thus, the only source of non-catenarity on qe schemes is that sometimes branches of different codimension on the same irreducible component can be glued on non-normal schemes. In particular, non-catenarity is close in nature to local non-equidimensionality.

(ii) If A is normal (or even regular) but not qe then it can happen that $\widehat{A}$ is not normal or even is not reduced. Also, there are normal but non-catenary not qe schemes.

Another danger with qe schemes is that even an excellent ring can be infinite dimensional (by famous Nagata's example). In particular, one cannot argue by induction on dimension and should use noetherian induction or induction on codimension instead.

Remark 2.7 The reader may wonder why these pathological examples are worth any discussion. I agree that the schemes from the above examples are curious but seem to be absolutely useless. However, the necessity to have them in mind seems to be very useful from my point of view. It makes one to argue correctly and allows to quickly reject approaches that could work for varieties but will not work for qe schemes (including the reasonable ones). For example, the main induction in our desingularization method will be by codimension. Also, the non-catenary example indicates that one must be extremely careful when dealing with non-equidimensional morphisms and schemes (including varieties!). We will discuss this caveat in §2.5.10.

2.3.5 Caveats with derivatives and bad DVR's

In all examples from §2.3.3(4) one establishes excellence by constructing a good theory of derivatives (algebraic or continuous). The latter does not have to exist on regular not qe schemes, and this can be interpreted as non-existence of global tangent space – the spaces m_x/m_x^2 do not glue to a nice sheaf. The source of the problem is that although the cotangent sheaf Ω^1_X is always quasi-coherent, it can be very large (e.g. $\Omega^1_{\mathbf{C}/\mathbf{Q}}$ is a

C-vector space of continual dimension), and then its dual sheaf $\mathrm{Der}_X = \mathcal{H}om_{\mathcal{O}_X}(\Omega^1_X, \mathcal{O}_X)$ can be arbitrarily bad (e.g. not quasi-coherent, or even a non-zero sheaf in a neighborhood of a point x but with zero stalk at x). Moreover, the following example shows that this can happen already for a qe trait, which is a very innocently looking scheme. (Recall that a trait X is the spectrum of a DVR, that is equivalent to X being regular, local and of dimension one.)

Example 2.8 Let k be a field and let $y = \sum_{i=0}^{\infty} a_i x^i \in k[[x]]$ be an element transcendental over $k[x]$. Then the field $K = k(x, y)$ embeds into $k((x))$ and $\mathcal{O} := k[[x]] \cap K$ is a DVR with fraction field K and completion $k[[x]]$.

(i) If $\mathrm{char}(k) = p$ and $y \in k[[x^p]]$ then $k((x)) = \widehat{K}$ is not separable over K because it contains $y^{1/p}$. In particular, the generic fiber of the completion homomorphism $\mathcal{O} \to k[[x]]$ is not geometrically reduced. This proves that the DVR $\mathcal{O}$ is not a qe ring.

(ii) Though one can show that $\mathcal{O}$ is excellent when $\mathrm{char}(k) = 0$, it still can have nasty differentials. For example, let us assume in addition that the derivative $y' = \sum_{i=1}^{\infty} i a_i x^{i-1}$ is transcendental over K. Consider the elements $y_i = x^{-n}(y - \sum_{i=0}^{n-1} a_i x^i) \in \mathcal{O}$. An easy computation shows that $\mathcal{O} = k[x, y_0, y_1, y_2, \dots]_{(x)}$ and $\Omega^1_{\mathcal{O}/k}$ is the $\mathcal{O}$-submodule of $\Omega^1_{K/k} \widetilde{\to} K dx \oplus K dy$ generated by $dx, dy_0, dy_1, \dots$. Then it follows that actually $\Omega^1_{\mathcal{O}/k} = \Omega^1_{K/k}$. (Note also that in the case when $y' \in K$, the same computation shows that $\Omega^1_{\mathcal{O}/k}$ is obtained from the free $\mathcal{O}$-module with generators dx and dy by adjoining the elements $x^{-n}(dy - y'dx)$ for all natural n.) In particular, we obtain that $\mathrm{Der}_k(\mathcal{O}, \mathcal{O}) = \mathrm{Hom}_{\mathcal{O}}(K^2, \mathcal{O}) = 0$ while $\mathrm{Der}_k(K, K) = \mathrm{Hom}_K(K^2, K) \widetilde{\to} K^2$ (and one shows similarly that $\mathrm{Der}_k(\mathcal{O}, \mathcal{O}) \widetilde{\to} \mathcal{O}$ when $y' \in K$). This shows that the sheaf of k-derivations on $\mathrm{Spec}(\mathcal{O})$ is not quasi-coherent and even has zero stalk at the closed point of $\mathrm{Spec}(\mathcal{O})$. Thus, there is no good theory of derivations on $\mathcal{O}$.

Remark 2.9 (i) We saw that Der_X can behave wildly even for a regular qe scheme X. Since all current desingularization algorithms over fields are based on derivatives, it is not clear if they can be extended to all qe schemes without serious modifications. On the other hand one might hope that they can be straightforwardly generalized to schemes that admit a closed immersion into a regular qe scheme "with good theory of derivations". See, [Tem2, 1.3.1(iii)] for a precise conjecture.

(ii) The ring $\mathcal{O}$ from Example 2.8(i) is a simplest example of a non-excellent ring. In addition, $\mathcal{O}$ is a very naively looking ring – it is a DVR, and so it is regular, local and of dimension one. As an additional demonstration of wildness of $\mathcal{O}$ we note that its normalization in $k(x, y^{1/p})$ is a DVR which is integral but not finite over $\mathcal{O}$.

2.4 Blow ups

2.4.1 Basics

Basic facts about blow ups can be found in [Tem1, §2.1] or in the literature cited there. Recall that the *blow up* $f : \mathrm{Bl}_V(X) \to X$ along a closed subscheme V is the universal morphism such that $V \times_X X'$ is a Cartier divisor. In particular, $f = \mathrm{Id}_X$ iff V is a Cartier divisor, $\mathrm{Bl}_X(X) = \emptyset$, and f is an isomorphism over $X \setminus V$. Also, $X \setminus V$ is dense in $\mathrm{Bl}_V(X)$ and so f is birational if V is nowhere dense in X. The blow up always exists and it is the projective morphism given by $\mathrm{Bl}_{\mathcal{I}}(X) \widetilde{\to} \mathrm{Proj}(\oplus \mathcal{I}^n)$ where $\mathcal{I} \subset \mathcal{O}_X$ is the ideal of V (we use our convention that X is locally noetherian and so $\mathcal{I}$ is locally finitely generated). Conversely, any projective modification is a blow up if X possesses an ample sheaf, and in any case, blow ups form a very large cofinal family among all modifications of a scheme (though the center of a typical blow up is highly non-reduced). A blow up $\mathrm{Bl}_{\emptyset}(X) \widetilde{\to} X$ is called *empty* or *trivial.*

Remark 2.10 Even empty blow ups play important role in functorial desingularization – they are responsible for synchronization.

In the sequel we adopt the convention of [Tem2] that a blow up of X consists of a morphism $f : \mathrm{Bl}_V(X) \to X$ and a center V, i.e. the blow up "remembers" its center. This approach is finer than the approach of [Tem1], where a blow up was defined as a morphism isomorphic to a morphism of the form $\mathrm{Bl}_V(X) \to X$ for some choice of V. As one may expect, we will see that the first approach is much better suited for studying functorial desingularization.

2.4.2 Operations with blow ups

Blow ups are compatible with flat morphisms $f : X' \to X$ in the sense that $\mathrm{Bl}_V(X) \times_X X' \widetilde{\to} \mathrm{Bl}_{V'}(X')$ where $V' = V \times_X X'$.

If $f : \mathrm{Bl}_V(X) \to X$ is a blow up and $Z \hookrightarrow X$ is a closed subscheme, then the scheme-theoretic preimage $Z \times_X \mathrm{Bl}_V(X)$ is called *full* or *total transform*, and we will denote it as $f^*(Z)$.

If $f : \mathrm{Bl}_V(X) \to X$ is a blow up, $Z \hookrightarrow X$ is a closed subscheme and $Z \setminus V$ denotes the open subscheme of Z obtained by removing V, then $Z \setminus V$ lifts to a subscheme in $\mathrm{Bl}_V(X)$ and its schematical closure is called the *strict transform* of Z under f and will be denoted $f^!(Z)$. For example, $f^!(Z) = \emptyset$ iff $|Z| \subset |V|$. Strict transforms are compatible with blow ups in the sense that $f^!(Z) \widetilde{\to} \mathrm{Bl}_{V|_Z}(Z)$.

If $i : U \hookrightarrow X$ is a locally closed immersion then any blow up $f : \mathrm{Bl}_W(U) \to U$ can be canonically *pushed forward* to a blow up $i_*(f) : \mathrm{Bl}_V(X) \to X$ where V is the schematic closure of W in X (W can be not reduced, so we must take the *schematical closure*, i.e. the minimal closed subscheme $V \hookrightarrow X$ with $V|_U = W$). Note that such extension is canonical because the blow up remembers its center. The restriction of $i_*(f)$ over U is f itself, i.e. $\mathrm{Bl}_W(U) \widetilde{\to} \mathrm{Bl}_V(X) \times_X U$ – this is obvious for open immersions and this follows from the properties of the strict transforms for a general locally closed immersion. Note that even if f is an isomorphism (i.e. U is a Cartier divisor) $i_*(f)$ does not have to be an isomorphism.

2.4.3 Blow up sequences

Although a composition of blow ups is known to be isomorphic to a blow up, it is not clear how to choose a center in a canonical way. If one ignores the centers one can study desingularizations by a single blow up, as it is done in [Tem1]. However, for the sake of a more explicit description of a desingularization one usually keeps all centers, i.e. considers whole *blow up sequences* $X' = X_n \to \cdots \to X_1 \to X_0 = X$ with the centers $V_i \hookrightarrow X_i$. Usually we will use the notation $X' \dashrightarrow X$ for blow up sequences. All operations with blow ups described above can be generalized to the blow up sequences straightforwardly (just iterate the construction step by step). We say that a blow up sequence $f = f_{n-1} \circ \cdots \circ f_1 \circ f_0$ is *Z-supported* for a closed subset $Z \subset X$ if all centers lie over Z.

2.5 Desingularization

2.5.1 Weak desingularization

A *weak desingularization* of an integral scheme X is a modification $f : X' \to X$ with regular source. If in addition $Z \times_X X'$ is monomial for a closed subscheme $Z \hookrightarrow X$ then we say that f is a weak desingularization of the pair (X, Z).

Remark 2.11 (i) Weak desingularization suffices to characterize qe schemes via [EGA, IV_2, §7.9.5].

(ii) Weak desingularization of varieties of characteristic zero can be proved by direct induction on dimension. One fibers X by curves and uses semi-stable modification theorem of de Jong and toroidal quotients, see [AdJ].

(iii) The essential weakness of weak desingularization is that it does not control the modification locus (i.e. the set of points of X over which the modification is not an isomorphism). In particular, a desingularization which modifies X_{reg} cannot be canonical.

(iv) The same result makes sense for any reduced scheme, but this generalization is not interesting since we can simply use normalization to separate the irreducible components.

2.5.2 Desingularization

By a (non-embedded) *desingularization* of a generically reduced scheme X we mean a blow up sequence $f : X' \dashrightarrow X$ with regular source and such that f is X_{sing}-supported. If all centers are regular then we say that f is a *strong desingularization.*

Remark 2.12 (i) Currently, all proofs of (not weak) desingularization go through embedding varieties into smooth ambient varieties and establishing an embedded desingularization, see §2.5.5. In its simplest form such approach leads to a non-strong desingularization, see [BM3, §8.2].

(ii) An additional strengthening of the notion of desingularization is to require that each blow up center is contained in the Hilbert-Samuel stratum of the largest order, where we naturally normalize the Hilbert-Samuel function by codimension (so that it becomes constant on regular schemes) and use the natural partial order on the set of all such functions. The methods of Hironaka, Bierstone-Milman and Villamayor provide such stronger desingularization, but currently it is not achieved for qe schemes.

2.5.3 Desingularization of pairs

By a *desingularization* of a pair (X, Z) we mean a $(Z \cup X_{\mathrm{sing}})$-supported blow up sequence $X' \dashrightarrow X$ with regular X' and monomial $Z' = Z \times_X X'$. Classically one splits desingularization of (X, Z) to usual desingularization $X'' \to X$ of X and subsequent embedded desingularization of $Z \times_X X'' \hookrightarrow X''$, but such splitting is not necessary and sometimes seems to be not natural, see §2.5.6.

2.5.4 Non-reduced schemes

As defined above, the desingularization is rather meaningless for generically non-reduced schemes since it just kills the generically non-reduced components. In particular, it can be easily obtained from desingularization of generically reduced schemes, and hence does not involve anything new. A "right" desingularization of such schemes is making them equisingular and it is usually achieved in the framework of strong desingularization. In particular, it was established in the works of Hironaka, Bierstone-Milman and Villamayor.

2.5.5 Embedded desingularization

Let X be a generically reduced variety of characteristic zero. Excluding special cases (e.g. low dimension), all known constructions of a desingularization $X' = X_n \dashrightarrow X_0 = X$ go by embedding X into a regular *ambient* variety M and successive blowing up M so that the strict transform of X becomes regular. Various embedded desingularization algorithms have many similar features which we only outline here.

(1) *The boundaries.* One has to take the history of the process into account, see for example [Kol, §3.6]. This is done by considering on each M_i a *boundary* E_i, which is the accumulated exceptional divisor of the blow up sequence $M_i \dashrightarrow M_0$. More concretely, E_i is an ordered set of divisors on M_i, which are called *components* of E_i and are numbered by the history function. The i-th boundary consists of the componentwise strict transform of the $(i-1)$-th boundary and the exceptional divisor of the blow up $M_i \to M_{i-1}$. The pair (M_i, E_i) is called the ambient variety with boundary, and the basic objects of the embedded desingularization are the triples (M_i, E_i, X_i). In classical embedded desingularization the boundary is always *snc*, that is, its components are regular and meet transversally. In applications one starts with $E_0 = \emptyset$ but any choice of an *snc* E_0 is fine (see below).

(2) *Permissible centers.* The center V_i of the blow up $M_{i+1} \to M_i$ is permissible in the sense that it is regular and has *simple normal crossings* with E_i, i.e. for any component $D \in E_i$ we have that locally at each point $x \in V_i \cap D$ either V_i is transversal to D or is contained in D. This ensures that each M_i is regular and each E_i is *snc*.

(3) *Principalization.* Probably, the main paradigm of embedded desingularization is to replace the desingularization problem with a very close problem of principalization of the ideal $\mathcal{I}_X \subset \mathcal{O}_M$ corresponding to X. Instead of the strict transform, one studies a *principal (controlled or weak) transform* of $\mathcal{I} = \mathcal{I}_X$ under a blow up $M' \to M$ along a permissible center. This transform is obtained from the full transform $\mathcal{I}\mathcal{O}_{M'}$ by dividing by an appropriate exceptional divisor. The ultimate aim of the principalization is to find a permissible blow up sequence $f : M' \dashrightarrow M$

such that the principal transform of $\mathscr{I}$ is $\mathcal{O}_{M'}$ and hence $\mathscr{I}' = \mathscr{I}\mathcal{O}_M$ is an exceptional divisor. In particular, f induces a desingularization of the pair (M, X). Embedded desingularization, is obtained from principalization by omitting the blow ups along components of the strict transform of X.

Remark 2.13 (i) The main advantage of the principalization is that it replaces a geometric problem with an algebraic one. In particular, it is much easier to compute principal transforms than the strict ones. In addition, all algorithms deform $\mathscr{I}$ severely in the process of principalization. This is done so that the ideal is replaced by an equivalent one, which has the same principalization. No geometric interpretation of this procedure is known so far. For example, X with an isolated singularity a is usually replaced with a highly non-reduced subscheme of M supported at a.

(ii) There are qe schemes that cannot be embedded into regular schemes (e.g. any non-catenary qe scheme). For this reason one should separately establish non-embedded and embedded desingularization of qe schemes. The first task was accomplished in [Tem2]. There are partial results on the embedded desingularization of qe schemes in [Tem3]. In particular, the centers are regular but not transversal to the boundary. Nevertheless, functorial desingularization of pairs is proved in [Tem3].

2.5.6 Non-embedded desingularization with boundary

For the sake of completeness we note that one can generalize boundaries to the non-embedded setting. A *boundary* E on a scheme X is an ordered set of locally principal closed subschemes of X (with possible repetitions). A finer form of desingularization of pairs is a desingularization of schemes with boundaries. The latter seems to be a very recent notion, which was studied only in [CJS] (for qe surfaces of all characteristics) and [Tem3] (for qe schemes over $\mathbf{Q}$). To argue why non-embedded desingularization with boundary might be a natural object to study we note that the embedded desingularization with boundary of (M, E, X) induces a non-embedded desingularization with boundary of $(X, E|_X)$ rather than just non-embedded desingularization of X.

2.5.7 Functorial desingularization

For the sake of concreteness we consider the non-embedded case in §§2.5.7–2.5.9. If $f : Y \to X$ is a regular morphism then desingularizations $g' : Y' \to Y$ and $g : X' \to X$ are *compatible* with respect to f if

g' is obtained from $g \times_X Y$ by skipping empty blow ups and, moreover, $g' = g \times_X X'$ whenever f is surjective (so, we even take the empty blow ups into account). If $\mathfrak{C}$ is a class of schemes (e.g. varieties, or qe schemes of characteristic zero) then by a *functorial desingularization on* $\mathfrak{C}$ we mean a rule $\mathscr{F}$ which to each $X \in \mathfrak{C}$ assigns a desingularization $\mathscr{F}(X) : \overline{X} \dashrightarrow X$ in a way compatible with all regular morphism between schemes from $\mathfrak{C}$, i.e. for any such morphism $f : X' \to X$ the desingularizations $\mathscr{F}(X)$ and $\mathscr{F}(X')$ are compatible with respect to f.

Remark 2.14 (i) Functoriality is a very strong property. It automatically implies desingularization in other categories including equivariant desingularization, desingularization of stacks, analytic spaces, etc. Moreover, in a seemingly paradoxical way it is usually easier to prove functorial desingularization since there is no problems with gluing local desingularizations.

(ii) When proving functorial desingularization one must be very careful with synchronizing various blow ups. For example, to construct $\mathscr{F}(X \sqcup Y)$ from $\mathscr{F}(X)$ and $\mathscr{F}(Y)$ we must compare the singularities of X and Y and decide which one is blown up earlier (or simultaneously). In other words, we amplify $\mathscr{F}(X)$ and $\mathscr{F}(Y)$ with synchronizing empty blow ups and then combine them into $\mathscr{F}(X \sqcup Y)$. This illustrates the role of the empty blow ups and explains why we worried for them in the definition of compatibility. See also [Tem2, Lem. 2.3.1, Rem. 2.3.2]. In addition, it is shown in [Tem2, Rem 2.3.4] how the idea of synchronization allows to represent any functorial desingularization as an algorithm governed by a desingularization invariant.

(iii) Since Hironaka's foundational work many improvements and simplifications were made, and one of the main achievements is that one obtains functorial desingularization. We try to outline (to some extent) the history of the subject in §2.5.8 below. Here we only note that in the recent papers [Wł], [Kol] and [BM3] one establishes functorial desingularization of varieties over a fixed field k of characteristic zero. Due to our convention from §2.5.7, this amounts to compatibility with all regular k-morphisms between k-varieties, which are precisely all smooth k-morphisms. As for the class of all varieties of characteristic zero and all regular morphisms between them, in addition to smooth morphisms these works only checked compatibility with the ground field extensions, i.e. with the regular morphisms of the form $X \otimes_k l \to X$ for a field extension l/k. It seems that full functoriality for varieties was established only in [BMT].

2.5.8 On the history of desingularization of varieties of zero characteristic

It is very difficult to present a complete history of the field. So, I will only describe three stages and will not even try to give all credits (including the very important contributions by Zariski and Giraud). The original Hironaka's proof in [Hir1] was purely existential. The proofs of the second generation started with the works [Vil1] and [Vil2] of Villamayor and [BM1] and [BM2] of Bierstone-Milman. The main focus in these (and many further) works is on constructing a canonical iterative desingularization algorithm (with history) controlled by an appropriate invariant (or a set of invariants). Canonicity of the algorithm was mainly understood as the fact that the constructed desingularization of a k-variety depended only on that variety and was compatible with open immersions, which simplified the proofs a lot. Note that the new methods heavily relied on some ideas (but not results) of Hironaka from [Hir1] an [Hir2], and, in addition, Villamayor used Hironaka's results on idealistic presentation of Hilbert-Samuel function to obtain strong desingularization. Starting with the recent work [Wł] of Wlodarczyk (who builds a self-contained non-strong desingularization algorithm but, again, heavily relies on the ideas of his predecessors), the main accent shifted to functoriality of the desingularization and to recursive description of the algorithm, sometimes making it unnecessary to introduce an invariant. In particular, it was shown in [BM3] how the algorithm of [BM2] can be rewritten in a recursive form, and it was checked that this algorithm is functorial with respect to all equidimensional smooth morphisms. Probably, all known canonical desingularization algorithms become functorial with respect to all smooth (or even regular) morphisms after minor adjustments (see, for example, §2.5.10(2)), but this was not checked for most of the algorithms yet.

2.5.9 Absolute desingularization of varieties

Intuitively it is clear that the functorial desingularization of varieties should be of absolute nature in the sense that a ground field k should not be taken into account. On the other hand, all known algorithms make extensive use of the sheaves of k-derivations $\mathrm{Der}_{X/k}$, and in principle this may be an obstacle. More concretely, the embeddings (or infinite localizations) like

$$\mathbf{A}^{n-1}_{\mathbf{Q}(x_n)} = \mathrm{Spec}(\mathbf{Q}(x_n)[x_1,\dots,x_{n-1}]) \hookrightarrow \mathrm{Spec}(\mathbf{Q}[x_1,\dots,x_n]) = \mathbf{A}^n_{\mathbf{Q}}$$

may be incompatible with the corresponding embedded desingularizations because we work with the $\mathbf{Q}(x_n)$-derivatives in the source and with all $\mathbf{Q}$-derivatives in the image. For example, for an ideal $I \subset A = \mathbf{Q}(x_n)[x_1,\dots,x_{n-1}]$ the derivative ideals $\mathrm{Der}_{A/\mathbf{Q}}(I)$ and $\mathrm{Der}_{A/\mathbf{Q}(x_n)}(I)$ used for these desingularizations are often different because the derivation along the "constant direction" $x_n = 0$ has a non-trivial effect. For this reason, it is not clear if all known algorithms are of absolute nature and are compatible with all regular morphisms (though, probably they are).

It was checked in [BMT] that the algorithm of Bierstone-Milman is of absolute nature. However, the proof used strong properties of the algorithm which are not known for some other algorithms. Moreover, it was shown in loc.cit. that the algorithm admits an absolute description if one replaces the k-derivations sheaves $\mathrm{Der}_{X/k}$ with the quasi-coherent absolute derivations sheaves $\mathrm{Der}_{X/\mathbf{Q}}$ (which can be very large). On the other hand, the following interesting result from [BMT] shows that any existing algorithm can be used to produce an absolute algorithm just by using only its "$\mathbf{Q}$-component". In other words, an absolute algorithm for varieties is the same as an algorithm for $\mathbf{Q}$-varieties.

Theorem 2.15 *Any functorial desingularization for $\mathbf{Q}$-varieties extends uniquely to a functorial desingularization of all varieties of characteristic zero, their localizations and henselizations.*

This slightly surprising Theorem is rather simple. The main idea is to use the approximation theory from [EGA, IV_2, §8] to approximate arbitrary varieties and regular morphisms between them with $\mathbf{Q}$-varieties and smooth morphisms between them. In general, such approximation is possible for any noetherian scheme over $\mathbf{Q}$, but, obviously, this is useless. So, the main observation about approximation of varieties was that each variety is a projective limit of $\mathbf{Q}$-varieties with *smooth* and affine transition morphisms. The latter smoothness condition is very special, and it reduces the problem to a standard juggling with references to [EGA, IV_2, §8]. Note that localizations and henselizations of varieties are also such special projective limits, and so we can treat them in the same Theorem. It is an interesting question if there are other natural schemes which can be represented as such limits.

2.5.10 Caveats with non-equidimensional schemes and morphisms

In the desingularization theory one should be very careful when dealing with non-equidimensional varieties and morphisms. We illustrate this by two examples.

(1) Usually, functorial embedded desingularization of X in M essentially depends on X and its codimension in M. For example, the resolution of (M, X) will run faster than that of $(\mathbf{A}^1_M, X)$ when we run them simultaneously (i.e. desingularize the disjoint union). Actually, for the algorithm from [Kol, Ch. 3] one can show that if $j : M \hookrightarrow M'$ is a closed immersion with regular M' then $\mathscr{E}(M', X)$ and $\mathscr{E}(M, X)$ induce the same desingularization of X only when j is of constant codimension. Probably, the same is true for many other algorithms.

(2) It was recently noted by O. Gabber that the algorithm of Bierstone-Milman in [BM3] is only functorial with respect to equidimensional smooth morphisms. A simple modification in the algorithm proposed in [BMT, §6.3] made the algorithm functorial with respect to all smooth (and regular) morphisms. Actually, one just adjusts the synchronization slightly (in a sense, one replaces synchronization by dimension with synchronization by codimension).

2.5.11 Strict desingularization and a caveat with non-monomial locus

By a *strict* desingularization of a pair (X, Z) in [Tem1] one means a desingularization $X' \to X$ that modifies only the non-monomial locus of (X, Z). This definition seemed natural to me but it turned out that it does not make much sense. A detailed analysis can be found in [Tem3, §A.1.3]. Here we only note that functorial strict desingularization does not exist even for varieties, and the assertion of [Tem1, Th. 2.2.11] should be corrected as explained in [Tem3, Rem. A.1.1].

2.6 Formal schemes

2.6.1 Quasi-excellent formal schemes

By a *formal variety* we mean a noetherian formal scheme $\mathfrak{X}$ whose special fiber $\mathfrak{X}_s$ (defined by the maximal ideal of definition) is a variety. An important stage of our method is desingularization of formal varieties of characteristic zero, so we will explain briefly how the desingularization setup extends to formal varieties. Formal varieties are *excellent* by results of Valabrega, see [Val]. That is, for any affine formal variety $\mathrm{Spf}(A)$ the ring A is excellent. Since everything applies to general qe formal schemes, we will work in such larger generality.

Remark 2.16 (i) To have a reasonable theory of qe formal schemes (other than formal varieties) one has to invoke Gabber's theorem from §2.3.3(iii). Otherwise, one does not even know that quasi-excellence is preserved by formal localizations. Also, it is Gabber's theorem that implies that (quasi-) excellence is preserved by formal completion.

(ii) The main intermediate progress towards Gabber's Theorem was done in the paper [NN] by Nishimura-Nishimura, where the same result was proved conditionally assuming weak resolution of singularities for local qe schemes. In particular, this settled the case of characteristic zero by using Hironaka's theorem (which covers local qe schemes). Alternatively, one can use the results of [Tem1] as the desingularization input.

(iii) Gabber strengthened the proof of [NN] so that desingularization of local qe schemes is replaced with a regular cover in the topology generated by alterations and flat quasi-finite covers. This argument is outlined in Gabber's letter to Laszlo. The existence of such a regular cover for any qe scheme is a subtle and important result by Gabber whose written version will (hopefully) be available soon. Actually, it is the only desingularization result established for all qe schemes.

2.6.2 Regularity for qe schemes

The underlying topological space of a formal scheme is too small to hold enough information even about reduced formal subschemes. For this reason we define the singular locus of a formal scheme $\mathfrak{X}$ as a closed formal subscheme rather than as a subset in $|\mathfrak{X}|$ (in particular, no regular locus is defined, though we remark for the sake of completeness that one could work set-theoretically at cost of considering also a generic fiber of $\mathfrak{X}$ in one of the non-archimedean geometries). If $\mathfrak{X} = \mathrm{Spf}(A)$ then we take for the singular (resp. non-reduced or non-equisingular) locus the ideal defining $\mathrm{Spec}(A)_{\mathrm{sing}}$ (or other loci), and it turns out that for qe formal schemes such definition is compatible with formal localizations and hence globalizes to general qe formal schemes. Obviously, we use here that formal localization morphisms are regular on qe formal schemes. Most probably, regularity and even reducedness does not make sense for general noetherian formal schemes. We say that $\mathfrak{X}$ is *regular* (resp. *reduced* or *equisingular*) if the singular (resp. non-reduced or non-equisingular) locus is empty. We say that $\mathfrak{X}$ is *rig-regular* if the singular locus is given by an open ideal, and hence is a usual scheme. Intuitively, the latter means that the generic fiber of $\mathfrak{X}$ is regular (and this makes precise sense in non-archimedean geometry).

Regular and reduced loci are preserved by formal completions. Also, one uses a similar definition to introduce the notion of regular morphisms between qe formal schemes and shows that the regular and reduced loci are compatible with regular morphisms similarly to the case of schemes.

2.6.3 Blow up sequences

The notion of the formal blow up $\widehat{\mathrm{Bl}}_{\mathfrak{V}}(\mathfrak{X})$ along a closed formal subscheme $\mathfrak{V}$ can be defined similarly to the case of schemes. Then the formal blow up sequences are defined obviously. These notions are compatible with formal completions, i.e. the $\mathscr{I}$-adic completion of $\mathrm{Bl}_V(X)$ is $\widehat{\mathrm{Bl}}_{\mathfrak{V}}(\mathfrak{X})$, where $\mathfrak{V}$ and $\mathfrak{X}$ are the $\mathscr{I}$-adic completions of V and X, respectively. All properties of usual blow ups are generalized straightforwardly to the formal case, see [Tem1, §2.1].

2.6.4 Formal desingularization

Since regular formal schemes and formal blow ups are defined, one defines desingularization of formal schemes similarly to desingularization of schemes (including embedded desingularization, etc.).

3 The method and the main results

3.1 Results

3.1.1 The non-embedded case

The main result of [Tem2] is that the class of all qe schemes over $\mathbf{Q}$ admits a strong non-embedded desingularization. Here is a detailed formulation of this result.

Theorem 3.1 *For any noetherian quasi-excellent generically reduced scheme $X = X_0$ over* $\mathrm{Spec}(\mathbf{Q})$ *there exists a blow up sequence $\mathscr{F}(X) : X_n \dashrightarrow X_0$ such that the following conditions are satisfied:*

(i) the centers of the blow ups are disjoint from the preimages of the regular locus X_{reg};

(ii) the centers of the blow ups are regular;

(iii) X_n is regular;

(iv) the blow up sequence $\mathscr{F}(X)$ is functorial with respect to all regular morphisms $X' \to X$, in the sense that $\mathscr{F}(X')$ is obtained from $\mathscr{F}(X) \times_X X'$ by omitting all empty blow ups.

Remark 3.2 An algorithm $\mathscr{F}$ will be constructed from an algorithm $\mathscr{F}_{\mathrm{Var}}$ for varieties, and we saw in §2.5.9 that $\mathscr{F}_{\mathrm{Var}}$ is completely defined by its restriction $\mathscr{F}_{\mathbf{Q}}$ onto the $\mathbf{Q}$-varieties. So, in some sense $\mathscr{F}$ is defined over $\mathbf{Q}$. Note, however, that $\mathscr{F}$ is obtained by "breaking $\mathscr{F}_{\mathrm{Var}}$ to pieces" and reassembling them into a new algorithm, so it differs from $\mathscr{F}_{\mathbf{Q}}$ even on $\mathbf{Q}$-varieties. This is necessary in order to have functoriality on all qe schemes.

3.1.2 The embedded case

Here is the main result of [Tem3] formulated in the language of embedded desingularization, see [Tem3, Th. 1.1.6]. Up to the non-embedded desingularization Theorem 3.1, this can be reformulated in the language of non-embedded desingularization with boundary. We do not discuss such approach in this survey and refer to [Tem3] for details.

Theorem 3.3 *For any quasi-excellent regular noetherian scheme X of characteristic zero with an snc boundary E and a closed subscheme $Z \hookrightarrow X$ there exists a blow up sequence $f = \mathscr{E}(X, E, Z) : X' \dashrightarrow X$ such that*

(i) X' is regular, the new boundary E' is snc and the strict transform $Z' = f^!(Z)$ is regular and has simple normal crossings with E',

(ii) each center of f is regular and for any point x of its image in X either Z is not regular at x or Z has not simple normal crossings with E at x,

(iii) $\mathscr{E}$ is functorial in exact regular morphisms; that is, given a regular morphism $g : Y \to X$ with $D = E \times_X Y$ and $T = Z \times_X Y$, the blow up sequence $\mathscr{E}(Y, D, T)$ is obtained from $g^*(\mathscr{E}(X, E, Z))$ by omitting all empty blow ups.

Remark 3.4 The main weakness of this result is that the functor $\mathscr{E}$ does not possess two important properties satisfied by classical embedded desingularization functors.

(1) The centers of $\mathscr{E}$ do not have to have normal crossings with the intermediate boundaries. In particular, intermediate boundaries can be not snc, and even the iterative definition of these boundaries given in §2.5.5(2) should be corrected by replacing strict transform with principal transform (see [Tem3, §2.2]).

(2) $\mathscr{E}$ does not resolve the principal transform of Z. In particular, this cannot be used to obtain a classical principalization of $\mathscr{I}_Z$ as defined in §2.5.5(3). However, if Z is a Cartier divisor (this situation is classically called "the hypersurface case") then $\mathscr{E}$ induces a principalization.

3.1.3 Desingularization of pairs

The strong principalization from §2.5.5(3) is not achieved for qe schemes so far. However, the functors $\mathscr{F}$ and $\mathscr{E}$ can be used to obtain a functorial desingularization of pairs.

Theorem 3.5 *For any quasi-excellent noetherian generically reduced scheme X of characteristic zero with a closed subscheme $Z \hookrightarrow X$ there exists a $(Z \cup X_{\mathrm{sing}})$-supported blow up sequence $\mathscr{P}(X,Z): X' \dashrightarrow X$ such that X' is regular, $Z \times_X X'$ is strictly monomial and $\mathscr{P}$ is functorial in exact regular morphisms.*

The proof is very simple. First we blow up X along Z achieving that the full transform of Z becomes a Cartier divisor (this is an obvious principalization). Set $X' = \mathrm{Bl}_Z(X)$ and $Z' = Z \times_X X'$. Then we apply $\mathscr{F}$ to desingularize X'. Note that this step is needed even if we started with regular X. Let $\mathscr{F}(X'): X'' \dashrightarrow X'$ and $Z'' = Z \times_X X''$. Finally, we apply $\mathscr{E}(X'', \emptyset, Z'')$ to monomialize Z. Note that a non-functorial desingularization of pairs is the main result of [Tem1], and Theorem 3.5 is a major strengthening of that result which was proved in [Tem3].

Remark 3.6 The main disadvantage of our construction is that even when X is regular $\mathscr{P}$ can blow up a non-regular center at the first step. This is only needed when Z is not a Cartier divisor.

3.1.4 Semi-stable reduction

In this section we just repeat the arguments from [KKMS, Ch. II, §3]. Assume that $\mathscr{O}$ is an excellent DVR of characteristic zero and $S = \mathrm{Spec}(\mathscr{O})$. Let η and s be its generic and closed points, respectively. Assume that X is a reduced flat S-scheme of finite type and with smooth generic fiber X_η. It is well known that using a desingularization $Y \to X$ of the pair (X, X_s) one can construct a DVR $\mathscr{O}'$ with a quasi-finite morphism $S' = \mathrm{Spec}(\mathscr{O}') \to S$ and a modification $X' \to X \times_S S'$ such that the morphism $X' \to S'$ is *semi-stable* (i.e. étale-locally X' is of the form $\mathrm{Spec}(\mathscr{O}'[t_1, \dots, t_n]/(t_1 \dots t_m - \pi'))$ for a non-zero $\pi' \in \mathscr{O}'$). Indeed, due to the assumption on the characteristic, Y is étale-locally of the form $\mathrm{Spec}(\mathscr{O}[t_1, \dots, t_n]/(t_1^{e_1} \dots t_m^{e_m} - \pi))$ for a non-zero $\pi \in \mathscr{O}$. Hence one can choose any $\mathscr{O}'$ whose ramification degree e over $\mathscr{O}$ is divided by all e_i and take X' to be the normalization of $Y \times_S S'$. Note that X' as above is regular if and only if π' is a uniformizer. A complicated but purely combinatorial method to achieve that X' is also regular is described in [KKMS]. It involves few blow ups along the strata of the reduction of

X'_s and its preimages (note that they are snc divisors) and an additional extension of the DVR. The algorithm is described in terms of the simplicial complex formed by the strata and the multiplicities of these strata in the closed fiber. In particular, although originally formulated in the context of varieties, it applies to our situation verbatim. This establishes the following Theorem.

Theorem 3.7 *Assume that $\mathcal{O}$ is an excellent DVR of characteristic zero, $S = \mathrm{Spec}(\mathcal{O})$ and X is an S-scheme of finite type and with smooth generic fiber. Then there exists a DVR $\mathcal{O}'$ with a quasi-finite morphism $S' = \mathrm{Spec}(\mathcal{O}') \to S$ and a modification $X' \to X \times_S S'$ such that X' is regular and the special fiber $X_{s'}$ is an snc divisor (in particular, X' is semi-stable over S').*

Remark 3.8 The first step in our construction used $\mathscr{P}$, so it is functorial. Functoriality of the whole construction depends, thereby, only on the functoriality of the combinatorial algorithm. The algorithm from [KKMS, Ch. III] seems to be not functorial (or canonical), but it seems very probable that functorial algorithms for this problem should exist. So, I expect that the ramification degree of $\mathcal{O}'/\mathcal{O}$ and the modification $X' \to X \times_S S'$ (for a fixed S' with correct ramification) can be chosen functorially.

3.2 The method

A very general idea of desingularizing qe schemes was discussed in Remark 2.4: one wants to pass to formal varieties by completion along subvarieties and desingularize the obtained formal varieties either by algebraization or by generalizing the algorithms for algebraic varieties. The technical background is provided by the following easy Lemma.

Lemma 3.9 *Let X be a qe scheme such that X_{sing} is contained in a closed subvariety $Z \hookrightarrow X$ (e.g. $Z = X_{\mathrm{sing}}$) and let $\mathfrak{X}$ be the formal completion of X along Z. Then $\mathfrak{X}$ is a rig-regular formal variety and the formal completion induces a bijective correspondence between desingularizations of X and $\mathfrak{X}$.*

The main point of the proof is that $\mathfrak{X}_{\mathrm{sing}}$ is given by an open ideal and hence any desingularization of $\mathfrak{X}$ blows up only open ideals. Since open ideals live on a nilpotent neighborhood of $\mathfrak{X}_s$ they algebraize to closed subschemes of X and hence the entire blow up sequence algebraizes as well. The Lemma (and few more similar claims) implies that any functorial desingularization $\widehat{\mathcal{F}}_{\mathrm{Var}}$ of formal varieties algebraizes uniquely to a functorial desingularization $\mathcal{F}_{\mathrm{small}}$ on the class of all qe schemes such that their singular locus is a variety. Now we can explain very generally what are the two main stages of our method, and we will describe them in more details in §§4–5.

Stage 1. *Algebraization.* The aim of this stage is to extend $\mathcal{F}_{\mathrm{Var}}$ to $\mathcal{F}_{\mathrm{small}}$. As we explained above this reduces to constructing a desingularization functor $\widehat{\mathcal{F}}_{\mathrm{Var}}$ for rig-regular formal varieties. The main tool is Elkik's theory which provides an algebraization of affine rig-regular formal schemes with a principal ideal of definition. It allows to easily extend desingularization of varieties to rig-regular formal varieties in a non-canonical way, see [Tem1, Th. 3.4.1]. In [Tem2], much more delicate arguments were used in order to make this construction partially functorial. For technical reasons related with Elkik's theory, $\widehat{\mathcal{F}}_{\mathrm{Var}}$ was only constructed for formal varieties with a fixed invertible ideal of definition and for regular morphisms that respect these ideals.

Remark 3.10 (i) It is difficult to control functoriality since the algebraization procedure is absolutely non-canonical. Much worse, as we observed in Remark 2.2(i) even the ground field of algebraization is not canonical. The latter turns out to be the main trouble since it is not easy to show that the existing desingularization algorithms for varieties essentially depend only on the formal completion viewed as an abstract formal scheme (i.e. without fixed morphism to a ground field). To illustrate the problem we note that our formal algorithm must be equivariant also with respect to the automorphisms not preserving any ground field, and such automorphisms do not have to be algebraizable by étale morphisms of varieties (compare with [Kol, 3.56] where all morphisms are defined over some ground field).

(ii) Most probably, $\widehat{\mathcal{F}}_{\mathrm{Var}}$ is fully functorial. The main problem of this stage is in establishing the properties of $\widehat{\mathcal{F}}_{\mathrm{Var}}$ rather than in constructing it.

Stage 2. *Localization* The aim of this stage is to reduce the general case to the case of schemes whose singular locus is a variety, that is, to construct a functorial desingularization $\mathcal{F}$ using a desingularization functor $\mathcal{F}_{\mathrm{small}}$ as an input. In a sense, we localize the desingularization problem at this stage. Although, we cannot reduce to the case of a local scheme with an isolated singularity, we will only use $\mathcal{F}_{\mathrm{small}}(X)$ for rather special schemes X such that X_{sing} is a variety. Namely, it will be enough to know $\mathcal{F}_{\mathrm{small}}(X)$ for a scheme X which can be represented as a blow up of a local scheme Y such that X_{sing} is contained in the preimage of the closed point of Y (and hence X_{sing} is a variety).

Remark 3.11 The algorithm $\mathcal{F}_{\text{small}}$ is an extension of $\mathcal{F}_{\text{Var}}$, i.e. both agree on varieties. During the localization stage a new algorithm $\mathcal{F}$ is produced from $\mathcal{F}_{\text{small}}$. We will see in §5.2 that $\mathcal{F}_{\text{Var}}$ and $\mathcal{F}$ differ already on algebraic curves. The construction of $\mathcal{F}_{\text{Var}}$ uses derivatives and embedded desingularization, so it seems that it cannot be generalized to all qe schemes. On the other hand, we have to build $\mathcal{F}$ for all qe schemes in a "uniform way". Thus, it seems almost unavoidable that $\mathcal{F}$ differs from $\mathcal{F}_{\text{Var}}$ on varieties.

4 Algebraization

Unless said to the contrary, we assume until the end of the paper that the characteristic is zero, i.e. all schemes are **Q**-schemes. The algebraization stage is rather subtle and technical and it is the bottleneck of the method. In particular, it is "responsible" for most of the cases that elude from our method, including generically non-reduced varieties, etc.

4.1 Non-embedded rig-regular case

Elkik's Theorem [Elk, Th. 7] implies that rig-regular formal varieties of characteristic zero with an invertible (or locally principal) ideal of definition are *locally algebraizable* in the sense that they are locally isomorphic to completions of varieties. For this reason we consider the pairs $(\mathfrak{X}, \mathfrak{I})$ where $\mathfrak{X}$ is a rig-regular formal variety and $\mathfrak{I}$ is an invertible ideal of definition. We will be only able to construct a formal desingularization $\widehat{\mathcal{F}_{\text{Var}}}(\mathfrak{X}, \mathfrak{I}) : \mathfrak{X}_n \dashrightarrow \mathfrak{X}$ associated to such a pair and functorial with respect to regular morphisms $\mathfrak{X}' \to \mathfrak{X}$ such that $\mathfrak{I}' = \mathfrak{I}\mathcal{O}_{\mathfrak{X}'}$. Most probably, $\widehat{\mathcal{F}_{\text{Var}}}$ is independent of $\mathfrak{I}$ and is fully functorial, but this was not proved. For shortness, let us say that $\mathfrak{X}$ is a *principal formal variety* if it is rig-regular, affine, and with fixed principal ideal of definition. Morphisms between such objects must be compatible with the fixed ideals.

Since we are going to establish functorial desingularization, it is enough to work locally. So, we can assume that $\mathfrak{X}$ is principal and then $\mathfrak{X}$ is algebriazable by [Elk, Th. 7] and [Tem1, 3.3.1], in the sense that $\mathfrak{X} = \widehat{X}$ and $\mathfrak{I} = \widehat{\mathcal{I}}$ for an affine variety X with a principal ideal $\mathcal{I}$. In particular, the desingularization $\mathcal{F}_{\text{Var}}(X)$ induces a desingularization $\widehat{\mathcal{F}_{\text{Var}}}(\mathfrak{X}, \mathfrak{I})$ of $\mathfrak{X}$. The only thing we should do is to check that $\widehat{\mathcal{F}_{\text{Var}}}(\mathfrak{X}, \mathfrak{I})$

is well defined (i.e. is independent of the choice of the algebraization) and functorial. The main idea beyond the argument is that all information about $\mathfrak{X}$ can be read off already from an infinitesimal neighborhood $X_n := (\mathfrak{X}, \mathcal{O}_{\mathfrak{X}}/\mathfrak{I}^n) = \mathrm{Spec}(\mathcal{O}_X/\mathcal{I}^n)$ with large n. This is based on [Tem2, 3.2.1] which is an easy corollary of Elkik's theory. Roughly speaking, this result states that if $\mathfrak{X}$ and $\mathfrak{X}'$ are principal formal varieties and $n = n(\mathfrak{X})$ is sufficiently large then any isomorphism $X'_n \widetilde{\to} X_n$ lifts to an isomorphism $\mathfrak{X}' \widetilde{\to} \mathfrak{X}$.

Thus, it is clear that all information about the desingularization of $\mathfrak{X}$ should be contained in some X_n, though it is not so easy to technically describe this; especially because we want to prove functoriality of the entire blow up sequence but only the first center is contained in X_n. We refer to [Tem2, §3.2] for a realization of this plan. To illustrate some technical problems that one has to solve we note that if X is an algebraization of $\mathfrak{X}$ and $f : X^{(p)} \dashrightarrow X^{(0)} = X$ is its desingularization then the sequence of n-th fibers $f_n : X_n^{(p)} \dashrightarrow X_n^{(0)}$ is not determined by $X_n^{(0)} = X_n$ (for any n). However, one can show that for sufficiently large numbers k, n with $n \gg kp$ the tower

$$\mathcal{F}_{n,k}(X_n^{(0)}, \mathcal{I}_n) : X_{n-kp}^{(p)} \to \cdots \to X_{n-k}^{(1)} \to X_n^{(0)}$$

is uniquely determined (up to a unique isomorphism) only by $X_n^{(0)}$ with the ideal $\mathcal{I}_n = \mathcal{I}\mathcal{O}_{X_n^{(0)}}$ and, moreover, is functorial in $X_n^{(0)}$ with respect to all regular morphisms. The above $\mathcal{F}_{n,k}$ is a functor of sequences of morphisms (not blow ups!) on certain non-reduced schemes with fixed principal ideal which are called Elkik fibers in [Tem2], and $\mathcal{F}_{n,k}$ is the heart of the technical proof that $\widehat{\mathcal{F}}_{\mathrm{Var}}$ is a well defined functor.

4.2 Limitations

The limitations of our algebraization method are related to the assumptions in Elkik's theory. For example, in the algebraization theorem [Elk, Th. 7] one assumes that the formal scheme is rig-smooth over the base and possesses a principal ideal of definition. In addition, no result for algebraization of a pair (X, Z), where $Z \hookrightarrow X$ is a closed subscheme, is known. Let us discuss what is the impact of these assumptions on our method.

4.2.1 Closed subschemes

I do not know if algebraization of pairs is possible under reasonable assumptions (say, $\mathfrak{X}$ is regular and $\mathfrak{Z} \hookrightarrow \mathfrak{X}$ is rig-regular). I hope that some progress in this direction is possible and this should be studied in the future. Currently, the lack of algebraization of pairs is the main reason that our embedded desingularization theorem is much weaker than its classical analog. In [Tem3] one only uses algebraization of pairs $(\mathfrak{X}, \mathfrak{Z})$, where $\mathfrak{X}$ is rig-regular and $\mathfrak{Z}$ is supported on the closed fiber (and hence is a scheme).

4.2.2 Rig-regularity

Simple examples show that rig-smoothness is necessary in order for algebraization to exist. Localization stage reduces desingularization of generically reduced schemes to desingularization of rig-regular varieties, and rig-smoothness is equivalent to rig-regularity in characteristic zero (see, for example, [Tem1, §3.3]). Thus, the algebraization stage as it is works only in characteristic zero (even if resolution of varieties would be known). Over $\mathbf{Q}$ the only limitation imposed by the rig-regularity assumption is that our method does not treat generically non-reduced schemes. To deal with the latter case on should desingularize *rig-equisingular* formal varieties (i.e. formal varieties whose non-equisingular locus is given by an open ideal), but the latter formal schemes are not locally algebraizable in general.

4.2.3 Principal ideal of definition

I do not know if this assumption in Elkik's theorem is necessary. It causes to certain technical difficulties in our proofs and extra-assumptions in intermediate results, but does not affect the final results. Step 1 in the localization stage (see §5) is needed only because of this assumption. Also, it is this assumption that makes us to define the functor $\widehat{\mathcal{F}}_{\mathrm{Var}}(\mathfrak{X}, \mathfrak{I})$ rather than a functor $\widehat{\mathcal{F}}_{\mathrm{Var}}(\mathfrak{X})$.

5 Localization

The localization stage is very robust, and can be adopted to work with almost all types of desingularization, including embedded desingularization, and desingularization of non-reduced schemes. Also, it is not sensitive to the characteristic. For simplicity, we will stick with the non-embedded case which is established in [Tem2, §4.3].

5.1 Construction of $\mathcal{F}$

Consider the category $\mathfrak{C}_{\mathrm{small}}$ whose elements are pairs (X, D) where X is a noetherian generically reduced qe scheme, $D \hookrightarrow X$ is a closed subvariety which is a Cartier divisor, and morphisms $(X', D') \to (X, D)$ are regular morphisms $f : X' \to X$ such that $D' = f^*(D)$. By the algebraization stage and Lemma 3.9, the original desingularization functor $\mathcal{F}_{\mathrm{Var}}$ extends to a desingularization functor $\mathcal{F}_{\mathrm{small}}$ on $\mathfrak{C}_{\mathrm{small}}$. The aim of the localization stage is to construct a desingularization $\mathcal{F}$ of qe schemes using $\mathcal{F}_{\mathrm{small}}$ as an input.

The construction of $\mathcal{F}$ goes by induction on codimension, i.e. we will construct inductively a sequence of blow up sequence functors $\mathcal{F}^d$ which desingularize X over $X^{\leq d}$, where the latter denotes the set of points of X of codimension at most d. Intuitively (and similarly to §2.4.2), each $\mathcal{F}^d(X)$ is the pushout of the desingularization $\mathcal{F}(X)|_{X^{\leq d}}$ under the embedding $X^{\leq d} \hookrightarrow X$, i.e. it is the portion of $\mathcal{F}^d(X)$ defined by the situation over $X^{\leq d}$. More specifically, each center of $\mathcal{F}(X)$ has a dense subset lying over $X^{\leq d}$ and $\mathcal{F}(X)$ is obtained from $\mathcal{F}^d(X)$ by inserting (in all places) few new blow ups whose centers lie over $X^{>d} := X \setminus X^{\leq d}$. Thus, the resulting algorithm works as follows. Take empty $\mathcal{F}^0(X)$. That is, start with X, which is the canonical desingularization of itself over $X^{\leq 0}$ (we use that X is generically reduced by our assumption). First we resolve the situation over the points of X of codimension one by a functor $\mathcal{F}^1$ (without caring for other points). Then we improve $\mathcal{F}^1$ over finitely many points of codimension two and leave the situation over the codimension one points unchanged. This gives a functor $\mathcal{F}^2$ which agrees with $\mathcal{F}^1$ over the codimension 1 points and resolves each generically reduced qe scheme X over $X^{\leq 2}$. We proceed similarly ad infinitum, but for each noetherian X the process stops after finitely many steps by noetherian induction.

Now let us describe how $\mathcal{F}^d$ is constructed from $\mathcal{F}^{d-1}$ and $\mathcal{F}_{\mathrm{small}}$. Given a blow up sequence $f : X' \dashrightarrow X$ by its *unresolved locus* f_{sing} we mean the set of points of X over which f is not a strong desingularization. In other words, f_{sing} is the union of the images of the singular loci of both X' and the centers of f. By the induction assumption, $\mathcal{F}^{d-1}(X)_{\mathrm{sing}}$ is of codimension at least d and hence it contains only finitely many points $x_1, \dots, x_m$ of exact codimension d. We should only improve $\mathcal{F}$ over these points, and the latter is done as follows.

Step 1. As a first blow up we insert the simultaneous blow up at all new points $x_1, \dots, x_m$ (we act simultaneously in order to ensure functoriality). Then the preimage of each x_i on any intermediate blow up of the sequence is a Cartier divisor (which will be needed later in order to use $\mathcal{F}_{\mathrm{small}}$).

Step 2. Next, we improve all centers of $\mathcal{F}^{d-1}(X)_{\mathrm{sing}}$ over x_i's by resolving the singularities of these centers over x_i's. We use here that these singularities are of codimension at most $d-1$ in the centers and so we can apply the functor $\mathscr{F}^{d-1}$. To summarize, before blowing up each center V_j of $\mathscr{F}^{d-1}$ we insert a blow up sequence which desingularizes V_j over x_i's.

Step 3. At the last step we obtain a sequence $X' \dashrightarrow X$ of blow ups whose centers are regular over $X^{\leq d}$, but X' may have singularities over x_i. Observe that the singular locus of the scheme

$$X_{x_i} = \mathrm{Spec}(\mathscr{O}_{X,x_i}) \times_X X'$$

is contained in the preimage of x_i, and the preimage of x_i is a Cartier divisor E_i (thanks to Step 1) which is a variety over $k(x_i)$. So, X_{x_i} can be resolved by a blow up sequence $f_i = \mathscr{F}_{\mathrm{small}}(X_{x_i}, E_i)$. It remains to extend all f_i's to a blow up of X' and to synchronically merge them into a single blow up sequence $X'' \dashrightarrow X'$. The composition $\mathscr{F}^d(X) : X'' \dashrightarrow X$ is a required desingularization of X over $X^{\leq d}$ which coincides with $\mathscr{F}^{d-1}$ over $X^{\leq d-1}$.

Remark 5.1 The center of a blow up is often reducible, and in Step 3 of the construction of $\mathscr{F}^d$ we often obtain a center with many components that are regular over x_i's but probably have non-empty intersections over $X^{>d}$. Thus, it is important that in Step 2 of the construction of $\mathscr{F}^d$ we are able to desingularize the reducible blow up centers inherited from $\mathscr{F}^i$ with $i < d$. In particular, even if we are only interested to desingularize integral schemes, we essentially use in our induction that the desingularization is constructed for all reduced schemes.

5.2 Examples

We will compare $\mathscr{F}$ and $\mathscr{F}_{\mathrm{Var}}$ in the case of few simple varieties. For $\mathscr{F}_{\mathrm{Var}}$ we take the desingularization functor of Bierstone-Milman, which is functorial in all regular morphisms by [BMT].

5.2.1 Plain curves

Assume that X is a generically reduced plain algebraic curve. A strong desingularization is uniquely defined up to synchronization because one has to blow up the singular points until the curve becomes smooth. On the other hand, synchronization of these blow ups depends on various choices. Note that the Hilbert-Samuel strata of X are the equimultiplicity strata because X embeds into a smooth surface. It follows easily that $\mathscr{F}_{\mathrm{Var}}$ is synchronized by the multiplicity. Namely, one blows up the points of maximal multiplicity at each step until all points are smooth. The synchronization of $\mathscr{F}$ is slightly different. Because of Step 1 in the localization stage we simultaneously blow up all singular points once. After that we skip Step 2 and use $\mathscr{F}_{\mathrm{Var}}$ at Step 3. To summarize, we blow up all singularities once, and then switch to the synchronization by multiplicity, similarly to $\mathscr{F}_{\mathrm{Var}}$.

Remark 5.2 The same description holds true for varieties with isolated singularities. At the first stage, $\mathscr{F}$ blows up all these singularities. One obtains a blow up $X' \to X$ and then simply applies $\mathscr{F}_{\mathrm{Var}}(X')$. Thus, $\mathscr{F}$ blows up the same centers as $\mathscr{F}_{\mathrm{Var}}$ but the synchronization can be different when X has more than one singular point.

5.2.2 Surfaces

Let us consider examples of the next level of complexity. Namely, let X be a surface such that $C = X_{\mathrm{sing}}$ is a curve with the set of generic points $\eta = \{\eta_1, \dots, \eta_n\}$. The functor $\mathscr{F}^1$ acts as follows. On the semi-local curve X_η, which is the semi-localization of X at η, $\mathscr{F}^1$ acts as was explained in §5.2.1. We extend the blow up sequence $\mathscr{F}^1(X_\eta) : X'_\eta \dashrightarrow X_\eta$ to a blow up sequence $\mathscr{F}^1(X) : X' \dashrightarrow X$ in the natural way (that is, the centers of $\mathscr{F}^1(X)$ are the Zariski closures of the centers of $\mathscr{F}^1(X_\eta)$). After that we produce $\mathscr{F}(X) = \mathscr{F}^2(X)$ by inserting new blow ups into $\mathscr{F}^1(X)$. This is done in three steps described in §5.1. All new blow ups will be inserted over the set $b = (b_1, \dots, b_m)$ such that $\mathscr{F}^1(X)$ is not a strong desingularization precisely over the points of b. In particular, the first blow up is along b.

Example 5.3 If $a \in X$ is a "generic point" of C then $\mathscr{F}(X) = \mathscr{F}^1(X)$ over a neighborhood of a. For example, this is the case of any surface of the form $X = \mathbf{A}^1_k \times Y$ for a curve Y. One easily sees that in this case $\mathscr{F}$ and $\mathscr{F}_{\mathrm{Var}}$ differ only by synchronization, as in the case of curves.

Next we consider two examples of a Cartier divisor in $\mathbf{A}^3_k = \mathrm{Spec}(k[x, y, z])$ with a non-isolated and "non-generic" singularity at the origin a. For the sake of comparison, we will show how $\mathscr{F}_{\mathrm{Var}}$ resolves the same examples. For reader's convenience, a brief explanation of how $\mathscr{F}_{\mathrm{Var}}$ can be computed in these examples will be given in §5.2.3.

Example 5.4 Whitney umbrella X is given by $y^2 + xz^2 = 0$. In this case C is the x-axis, and blowing it up resolves all singularities. So, $\mathscr{F}(X) = \mathscr{F}^2(X) = \mathscr{F}^1(X)$ just blows up C. However, a is not a "generic point" of the singular locus and other algorithms feel this. In particular, $\mathscr{F}_{\mathrm{Var}}$ first blows up a. The blow up $X_1 = \mathrm{Bl}_a(X)$ is covered by two charts: the x-chart X_{1x} and the z-chart X_{1z}. Since X_{1x} is defined by $y_1^2 + x_1 z_1^2 = 0$ for $x_1 = x$, $y_1 = y/x$ and $z_1 = z/x$, we see that it has the same singularity as X. Namely, the singular locus C_1 is the line $y_1 = z_1 = 0$ (which is the strict transform of C). Since X_{1z} is defined by $y_1^2 + x_1 z_1 = 0$ for $x_1 = x/z$, $y_1 = y/z$, $z_1 = z$ (for simplicity we denote the local coordinates by the same letters as earlier), the singularity of X_{1z} is the isolated orbifold point c_1 given by $x_1 = y_1 = z_1 = 0$. Because of a synchronization issue, C_1 is dealt with first (a non-monomial ideal $\mathcal{N}$ has order two along C_1 and order one at c_1, see §5.2.3 for a similar computation). The second blow up is, again, at the pinch point $a_1 \in C_1$. The same computation as above shows that the singular locus of X_2 consists of C_2, which is the strict transform of C (and C_1), an orbifold point c_2', which sits over a_1 and an orbifold point c_2 which is the preimage of c_1. The third blow up is along C_2, so the singular locus of X_3 consists of two orbifold points c_3 and c_3' sitting over c_2 and c_2', and the last blow up is along $\{c_3, c_3'\}$. This resolves X completely. Note also that for $0 \le i \le 2$ the local structure of X_i along C_i is similar. Nevertheless, the algorithm blows up the pinch point twice and then decides to blow up the whole singular line C_2 because of the history of the process.

Example 5.5 Let Y be given by $y^2 + xz^3 = 0$. The entire resolution process is too messy, so we will only work with a sequence of affine charts. As we saw in the previous Example, there might be (and there are) blow ups on the other charts that may (but do not have to) be simultaneous with the blow ups on our charts. In this case, we also have that $C = Y_{\rm sing}$ is the x-axis and $\mathcal{F}^1(Y) : \widetilde{Y} \to Y$ is the blow up along C. Let us describe an *affine chart* $Y_n \dashrightarrow Y_0 = Y$ of $\mathcal{F}(Y) : \overline{Y}_{\overline{n}} \dashrightarrow \overline{Y}_0 = Y$. It is obtained by choosing each time an appropriate affine chart of the blow up and restricting the remaining sequence over that chart. In particular, $n < \overline{n}$ because of the synchronization with other charts, which we ignore for simplicity.

Note that $\widetilde{Y}$ has an isolated orbifold singularity above a locally given by $y_1^2 + x_1 z_1 = 0$, where $x_1 = x$, $z_1 = z$ and $y_1 = y/z$. Thus, $\mathcal{F}(Y) \neq \mathcal{F}^1(Y)$, $b = \{a\}$ and Step 1 inserts the blow up at a as the first blow up. So, $\overline{Y}_1 = \mathrm{Bl}_a(Y)$ and we will study how $\mathcal{F}$ proceeds on the (most interesting) x-chart Y_1 defined by the equation $y_1^2 + x_1^2 z_1^3 = 0$, where $x_1 = x$, $y_1 = y/x$ and $z_1 = z/x$. Note that the strict transform of C is the x_1-axis, which we denote by C_1. Since C_1 is regular, no blow up is inserted at Step 2, and so the second blow up is along C_1. Thus, $\mathrm{Bl}_{C_1}(Y_1)$ has only one chart Y_2 given by $y_2^2 + x_2^2 z_2 = 0$, where $x_2 = x_1$, $y_2 = y_1/z_1$ and $z_2 = z_1$. In particular, Y_2 is a Whitney umbrella, and $(Y_2)_{\rm sing}$ is contracted to the point a by the projection $Y_2 \to Y$. The singularity of Y_2 is resolved at Step 3 by applying $\mathcal{F}_{\rm Var}(Y_2)$, which is the same as $\mathcal{F}_{\mathrm{Var}}(X)$ from Example 5.4. So, the pinch point of Y_2 is blown up twice and the fifth blow up blows up the entire line, which is the strict transform of $(Y_2)_{\mathrm{sing}}$ (note that this singular line appeared for the first time as the preimage of a under $Y_1 \to Y$).

Now, let us describe $\mathcal{F}_{\mathrm{Var}}(Y) : Z_n \dashrightarrow Z_0 = Y$. The first two blow ups are at the origin (similarly, to Whitney umbrella). Thus $Z_1 = Y_1$ and for Z_2 we take the (most interesting) x-chart, so Z_2 is given by $y_2^2 + x_2^3 z_2^3 = 0$, where $x_2 = x_1$, $y_2 = y_1/x_1$ and $z_2 = z_1/x_1$. Note that the singularities of Y_2 and Z_2 are different. The third blow up is along the line $y_2 = z_2 = 0$ (which is the strict transform of the original singular line C), hence its only chart is Z_3 given by $y_3^2 + x_2^3 z_2 = 0$, where $x_3 = x_2$, $y_3 = y_2/z_2$ and $z_3 = z_2$. The next blow up is along the line $x_3 = y_3 = 0$ and Z_4 is given by $y_4^2 + x_4 z_4 = 0$. Finally, the last blow up is at the point $x_4 = y_4 = z_4$.

We summarize by saying that (up to our matching of affine charts, which is sort of informal) $\mathcal{F}$ blows up point, old line (the strict transform of C), point, point, new line, while $\mathcal{F}_{\mathrm{Var}}$ blows up point, point, old line, new line, point.

Remark 5.6 (i) We saw that when the singularities are not isolated, the two algorithms can blow up different centers. Unlike the synchronization issues, this makes the algorithms very different because they deal with different singularities after the first choice of different centers. In particular, it is unclear to me how one can compare the algorithms in general (or match affine charts).

(ii) Although these two examples do not give enough intuition for deciding which algorithm is faster, I would expect that in general $\mathcal{F}$ produces more complicated desingularizations than $\mathcal{F}_{\mathrm{Var}}$. Nevertheless, the tendency of $\mathcal{F}$ to blow up curves (and other higher dimensional centers) at first occasion can shorten the desingularization process in some cases.

5.2.3 Computing $\mathcal{F}_{\mathrm{Var}}$

Let X and Y be as in Examples 5.4 and 5.5. We will compute $\mathcal{F}_{\mathrm{Var}}(Y)$ and $\mathcal{F}_{\mathrm{Var}}(X)$ can be found similarly. We apply the algorithm from [BM3, §5], so the reader is advised to consult [BM3] for more details. Since Y is a hypersurface in $M = \mathbf{A}^3$, we can take the equimultiple stratum of $Y \hookrightarrow M$ as a presentation of the maximal Hilbert-Samuel stratum of Y. Thus, it suffices to resolve the marked ideal $(y^2 + xz^3)$ with $d = 2$ on M. We can take the (xz) plane P given by $y = 0$ as the hypersurface of maximal contact. Restricting the coefficient ideal $\mathscr{C}_{\emptyset}^{d-1} = (y^2 + xz^3) + (2y, 3zx^2, z^3)^2$ on P gives the ideal $\mathscr{I} = (z^6, xz^3)$ with $d_{\mathscr{I}} = d = 2$ (we refer to [BM3] and especially its §3.4, §4 and Case A of Step II in §5). To resolve $(y^2 + xz^3)$ on M is the same as to resolve $\mathscr{I}$ on P, and for the latter we go to Case B of Step II in [BM3, §5]. The monomial part $\mathscr{M}$ of $\mathscr{I}$ is trivial (no history) and the non-monomial part is $\mathscr{N} = (z^6, xz^3)$ with $d_{\mathscr{N}} = 4$. So, the companion ideal $\mathscr{G}$ is $\mathscr{N}$ with $d_{\mathscr{G}} = 4$ and it easy to see (via further maximal contact reduction) that $\mathscr{G}$ is resolved by blowing up the origin. Hence the first blow up in the resolution of $\mathscr{I}$ is along the origin a (and similarly for the resolution of Y). Consider the x-chart P_1 of $\mathrm{Bl}_a(P)$. The principal transform of $\mathscr{I}$ on this chart is $\mathscr{I}_1 = (x_1^2 z_1^3, x_1^4 z_1^6) = (x_1^2 z_1^3)$ (it is obtained from the full transform by dividing by $z_1^d = z_1^2$). Now, $\mathscr{I}_1 = \mathscr{M}_1 \mathscr{N}_1$ for monomial $\mathscr{M}_1 = (x_1^2)$ and non-monomial $\mathscr{N}_1 = (z_1^3)$. It follows that $\mathscr{G}_1 = (z_1^3)$ with the exceptional divisor $V(x_1)$. Hence $\mathscr{G}_1$ is resolved by two blow ups: first we blow up the point $x_1 = z_1 = 0$ due to Case B of Step I in [BM3, §5] in order to separate the singular locus $V(z_1)$ from the old boundary. This gives $\mathscr{G}_2 = (z_2^3)$ on the x-chart, and then we blow up the line $V(z_2)$. Tracking the effect on the principal transforms of $\mathscr{I}_1$ under these two blow ups we see that $\mathscr{I}_2 = x_2^{-2}(x_2^5 z_2^3) = (x_2^3 z_2^3)$ and $\mathscr{I}_3 = (x_3^3 z_3)$. The latter ideal is monomial because the exceptional divisor is $V(x_3 z_3)$ at this stage. So, Step II in [BM3, §5] deals with it by Case A. It is easy to see that $\mathscr{I}_3$ is resolved by two additional blow ups. First one blows up the line $V(x_3)$, obtaining $\mathscr{I}_4 = (x_4 z_4)$, and then one blows up the remaining singular point $V(x_4, z_4)$, resolving $\mathscr{I}$ completely. Thus, $\mathscr{I}$ is locally resolved by an affine chart of a blow up sequence of length five, and its centers are as follows: the point $V(x, z)$, the point $V(x_1, z_1)$, the line $V(z_2)$, the line $V(x_3)$ and the point $V(x_4, z_4)$. Therefore the centers of the corresponding "affine chart" of $\mathscr{F}_{\mathrm{Var}}(Y)$ are as follows: the point $V(x, y, z)$, the point $V(x_1, y_1, z_1)$, the line $V(y_2, z_2)$, the line $V(x_3, y_3)$ and the point $V(x_4, y_4, z_4)$.

6 Desingularization in other categories

We will show that the main Theorems 3.1, 3.3 and 3.5 imply analogous desingularization theorems in many other categories, including qe stacks, formal schemes and various analytic spaces in characteristic zero. Also, we will show that desingularization of non-compact objects follows as well.

6.1 Stacks

Let $\mathfrak{X}$ be an Artin stack with a smooth atlas $p_{1,2} : R \rightrightarrows U$, so $U \to \mathfrak{X}$ is a smooth covering and $R = U \times_{\mathfrak{X}} U$. If $\mathfrak{V} \hookrightarrow \mathfrak{X}$ is a closed substack then we define the blow up $\mathrm{Bl}_{\mathfrak{V}}(\mathfrak{X})$ using the chart $\mathrm{Bl}_W(R) \rightrightarrows \mathrm{Bl}_V(U)$ where $V = \mathfrak{V} \times_{\mathfrak{X}} U$ and $W = \mathfrak{V} \times_{\mathfrak{X}} R$ (we use here that the blow ups are compatible with flat morphisms and so $\mathrm{Bl}_W(R) = \mathrm{Bl}_V(U) \times_{U,p_i} R$ for $i = 1, 2$). We say that a stack is regular or qe if it admits a smooth cover by such a scheme. A strong desingularization is now defined as in the case of schemes.

Theorem 6.1 *The blow up sequence functors $\mathcal{F}$, $\mathcal{E}$ and $\mathcal{P}$ extend uniquely to noetherian qe stacks over* $\mathbf{Q}$.

To prove this Theorem for $\mathcal{F}$ we take any stack $\mathfrak{X}$ as above and find its smooth atlas $R \rightrightarrows U$. Then $\mathcal{F}(R) \rightrightarrows \mathcal{F}(U)$ is an atlas of a blow up sequence $\mathcal{F}(\mathfrak{X})$. An interesting case is when $\mathfrak{X} = X/G$ for an S-scheme X acted on by an S-smooth group scheme G. Then the above Theorem actually states that X admits a G-equivariant desingularization. Other functors are dealt with similarly.

6.2 Formal schemes and analytic spaces

In the categories of qe formal schemes, complex analytic spaces, rigid analytic spaces and analytic k-spaces of Berkovich the notions of regularity and blow ups are defined. So, one can define strong desingularization similarly to the case of schemes. Hironaka proved desingularization of complex analytic spaces, but this required to insert major changes in his method (and the main reason is that his method is not canonical). The new algorithms are known to work almost verbatim for complex analytic spaces, though strictly speaking, one should repeat the entire proof word by word. The desingularization of affine formal schemes was deduced in [Tem1] from non-functorial desingularization of affine schemes, but this approach did not yield global desingularization of formal schemes.

It turns out that functorial desingularization of qe schemes is so strong that it rigorously implies functorial desingularization of all above objects. The strategy is always the same, so let us stick with the non-embedded desingularization. We cover a generically reduced object X (i.e. the non-reduced locus is nowhere dense) by compact local subobjects $X_1, \dots, X_n$ (e.g. affinoid subdomains, affine formal subschemes or semi-algebraic Stein compacts, e.g., closed subspaces of a closed polydisk) and observe that $A_i = \mathcal{O}_X(X_i)$ are qe rings and for any smaller object $X_{ijk} \subset X_i \cap X_j$ the localization homomorphisms $A_i \to \mathcal{O}_X(X_{ijk})$ are regular (e.g. formal localization is regular on qe formal schemes). Thus, completion/analytification of the desingularization $\mathcal{F}(\mathrm{Spec}(A_i))$ yields a desingularization of X_i, and these local desingularizations glue together because $\mathcal{F}$ is compatible with all regular morphisms.

Remark 6.2 (i) Recall that $\mathcal{F}$ is of absolute nature, and actually it is constructed from a functor $\mathcal{F}_{\mathbf{Q}}$ on $\mathbf{Q}$-varieties. Thus, the obtained desingularization of all above objects is algebraic, and even defined over $\mathbf{Q}$ in some sense. The latter might look surprising since there are non-algebraizable analytic singularities, so we illustrate below the differences between our method and "naive algebraization".

(ii) First of all, thanks to the localization stage we only have to algebraize rather special classes of singularities, which generalize in some sense the isolated singularities. Furthermore, even when $x \in X$ is an isolated complex singularity, we do not algebraize a complex neighborhood of x but only its formal neighborhood $\widehat{X}_x = \mathrm{Spf}(\widehat{\mathcal{O}}_{X,x})$. This operation is "too local" at x, so it does not have to extend to an analytic neighborhood of x. Moreover, for the sake of functoriality we had to study all algebraizations of $\widehat{X}_x$, including those that induce embeddings $\mathbf{C} \hookrightarrow \widehat{\mathcal{O}}_{X,x}$ not landing in $\mathcal{O}_{X,x}$.

6.3 Non-compact objects and hypersequences

Because of functoriality of the algorithms from §6.2, one immediately obtains a functorial desingularization of non-compact qe schemes, formal schemes, and various analytic spaces. However, this time the desingularization is just a proper morphism $X' \to X$ because its functorial splitting into a sequence of blow ups can be infinite. One can often perform few blow ups simultaneously to obtain a finite splitting $X' = X_n \dashrightarrow X_0 = X$, or an infinite splitting

$$\cdots \to X_n \to \cdots \to X_1 \to X_0 = X$$

that reduces to a finite sequence (and infinitely many empty blow ups) over any relatively compact subspace of X, but this is not functorial. Moreover, despite some claims in the literature, even such an infinite splitting does not always exist.

Example 6.3 For concreteness, let us work complex-analytically and fix a strong desingularization algorithm $\mathcal{F}$. It is easy to construct a complex surface S with an irreducible curve $C \subset S_{\text{sing}}$ and a point $a_n \in S$ such that the following conditions hold. At a generic point $\eta \in C$ one can locally describe S by the equation $f(x,y,z) = y^2+z^2 = 0$ (i.e. S consists of two smooth branches meeting transversally along C), but a_n is a so special point that the resolution $\mathcal{F}(S)$ on the appropriate charts over a_n looks as follows: at least n times one blows up the preimage of a_n on the strict transforms of C, and only then one blows up the strict transform of C (thus resolving the generic points of C). For example, an easy computation shows that both for our algorithm $\mathcal{F}$ and for the algorithm $\mathcal{F}_{\text{Var}}$ of Bierstone-Milman one can define the germ of S at a_n by $y^2+z^2+x^{2n+3} = 0$. Clearly, we can construct a non-compact surface with a curve C and an infinite sequence of points $a_1, a_2, \dots$ as above (use an infinite pasting procedure). If a factorization of $\mathcal{F}(S)$ into an infinite sequence $\dots \to S_1 \to S_0 = S$ would exist, then the strict transform of C would be a component of the center of some blow up, say $S_{n+1} \to S_n$. And this would contradict the assumption that the composition is $\mathcal{F}(S)$ over a_{n+1}.

Nevertheless, there is a functorial way to split the desingularization. Instead of infinite blow up sequences ordered by **N**, one should consider their generalization, which is called blow up *hypersequences* in [Tem2]. The latter are sequences ordered by a countable ordered set (the set of invariants of the algorithm in this case) and such that over each relatively compact subobject $Y \hookrightarrow X$ the hypersequence reduces to the finite blow up sequence $\mathcal{F}(Y)$ saturated with infinitely many empty blow ups. Existence of such a splitting is more or less a tautology and we refer to [Tem2, §5.3] for details.

References

[AdJ] Abramovich, D.; de Jong, A.J.: *Smoothness, semistability, and toroidal geometry.*, J. Alg. Geom. **6** (1997), no. 4, 789–801.

[BM1] Bierstone, E.; Milman, P.: *A simple constructive proof of canonical resolution of singularities*, Effective methods in algebraic geometry (Castiglioncello, 1990), 11–30, Progr. Math., 94, Birkhäuser Boston, Boston, MA, 1991.

[BM2] Bierstone, E.; Milman, P.: *Canonical desingularization in characteristic zero by blowing up the maximum strata of a local invariant*, Invent. Math. **128** (1997), no. 2, 207–302.

[BM3] Bierstone, E.; Milman, P.: *Functoriality in resolution of singularities*, Publ. Res. Inst. Math. Sci. **44** (2008), 609–639.

[BMT] Bierstone, E.; Milman, P.; Temkin M.: **Q**-*universal desingularization*, preprint, arXiv:[0905.3580], to appear in Asian Journal of Mathematics.

[CJS] Cossart, V.; Jannsen, U.; Saito, S.: *Canonical embedded and non-embedded resolution of singularities for excellent two-dimensional schemes*, preprint, arXiv:[0905.2191].

[EGA] Dieudonné, J.; Grothendieck, A.: *Éléments de géométrie algébrique*, Publ. Math. IHES, **4, 8, 11, 17, 20, 24, 28, 32**, (1960-7).

[EH] Encinas, S.; Hauser, H.: *Strong resolution of singularities in characteristic zero*, Comment. Math. Helv. **77** (2002) 821–845.

[Elk] Elkik, R.: *Solution d'équations à coefficients dans un anneau hensélien*, Ann. Sci. Ecole Norm. Sup. (4) **6** (1973), 553–603.

[FEM] de Fernex, T.; Ein, L.; Mustata, M.: *Shokurov's ACC Conjecture for log canonical thresholds on smooth varieties*, arXiv:[0905.3775], to appear in Duke Math. J.

[Hir1] Hironaka, H.: *Resolution of singularities of an algebraic variety over a field of characteristic zero. I, II*, Ann. of Math. **79** (1964), 109–203.

[Hir2] Hironaka, H.: *Idealistic exponents of singularity*, Algebraic geometry (J. J. Sylvester Sympos., Johns Hopkins Univ., Baltimore, Md., 1976), pp. 52–125. Johns Hopkins Univ. Press, Baltimore, Md., 1977.

[Ked] Kedlaya, K.: *Good formal structures for flat meromorphic connections, II: Excellent schemes*, arXiv:[1001.0544].

[KKMS] Kempf, G.; Knudsen, F.; Mumford, D.; Saint-Donat, B.: *Toroidal embeddings*, Lecture Notes in Mathematics, Vol. 339. Springer-Verlag, Berlin-New York, 1973. viii+209 pp.

[Kol] Kollár, J.: *Lectures on resolution of singularities*, Annals of Mathematics Studies, 166. Princeton University Press, Princeton, NJ, 2007. vi+208 pp.

[Mat] Matsumura, H.: *Commutative ring theory*, Translated from the Japanese by M. Reid. Second edition. Cambridge Studies in Advanced Mathematics, 8. Cambridge University Press, Cambridge, 1989.

[Nic] Nicaise, J.: *A trace formula for rigid varieties, and motivic Weil generating series for formal schemes*, Math. Ann. **343** (2009), 285–349.

[NN] Nishimura, J.; Nishimura, T. *Ideal-adic completion of Noetherian rings. II*, Algebraic geometry and commutative algebra, Vol. II, 453–467, Kinokuniya, Tokyo, 1988.

[Tem1] Temkin, M.: *Desingularization of quasi-excellent schemes in characteristic zero*, Adv. Math., **219** (2008), 488-522.

[Tem2] Temkin, M.: *Functorial desingularization of quasi-excellent schemes in characteristic zero: the non-embedded case*, preprint, arXiv:[0904.1592].

[Tem3] Temkin, M.: *Functorial desingularization over* **Q**: *boundaries and the embedded case*, preprint, arXiv:[0912.2570].

[Val] Valabrega, P.: *A few Theorems on completion of excellent rings*, Nagoya Math. J. **61** (1976), 127–133.

[Vil1] Villamayor, O.: *Constructiveness of Hironaka's resolution*, Ann. Sci. Ecole Norm. Sup. (4) **22** (1989), 1–32.

[Vil2] Villamayor, O.: *Patching local uniformizations*, Ann. Sci. École Norm. Sup. (4) **25** (1992), 629–677.

[Wł] Włodarczyk, J.: *Simple Hironaka resolution in characteristic zero*, J. Amer. Math. Soc. **18** (2005), no. 4, 779–822 (electronic).

Printed in the United States
by Baker & Taylor Publisher Services